Babak Fotouhi

EVOLUTIONARY DYNAMICS

EVOLUTIONARY DYNAMICS

EXPLORING THE EQUATIONS OF LIFE

MARTIN A. NOWAK

THE BELKNAP PRESS OF HARVARD UNIVERSITY PRESS
CAMBRIDGE, MASSACHUSETTS, AND LONDON, ENGLAND 2006

Printed in Canada

Library of Congress Cataloging-in-Publication Data

Nowak, M. A. (Martin A.)
Evolutionary dynamics : exploring the equations of life /
Martin A. Nowak.
p.   cm.
Includes bibliographical references and index.
ISBN-13: 978-0-674-02338-3 (alk. paper)
ISBN-10: 0-674-02338-2 (alk. paper)
1. Evolution (Biology)—Mathematical models.   I. Title.

QH371.3.M37N69   2006
576.801′5118—dc22   2006042693

Designed by Gwen Nefsky Frankfeldt

To my parents,
in gratitude and love

CONTENTS

PREFACE

Evolutionary Dynamics presents those mathematical principles according to which life has evolved and continues to evolve. Since the 1950s biology, and with it the study of evolution, has grown enormously, driven by the quest to understand the world we live in and the stuff we are made of. Evolution is the one theory that transcends all of biology. Any observation of a living system must ultimately be interpreted in the context of its evolution. Because of the tremendous advances over the last half century, evolution has become a discipline that is based on precise mathematical foundations. All ideas regarding evolutionary processes or mechanisms can, and should, be studied in the context of the mathematical equations of evolutionary dynamics.

The original formulation of evolutionary theory and many of the investigations of its first hundred years dealt with the genetic evolution of the origin and adaptation of species. But more recently evolutionary thinking has expanded to all areas of biology and many related disciplines of the life sciences. Wherever information reproduces, there is evolution. Mutations are caused by errors in information transfer, resulting in different types of messages. Selection among types emerges when some messages reproduce faster than others.

Mutation and selection make evolution. Mutation and selection can be described by exact mathematical equations. Therefore evolution has become a mathematical theory.

The life sciences in general, and biology in particular, are on the brink of an unprecedented theoretical expansion. Every university is currently aiming to establish programs in mathematical biology and to offer its students an interdisciplinary education that spans fields as diverse as mathematics and molecular biology, linguistics and computer science. At the borders of such disciplines, progress occurs. Whenever the languages of two disciplines meet, two cultures interact, and something new happens.

In this book, the languages of biology and mathematics meet to talk about evolution. *Evolutionary Dynamics* introduces the reader to the fascinating and simple laws that govern the evolution of living systems, however complicated they may seem. I will start with the basics, avoid unnecessary complications, and reach cutting-edge research problems within a few steps.

The book grew out of a course I taught at Harvard University in 2004 and 2005. The students in my first class were Blythe Adler, Natalie Arkus, Michael Baym, Paul Berman, Illya Bomash, Nathan Burke, Chris Clearfield, Rebecca Dell, Samuel Ganzfried, Michael Gensheimer, Julia Hanover, David Hewitt, Mark Kaganovich, Gregory Lang, Jonathan Leong, Danielle Li, Alex Macalalad, Shien Ong, Ankit Patel, Yannis Paulus, Jura Pintar, Esteban Real, Daniel Rosenbloom, Sabrina Spencer, and Martin Willensdorfer, and the teaching fellows Erez Lieberman, Franziska Michor, and Christine Taylor. I have learned much from you. Your questions were my motivation. I wrote this book for you.

I am indebted to many people. Most of all I would like to thank May Huang and Laura Abbott, who helped me to prepare the final manuscript and index. They turned chaos into order. I could not have finished without them. I also thank the excellent editors of Harvard University Press, Elizabeth Gilbert and Michael Fisher.

I thank Ursula, Sebastian, and Philipp for their patience and for their burning desire to understand everything that can be understood.

I would like to express my gratitude to my teachers, Karl Sigmund and Robert May. Both of them are shining examples of how scientists should be. They have again and again impressed me with their superior judgment, insight, and generosity. I also appreciate the work and friendship of the many people with whom I have had the honor of collaborating and whose enthusiasm for science is woven into the ideas presented here: Roy Anderson, Rustom Antia, Ramy Arnaout, Charles Bangham, Barbara Bittner, Baruch Blumberg, Maarten Boerlijst, Sebastian Bonhoeffer, Persephone Borrow, Reinhard Bürger, Michael Doebeli, Peter Doherty, Andreas Dress, Ernst Fehr, Steve Frank, Drew Fudenberg, Beatrice Hahn, Christoph Hauert, Tim Hughes, Lorens Imhof, Yoh Iwasa, Vincent Jansen, Paul Klenerman, Aron Klug, Natalia Komarova, David Krakauer, Christoph Lengauer, Richard Lenski, Bruce Levin, Erez Lieberman, Jeffrey Lifson, Marc Lipsitch, Alun Lloyd, Joanna Masel, Erick Matsen, Lord May of Oxford (Defender of Science), John Maynard Smith, Angela McLean, Andrew McMichael, Franziska Michor, Garrett Mitchener, Richard Moxon, Partha Niyogi, Hisashi Ohtsuki, Jorge Pacheco, Karen Page, Robert Payne, Rodney Phillips, Joshua Plotkin, Roland Regoes, Ruy Ribeiro, Akira Sasaki, Charles Sawyers, Peter Schuster, Anirvan Sengupta, Neil Shah, George Shaw, Karl Sigmund, Richard Southwood, Ed Stabler, Dov Stekel, Christine Taylor, David Tilman, Peter Trappa, Arne Traulsen, Bert Vogelstein, Lindi Wahl, Martin Willensdorfer, and Dominik Wodarz.

I thank Jeffrey Epstein for many ideas and for letting me participate in his passionate pursuit of knowledge in all its forms.

EVOLUTIONARY DYNAMICS

1 INTRODUCTION

IN 1831, at the age of twenty-two, Charles Darwin embarked on his journey around the world. He gazed at the breath-taking diversity of tropical flora and fauna, collected creepy-crawlies from the vast oceans that he traversed, was hopelessly seasick, saw slavery in Brazil, witnessed genocide in Argentina, and was underwhelmed by the naked humanity at Tierra del Fuego. He experienced the effects of a devastating earthquake in Chile that raised the South American continent. He led an expedition into the Andes and discovered marine fossils at high altitude. He paid little attention to which finches came from which islands in the Galápagos and ate most of the delicious turtles he had gathered on his way home across the Pacific. He saw Tahiti and the economic rise of Australia. He visited John Hershel, England's leading physicist of the time, in South Africa; Hershel told him that "the mystery of mysteries" was the as yet unknown mechanism that gave rise to new species. Darwin returned to England's shores after five years, having collected six thousand specimens that would require decades of analysis by an army of experts.

His own observations in geology and the theory of his mentor, Sir Charles Lyell, that mountains were not lifted up in one day, but rose slowly over unimaginable periods of time, led Darwin to a key idea: given enough time everything can happen.

Charles Darwin did not invent the concept of evolution. When he was a student in Edinburgh in the late 1820s, evolution was already the talk of the town. But evolution was rejected by the establishment. Those who adhered to evolutionary thinking were called Lamarckists, after the French scientist Jean-Baptiste Lamarck, who was the first to propose that species are not static, but change over time and give rise to new species. Lamarck had offered this perspective in a book published in 1809. He did not, however, propose a correct mechanism for how species change into each other. This mechanism was discovered first by Charles Darwin and independently by Alfred Russel Wallace.

From reading the economist Thomas Malthus, Darwin was aware of the consequences of exponentially growing populations. Once resources become limiting only a fraction of individuals can survive. Darwin was also a keen observer of animal breeders. He analyzed their methods and studied their results. Slowly he understood that nature acted like a gigantic breeder. This was the first time that natural selection materialized as an idea, a scientific concept in a human mind. Darwin was thirty-three years old.

The one problem that Darwin did not solve concerned the mechanism that could maintain enough diversity in a population for natural selection to operate. Darwin was unaware of the Austrian monk and botanist Gregor Mendel and his experiments on plant heredity. Mendel's work had already been published but was hidden, gathering dust in the *Annals* of the Brno Academy of Sciences.

Darwin once remarked, "I have deeply regretted that I did not proceed far enough at least to understand something of the great leading principles of mathematics; for men thus endowed seem to have an extra sense." The engineer Fleeming Jenkins, who reviewed Darwin's *On the Origin of Species*, published in 1859, had raised a fundamental and seemingly intractable objection to Darwin's theory: if offspring inherit a blend of the parents' characteristics, then variability diminishes in successive generations. Several decades later a simple mathematical equation, independently found by the famous British mathematician G. H. Hardy and the German physician Wilhelm Weinberg, showed that Mendelian (particulate) inheritance does lead to a maintainance of genetic diversity under random mating. The Hardy-Weinberg law is one of the fundamental principles of evolution under sexual reproduction.

Mendelian genetics and Darwinian evolution were unified in the new discipline of mathematical biology, which developed from the seminal investigations of Ronald Fisher, J. B. S. Haldane, and Sewall Wright in the 1920s and 1930s. Through their work, fundamental concepts of evolution, selection, and mutation were embedded in a precise mathematical framework. This line of mathematical analysis was taken up in the 1950s by Motoo Kimura, who formulated the neutral theory of evolution. Kimura realized that most genetic mutations do not affect fitness and are fixed in populations only by random drift.

Other milestones of evolutionary dynamics include William Hamilton's discovery in 1964 that selection of "selfish genes" can favor altruistic behavior among relatives and John Maynard Smith's invention of evolutionary game theory in 1973. In the mid-1970s Robert May revolutionized the mathematical approaches to ecology and epidemiology. Manfred Eigen and Peter Schuster formulated quasispecies theory, which provides a link between genetic evolution, physical chemistry, and information theory. Peter Taylor, Josef Hofbauer, and Karl Sigmund studied the replicator equation, the foundation of evolutionary game dynamics.

This very brief and incomplete account of the evolution of evolutionary dynamics brings us to the present book. It has fourteen chapters. Although there is some progression of complexity, the chapters are largely independent. Therefore, if you know something about the subject, you can read the book in whatever order you like. My aim has been to keep things as simple as possible, as linear as possible, and as deterministic as possible. I will start with the basics and in a few steps lead you to some of the most interesting and unanswered research questions in the field. Having read the book, you will know what you need to embark on your own journey and make your own discoveries.

This book represents an introduction to certain aspects of mathematical biology, but it is not comprehensive. Mathematical biology includes many topics, such as theoretical ecology, population genetics, epidemiology, theoretical immunology, protein folding, genetic regulatory networks, neural networks, genomic analysis, and pattern formation. The field is too diverse for any one book to represent it without running the risk of becoming as entertaining as a telephone directory. I have chosen those topics that I know well and where my explanation can be brief and effective. I have concentrated on evolution because it is the one unifying principle of all of biology.

It might seem surprising that a book on evolutionary dynamics is not primarily about population genetics. Nevertheless the ideas and concepts of this fascinating field stand behind many of my explorations: the basic mathematical formulations of selection, mutation, random drift, fitness landscapes, and frequency-dependent selection as well as of evolution in structured populations have originated in population genetics. Several major themes of population genetics, however, such as sexual reproduction, sexual selection, recombination, and speciation, are not discussed here. In contrast, classical population genetics does not deal with evolutionary dynamics of infectious agents, the somatic evolution of cancer, evolutionary game theory, or the evolution of human language, all of which are subjects that I do explore.

The main ingredients of evolutionary dynamics are reproduction, mutation, selection, random drift, and spatial movement. Always keep in mind that the population is the fundamental basis of any evolution. Individuals, genes, or ideas can change over time, but only populations evolve.

The structure of the book is as follows. After this introduction, in Chapter 2 I will discuss populations of reproducing individuals and the basic ideas of natural selection and mutation. Simple models of population dynamics can lead to an exponential explosion, to a stable equilibrium, or to oscillations and chaos. Selection emerges whenever two or more individuals reproduce at different rates. Mutation means that one type can change into another. There are models of population growth that lead to the survival of whoever reproduces fastest ("survival of the fittest"). Other models lead to the survival of the first or the coexistence of all.

In Chapter 3, quasispecies theory is introduced. Quasispecies are populations of reproducing genomes subject to mutation and selection. They live in sequence space and move over fitness landscapes. An important relationship between mutation rates and genome length is called the "error threshold": adaptation on most fitness landscapes is possible only if the mutation rate per base is less than one over the genome length, measured in bases.

In Chapter 4, we study evolutionary game dynamics, which arise whenever the fitness of an individual is not constant but depends on the relative abundance (= frequency) of others in the population. Thus evolutionary game theory is the most comprehensive way to look at the world. People who do not engage in evolutionary game theory restrict themselves to the rigidity of constant selection, where the fitness of one individual does not depend on others. The replicator equation is a nonlinear differential equation that describes frequency-dependent selection among a fixed number of strategies. We will encounter the Nash equilibrium and evolutionarily stable strategies. Evolutionary game theory and ecology are linked in an important way: the replicator equation is equivalent to the Lotka-Volterra equation of ecological systems, which describes the interation between predator and prey species.

Chapter 5 is dedicated to the best game in town, the Prisoner's Dilemma. The cooperation of reproducing entities is essential for evolutionary progress. Genes cooperate to form a genome. Cells cooperate to produce multicellular organisms. Individuals cooperate to form groups and societies. The emergence of human culture is a cooperative enterprise. The very problem of how to obtain cooperation by natural selection is described by the Prisoner's Dilemma. In the absence of any other assumption, natural selection favors defectors over cooperators. Cooperation has a chance, however, if there are repeated interactions between the same two individuals. We will encounter the strategy Tit-for-tat, which is defeated first by Generous Tit-for-tat and then by Win-stay, lose-shift.

In Chapter 6 we move to a stochastic description of finite populations. Neutral drift is a crucial aspect of evolutionary dynamics: if a finite population consists of two types of individuals, red and blue, and if both individuals have identical fitness, then eventually the population will be either all red or all blue. Even in the absence of selection, coexistence is not possible. If there is a fitness difference, then the fitter type has a greater chance of winning, but no certainty. We calculate the probability that the descendants of one individual will take over the whole population. This so-called fixation probability is important for estimating the rate of evolution.

Chapter 7 is about games in finite populations. Most of evolutionary game theory has been formulated in terms of deterministic dynamics describing the limit of infinitely large populations. Here we move game theory to finite populations and make surprising observations. Neither a Nash equilibrium, nor an evolutionarily stable strategy, nor a risk-dominant strategy is protected by natural selection. There can be advantageous mutants against all three. In a bistable situation between two strategies, there is a simple "1/3 rule" that determines whether a strategy is favored by natural selection.

In Chapter 8, the individuals of a population are represented by the vertices of a graph. The edges of the graph specify who interacts with whom. The graph can denote spatial relationships or social networks. The first observations of "evolutionary graph theory" are reported. The classical homogeneous population is defined by the complete graph, where all vertices are connected. We will see that circulations have the same evolutionary behavior as the complete graph in terms of fixation probability under constant selection, and therefore represent a particular balance between drift and selection. Graphs that enhance drift act as suppressors of selection. Graphs that reduce drift act as amplifiers of selection. In the limit of large population size, there exist graphs that guarantee the fixation of any advantageous mutant and the extinction of any disadvantageous mutant. Games on graphs are also studied in this chapter. There is a remarkably simple rule for the evolution of cooperation.

Chapter 9 gives an account of evolutionary game dynamics on spatial grids. The primary approach will be deterministic, discrete in time, and discrete in space. This approach brings together game theory and cellular automata. We will observe evolutionary kaleidoscopes, dynamic fractals, and spatial chaos. There is all the complexity one could ever wish for—making it unnecessary for God to play dice. Moreover, cooperation can evolve on spatial grids. This is the concept of "spatial reciprocity."

In Chapter 10, we study the evolutionary dynamics of virus infections. I will argue that the mechanism of disease progression caused by the human immunodeficiency virus (HIV) is an evolutionary one. The immune system constantly attacks the virus, but the virus continuously evolves away, appears elsewhere in sequence space, and eventually overpowers the immune system. The resulting "diversity threshold theory" can explain why people succumb to the fatal immunodeficiency disease AIDS after a long and variable infection with HIV.

Chapter 11 discusses the evolution of infectious agents, their attempts to infect new hosts, and the selection pressures that determine the level of virulence. The conventional wisdom is that well-adapted parasites are harmless to their hosts. This perspective is revised in the context of evolutionary dynamics. Competition between different mutants of a parasite maximizes its basic reproductive ratio. Superinfection takes into account that parasites compete on two levels of selection: within an infected host and in the population of hosts. Superinfection holds many surprising aspects, including the shortsighted evolution of higher and higher virulence beyond what would be optimum for the parasite.

Chapter 12 explores the evolutionary dynamics of human cancer. Cancer arises when cooperation among cells breaks down. The mutated cells revert to their primitive program of uncontrolled replication. We calculate the rate of activation of oncogenes and inactivation of tumor suppressor genes. We analyze the impact of mutations that trigger "genetic instability." We outline the conditions necessary for "chromosomal instability" to initiate cancer progression.

Chapter 13 is devoted to the evolutionary dynamics of the one trait that is truly our own invention and that is arguably the one interesting thing that has happened in the last six hundred million years on earth. Bacteria invented all the biochemistry of life. Eukaryotes invented some advanced genetics and how to build complicated multicellular plants and animals. Humans will be remembered for language.

Chapter 14 summarizes and concludes. Further readings will be found at the back of the book.

All the diverse topics of this book are unified by evolutionary dynamics. The mathematical description of evolution has moved from the study of purely genetic systems to any kind of process where information is being reproduced in a noisy (= natural) environment. What you will encounter in this book is responsible for shaping life around you. Every living system, and everything that arises as a consequence of living systems, is a product of evolutionary dynamics.

WHAT EVOLUTION IS 2

THIS CHAPTER introduces three basic building blocks of evolutionary dynamics: replication, selection, and mutation. These are the fundamental and defining principles of biological systems. They apply to any biological organization anywhere in our or other universes and do not depend on the particular details of which chemistry was recruited to embody life. Any living organism has arisen and is continually modified by these three principles.

Evolution requires populations of reproducing individuals. In the right environment, biological entities, such as viruses, cells, and multicellular organisms can make copies of themselves. The blueprint that determines their structure, the genomic material in form of DNA or RNA, is replicated and passed on to the offspring. Selection results when different types of individuals compete with each other. One type may reproduce faster and thereby outcompete the others. Reproduction is not perfect, but involves occasional mistakes, or mutations. Mutation is responsible for generating different types that can be evaluated in the selection process, and thus results in biological novelty and diversity. Selection will choose to maintain some innovations and dismiss others, and can favor or oppose genetic diversity.

At the end of this chapter we will focus on the Hardy-Weinberg law of random mating. This discussion will be our only venture into the mathematics of sexual reproduction. In subsequent chapters we will encounter additional principles of evolutionary dynamics, such as random drift and spatial movement.

2.1 REPRODUCTION

Imagine a single bacterial cell in a perfect environment that contains all the nutrients required for growth and happiness. In this bacterial heaven, the fortunate cell and all its offspring divide every 20 minutes, which is the known world record for bacterial cell division in an ideal lab setting. After 20 minutes the cell has given rise to 2 daughter cells. After 40 minutes there are 4 granddaughters, and after one hour there are 8 great granddaughters. How many cells will there be after three days?

After t generations there are 2^t cells. In three days there are 216 generations. Hence we expect $2^{216} = 10^{65}$ cells. The total mass of these cells would exceed the mass of the earth by many orders of magnitude.

The growth law for this overwhelming expansion can be written as a recursive equation

$$x_{t+1} = 2x_t. \tag{2.1}$$

Here x_t is the number of cells at time t, and x_{t+1} is the number of cells at time $t+1$. The equation means that at time $t+1$ there are twice as many cells as at time t. Time is measured in numbers of generations.

The number of cells at time 0 is given by x_0. With this initial condition, the solution of equation (2.1) can be written as

$$x_t = x_0 2^t. \tag{2.2}$$

Equation (2.1) is a so-called difference equation, because time is measured in discrete steps.

We can also formulate a differential equation for exponential growth that measures time as a continuous quantity. Let $x(t)$ denote the abundance of cells at time t. Suppose that cells divide at rate r. More precisely, we assume that the time for cell division follows an exponential distribution with average $1/r$. We can write the differential equation

$$\dot{x} = \frac{dx}{dt} = rx. \tag{2.3}$$

Throughout this book, I will use the standard notation $\dot{x}$ to refer to differentiation (of x) with respect to time. If the abundance of cells at time 0 is given by x_0 then the solution of the differential equation (2.3) is

$$x(t) = x_0 e^{rt}. \tag{2.4}$$

Let us reconsider our bacterial supernova. If we measure time in units of days, then $r = 72$ means that the time for a cell cycle requires, on average, 20 minutes (calculated by dividing the total number of minutes in a day, 1,440, by 72). Hence there are 72 cell divisions in one day. After three days, one bacterial cell has generated e^{216} cells which is approximately 6×10^{93} cells.

The discrepancy between the differential equation and the difference equation is a consequence of the varying assumptions for the distribution of the generation time. The difference equation assumes that each cell division occurs after exactly 20 minutes. The differential equation assumes that each cell division occurs after a time which is exponentially distributed around an average of 20 minutes. The exponential distribution is defined as follows: the probability that cell division occurs between time 0 and τ is given by $1 - e^{-r\tau}$. On average, cells divide after $1/r$ time units.

So far we have ignored cell death. Let us now suppose that cells die at rate d, which means that they have an exponentially distributed lifespan with an average of $1/d$. The differential equation becomes

$$\dot{x} = (r - d)x. \tag{2.5}$$

The effective growth rate is the difference between the birth rate, r, and the death rate, d. If $r > d$, then the population will expand indefinitely. If $r < d$, then the population will converge to zero and become extinct. If $r = d$, then the population size remains constant, but this situation is unstable: small deviations from absolute equality between birth and death will lead to either exponential expansion or decline. It is important to note that setting $r = d$

in equation (2.5) does not constitute a mechanism for maintaining a stable constant population size.

The simple equation (2.5) allows us to introduce an extremely important concept in evolution, ecology and epidemiology: the basic reproductive ratio, r/d. This ratio denotes the expected number of offspring that come from any one individual. The average lifetime of a cell is $1/d$. The rate of producing offspring cells is given by r. If each cell produces on average more than one offspring, $r/d > 1$, then an exponential expansion will follow. A basic reproductive ratio greater than one is a necessary condition for population expansion.

We have observed that ongoing exponential growth can lead to unreasonably high numbers in a very short time. In a realistic environment, the expanding population will hit constraints that prevent further expansion. For example, the population might run out of nutrients or physical space.

A model for population expansion with a maximum carrying capacity is given by the logistic equation

$$\dot{x} = rx(1 - x/K). \tag{2.6}$$

As before, the parameter r refers to the rate of reproduction in the absence of density regulation, when the population size, x, is much smaller than the carrying capacity K. As x increases, the rate of growth slows down. When x reaches the carrying capacity, K, then the population expansion ceases. For the initial condition x_0, the solution of equation (2.6) is given by

$$x(t) = \frac{Kx_0e^{rt}}{K + x_0(e^{rt} - 1)}. \tag{2.7}$$

In the limit of infinite time, $t \to \infty$, the population size converges to the equilibrium $x^* = K$. Throughout the book we will use a superscript asterisk to denote a quantity at equilibrium.

2.1.1 Deterministic Chaos

We can also study a logistic difference equation. Without loss of generality, let us rescale the population abundance in such a way that the maximum carrying capacity is given by $K = 1$. We have

$$x_{t+1} = ax_t(1 - x_t). \quad (2.8)$$

Note that the growth rate in the difference equation, a, is analogous to $1 + r$ in the differential equation (2.6). In contrast to the differential equation, the logistic difference equation (2.8) has many surprises. The behavior of this equation is so rich that many papers and even books have been written about it, and it has generously awarded glorious careers to some scientists who have studied it.

The abundance of the population, x, is given by a number between 0 and 1. The growth rate, a, can vary between 0 and 4. If $a < 0$ or $a > 4$, then negative x values will be generated, which are not biologically meaningful.

The point $x = 0$ is always an equilibrium. If $a < 1$, then the only stable equilibrium of the system is given by $x^* = 0$. This means the population will die out. If $1 < a < 3$, then the only stable equilibrium is given by $x^* = (a - 1)/a$. All trajectories starting from any initial condition x_0 (greater than 0 and less than 1) will converge to this value. The point x^* is a global attractor for the open interval (0, 1).

If $a > 3$, then the point x^* becomes unstable. For a values slightly above 3, we find a stable oscillation of period two. As a increases, the period two oscillator is replaced by period four, then by eight, and so on. For $a = 3.57$ there are infinitely many even periods. For $a = 3.6786$ the first odd periods appear. For $3.82 < a \leq 4$ all periods occur.

The logistic map with $a = 4$ is a simple and most illuminating example for studying deterministic chaos. For any value x_t it is straightforward to compute the population size in the subsequent generation, x_{t+1}. Yet the dynamics are unpredictable in the following sense. Suppose the value of x_t is only known subject to a small uncertainty. It may not be clear whether $x_t = 0.3156$ or 0.3157. After ten generations, however, the trajectories starting from these two initial values will have diverged completely. Hence prediction is impossible. Anything can happen.

We conclude that simple rules can generate complicated behavior. Much of the apparent complexity and unpredictability of biological time series, such as the population size of birds in a particular habitat, the number of measles cases in New York City, or the price fluctuations of stocks and bonds, could in principle be the consequence of deterministic laws.

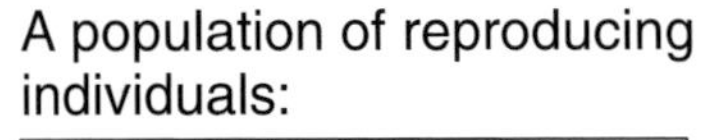

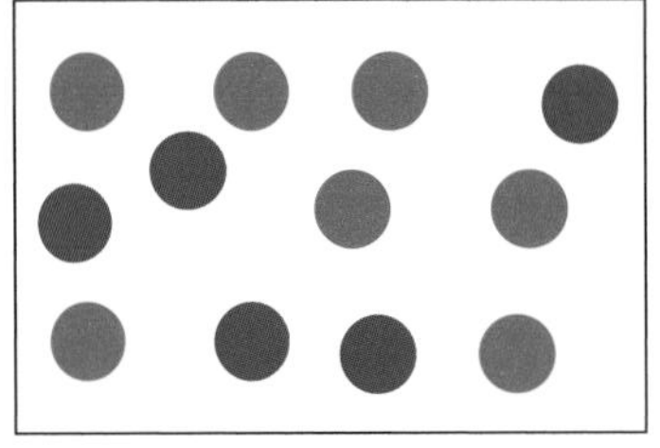

Reproduction:

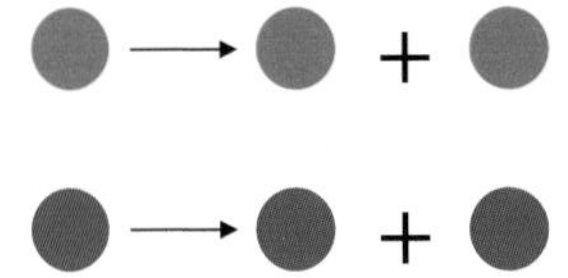

Figure 2.1 Evolution requires a population of reproducing individuals. Strictly speaking, neither genes, nor cells, nor organisms, nor ideas evolve. Only populations can evolve.

2.2 SELECTION

Selection operates whenever different types of individuals reproduce at different rates. At the very least we need two types (Figure 2.1). Let us call them A and B. Type A individuals reproduce at rate a. Type B individuals reproduce at rate b. The rate of reproduction is interpreted as fitness. Therefore the fitness of A is a, the fitness of B is b. Denote by $x(t)$ the number of A individuals at time t. Denote by $y(t)$ the number of B individuals at time t. At time $t = 0$, the numbers of A and B are respectively given by x_0 and y_0. The A and B subpopulations grow according to the differential equations

$$\begin{aligned} \dot{x} &= ax \\ \dot{y} &= by \end{aligned} \tag{2.9}$$

Equation (2.9) is a system of two ordinary, linear differential equations. The analytical solution is given by

$$\begin{aligned} x(t) &= x_0 e^{at} \\ y(t) &= y_0 e^{bt} \end{aligned} \tag{2.10}$$

Hence the A and B subpopulations grow exponentially at rates a and b, respectively. The doubling time for A is $\log 2/a$. The doubling time for B is $\log 2/b$. If a is greater than b, then A reproduces faster than B: after some time, there will be more A than B individuals.

Denote by $\rho(t) = x(t)/y(t)$ the ratio of A over B at time t. We have

$$\dot{\rho} = \frac{\dot{x}y - x\dot{y}}{y^2} = (a - b)\rho. \tag{2.11}$$

The solution of this differential equation, for the initial condition $\rho_0 = x_0/y_0$, is given by

$$\rho(t) = \rho_0 e^{(a-b)t}. \tag{2.12}$$

Hence if $a > b$ then ρ tends to infinity. In this case A will outcompete B, which means selection favors A over B. If, on the other hand, $a < b$, then ρ tends to zero. In this case B will outcompete A, which means that selection favors B over A.

Let us now consider a situation in which the total population size is held constant. This situation can arise, for example, when an ecosystem has a constant maximum carrying capacity. Let $x(t)$ denote the relative abundance of A at time t. Instead of "relative abundance" we can also say "frequency." Let $y(t)$ denote the frequency of B. Since there are only A and B individuals in the population, we have $x + y = 1$. As before, A and B individuals reproduce, respectively, at rates a and b.

We have the system of equations

$$\begin{aligned} \dot{x} &= x(a - \phi) \\ \dot{y} &= y(b - \phi) \end{aligned} \tag{2.13}$$

The term ϕ ensures that $x + y = 1$. This is only possible if $\phi = ax + by$. Observe that ϕ is the average fitness of the population.

The system (2.13) describes only a single differential equation, because y can be replaced by $1 - x$. We obtain

$$\dot{x} = x(1 - x)(a - b). \tag{2.14}$$

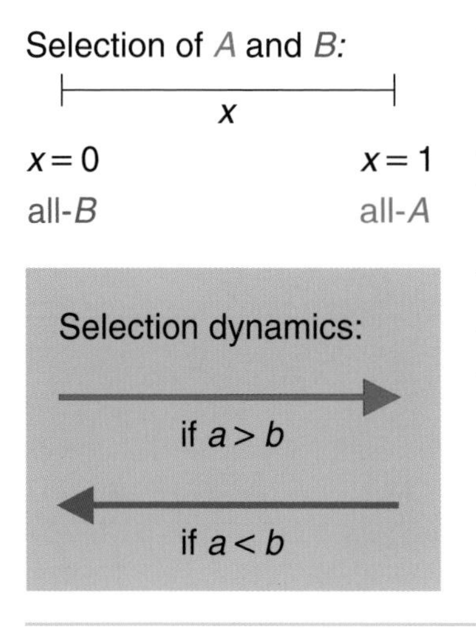

Figure 2.2 Selection arises if two types, A and B, have different rates of reproduction, a and b. If A reproduces faster than B, which means $a > b$, then A will become more abundant than B. Eventually A will take over the entire population; B will become extinct. Denote by x the relative abundance (= frequency) of type A. The quantity x is a number between 0 and 1. Therefore selection dynamics are defined on the closed interval $[0, 1]$.

This differential equation has two equilibria, one for $x = 0$ and the other for $x = 1$. At these two points, we have $\dot{x} = 0$. This observation makes sense: if $x = 1$ then the system consists only of A individuals and nothing more can happen; if $x = 0$, then the system consists only of B individuals and again nothing more can happen.

We can, however, make an additional observation. If $a > b$, then $\dot{x} > 0$ for all values of x that are strictly greater than 0 and strictly smaller than 1. This means that for any mixed system (consisting of some A and some B individuals) the fraction of A will increase if the fitness of A is greater than the fitness of B. In this case, the fraction of B will converge to 0, while the fraction of A converges to 1. We have encountered the concept of "survival of the fitter" (Figure 2.2).

2.2.1 Survival of the Fittest

The model can be extended to describe selection among n different types. Let us label them $i = 1, \ldots, n$. Denote by $x_i(t)$ the frequency of type i. The structure of the population is given by the vector $\vec{x} = (x_1, x_2, \ldots, x_n)$.

Denote by f_i the fitness of type i. As before, fitness is a non-negative real number and describes the rate of reproduction. The average fitness of the

The simplex is the set of all points whose coordinates are not negative and add up to one

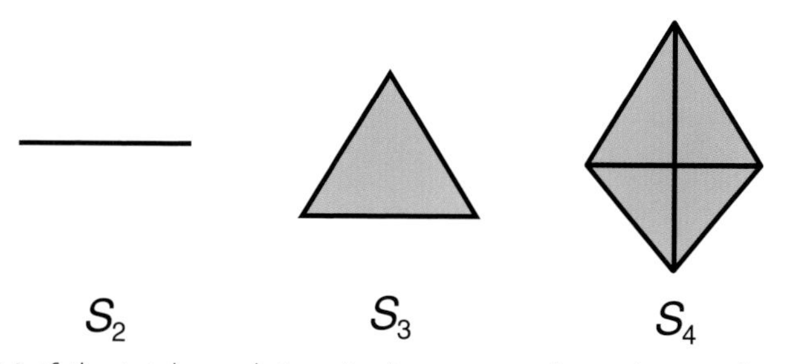

Figure 2.3 If the total population size is constant, then selection dynamics can be formulated in terms of relative abundance (= frequency). Suppose there are n different types, $i = 1, \ldots, n$. Type i has frequency x_i. The sum over all x_i is one. The set of all points, $(x_1, \ldots, x_n)$ with the property $\sum_{i=1}^{n} x_i = 1$, is called the simplex S_n. Selection dynamics occur on the simplex S_n. The figure shows S_2, S_3, and S_4. The simplex S_n is an $n - 1$ dimensional structure embedded in an n-dimensional Euclidian space. The simplex S_n has n faces that each consist of the simplex S_{n-1}.

population is given by

$$\phi = \sum_{i=1}^{n} x_i f_i. \tag{2.15}$$

Selection dynamics can be written as

$$\dot{x}_i = x_i(f_i - \phi) \qquad i = 1, \ldots, n \tag{2.16}$$

The frequency of type i increases, if its fitness exceeds the average fitness of the population. Otherwise it will decline. The total population size remains constant: $\sum_{i=1}^{n} x_i = 1$ and $\sum_{i=1}^{n} \dot{x}_i = 0$.

The set of points with the property $\sum_{i=1}^{n} x_i = 1$ is called the simplex S_n (Figure 2.3). Each point in the simplex refers to a particular structure of the population. The interior of the simplex is the set of points $\vec{x}$ with the property that $x_i > 0$ for all $i = 1, \ldots, n$. The face of the simplex is the set of points $\vec{x}$ with the property that $x_i = 0$ for at least one i. The vertices of the simplex

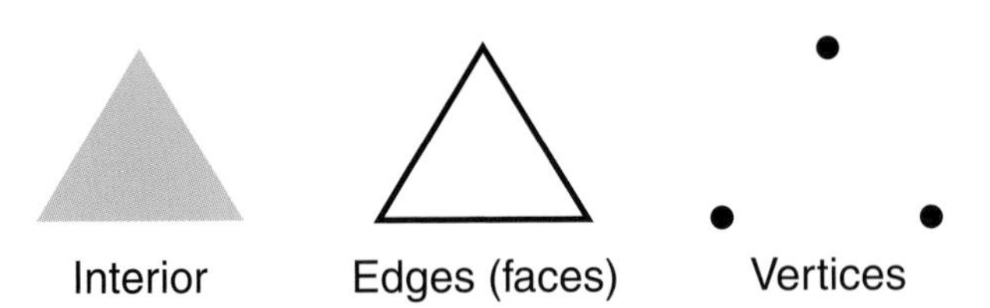

Figure 2.4 The *interior* of a simplex is the set of all points where all coordinates are strictly positive; this means no type has become extinct. The *faces* are the sets of points where at least one coordinate is zero; this means at least one type has become extinct. The *vertices* describe pure populations, where all but one type have become extinct.

are the corner points where exactly one type is present, $x_i = 1$, while all other types are extinct, $x_j = 0$ for all $j \neq i$ (Figures 2.4 and 2.5).

The simplex S_2 is given by the closed interval [0, 1]. The notation [0, 1] refers to all numbers which are greater than or equal to 0 and less than or equal to 1. In contrast, (0, 1) is the open interval; it contains all numbers that are strictly greater than 0 and strictly less than 1. The open interval (0, 1) is the interior of the closed interval [0, 1] and, therefore, is also the interior of the simplex S_2.

Equation (2.16) contains a single globally stable equilibrium. Starting from any initial condition in the interior of the simplex, the population will converge to a corner point where all but one type have become extinct. The winner, k, enjoys a well-deserved victory because it has the property of having the largest fitness, f_k. Thus $f_k > f_i$ for all $i \neq k$. The system shows competitive exclusion: the fittest type will outcompete all others. This is the concept of "survival of the fittest."

2.2.2 Survival of the First, Survival of All

Let us return to the selection of two types, A and B, but without making the assumption that their growth rates are linear functions of their frequencies. Instead consider the equation

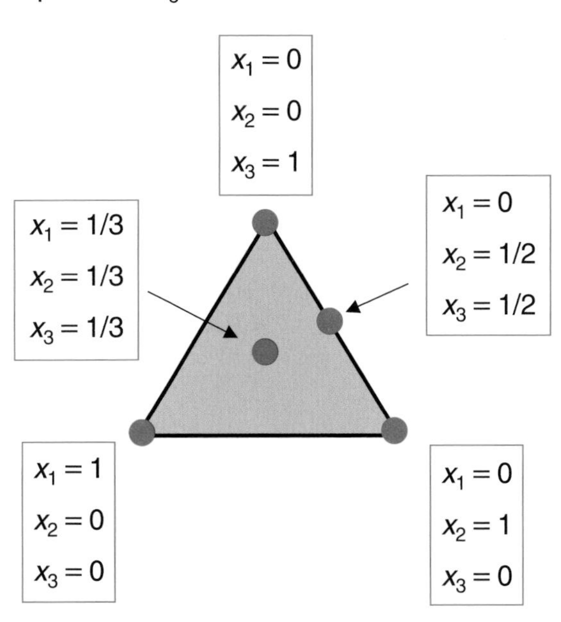

Figure 2.5 Five points on the simplex S_3. In the center, $(1/3, 1/3, 1/3)$, all three types have the same frequency. There are three faces. The center of one particular face is given by $(0, 1/2, 1/2)$; one type has become extinct. The corner points (vertices) indicate populations that consist of only one type. S_3 has three corners: $(1, 0, 0)$, $(0, 1, 0)$, and $(0, 0, 1)$.

$$\begin{cases} \dot{x} = ax^c - \phi x \\ \dot{y} = by^c - \phi y \end{cases} \tag{2.17}$$

As before, a and b denote the fitness values of A and B, respectively. If $c = 1$, we are back to equation (2.13). If $c < 1$, then growth is subexponential. In the absence of the density limitation, ϕ, the growth curve of the two types would be slower than exponential.

In contrast, if $c > 1$, then growth is superexponential. In the absence of the density limitation, ϕ, the growth curve of the two types would be faster than exponential (hyperbolic). To maintain a constant population size, $x + y = 1$, we set $\phi = ax^c + by^c$. Equation (2.17) reduces to

$$\dot{x} = x(1 - x)f(x) \tag{2.18}$$

where

$$f(x) = ax^{c-1} - b(1 - x)^{c-1}. \tag{2.19}$$

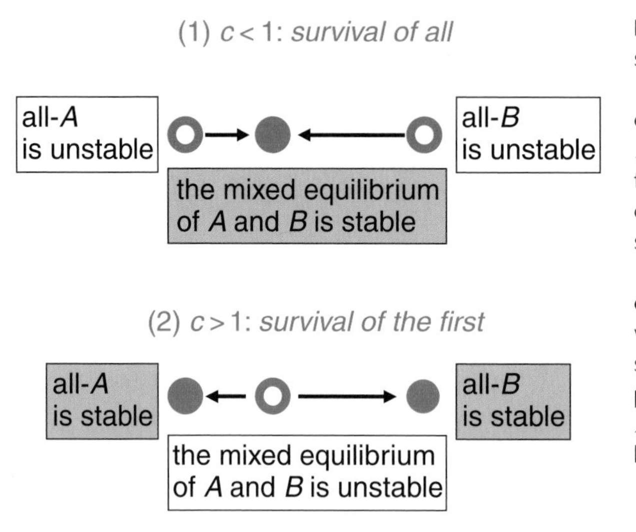
(1) c < 1: survival of all
all-A is unstable
all-B is unstable
the mixed equilibrium of A and B is stable
(2) c > 1: survival of the first
all-A is stable
all-B is stable
the mixed equilibrium of A and B is unstable

Figure 2.6 Survival of all: for subexponential growth ($c < 1$), there is a stable mixed equilibrium between A and B, even if one type has a faster growth rate than the other. Survival of the first: for superexponential growth ($c > 1$), there is an unstable mixed equilibrium between A and B, while the pure populations are stable. For example, if the whole population consists of B, then A cannot invade even if it has a higher growth rate.

This equation always has fixed points for $x = 0$ and $x = 1$. For $c \neq 1$ there exists exactly one other fixed point between 0 and 1. It is given by

$$x^* = \frac{1}{1 + \sqrt[c-1]{a/b}}. \tag{2.20}$$

If $c < 1$, then the boundary fixed points, $x = 0$ and $x = 1$, are always unstable; the interior fixed point, x^*, is globally stable. Hence there is survival of both A and B. Surprisingly, even if A is fitter than B, $a > b$, then a small amount of B can invade an A population.

If $c > 1$, then the boundary fixed points, $x = 0$ and $x = 1$, are always stable; the interior fixed point, x^*, is unstable. If $x > x^*$, then A will outcompete B. If $x < x^*$, then B will outcompete A. Again this observation is remarkable. Even if A is fitter than B in the sense that $a > b$, a B population cannot be invaded by an A mutant.

We conclude that superexponential growth favors whoever was there first (survival of the first) whereas subexponential growth leads to the survival of all (Figure 2.6).

What is the intuition behind this observation? An extreme form of subexponential growth is "immigration," $c = 0$. The growth rate does not depend on x or y at all. We have

$$\begin{aligned} \dot{x} &= a - \phi x \\ \dot{y} &= b - \phi y \end{aligned} \tag{2.21}$$

with $\phi = a + b$. This equation can be interpreted as the immigration of A and B into the population from some other place. It is clear that these immigration dynamics lead to coexistence. A value of c between 0 and 1 is a mixture between immigration and linear growth and retains the property of coexistence.

If $c > 1$, on the other hand, then A cannot invade B even if $a > b$. ("Invasion" means that an infinitesimally small fraction of A individuals can increase in abundance in a population where almost all individuals are of type B.) The intuitive explanation is as follows: we can think of the case $c = 2$ as implying that two individuals of the same type have to meet in order to reproduce. If there is only an infinitesimally small fraction of A individuals, then two A individuals will never meet and hence A will not reproduce. If $c = 3$ then three individuals of the same type have to meet in order to reproduce. Again arbitrarily small fractions of a type can never increase. The same intuition holds for all values $c > 1$.

The case $c = 2$ can also be interpreted as an evolutionary game between two strategies, A and B, that are strict Nash equilibria. Neither strategy can invade the other. We will encounter these concepts in Chapter 4.

2.3 MUTATION

Life takes advantage of mistakes. Replication of DNA or RNA can lead to slightly modified sequences that represent novel variants. Errors during reproduction are called mutations. In this section, we study the simplest possible differential equations that describe mutation (Figure 2.7).

Let us again consider just two types, A and B. Denote by u_1 the mutation rate from A to B: u_1 is the probability that the reproduction of A leads to B.

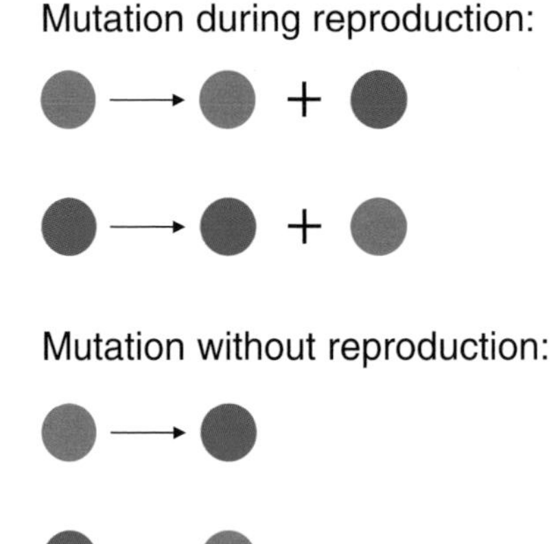

Figure 2.7 Mutation can occur during reproduction: type A produces an offspring that is type B. Mutation can also occur in the absence of reproduction: type A changes into type B. Many genetic mutations occur when the genomic material of a cell is being copied. But mutagens can also change the genetic material of a cell when it is not dividing.

Conversely, denote by u_2 the mutation rate from B to A. As before, let x and y denote the frequencies of A and B, respectively. We have

$$\begin{aligned} \dot{x} &= x(1-u_1) + yu_2 - \phi x \\ \dot{y} &= xu_1 + y(1-u_2) - \phi y \end{aligned} \tag{2.22}$$

Since A and B have the same fitness ($a = b = 1$), the average fitness of the population is constant and given by $\phi = 1$. Taking into account $x + y = 1$, system (2.22) reduces to the differential equation

$$\dot{x} = u_2 - x(u_1 + u_2). \tag{2.23}$$

The frequency of A converges to the stable equilibrium

$$x^* = \frac{u_2}{u_1 + u_2}. \tag{2.24}$$

Hence mutation leads to coexistence between A and B. The relative proportion of A and B at equilibrium depends on the mutation rates. At equilibrium, the ratio of A to B is given by $x^*/y^* = u_2/u_1$. If the mutation rates are the same, $u_1 = u_2$, and then $x^* = y^*$.

Sometimes the mutation rate in one direction is much larger than in the other direction. In these cases, it often makes sense to ignore mutation in the other direction altogether. Let $u_2 = 0$. We have

$$\dot{x} = -xu_1. \tag{2.25}$$

Therefore the frequency of A declines over time as

$$x(t) = x_0 e^{-u_1 t}. \tag{2.26}$$

The frequency of B increases as

$$y(t) = 1 - (1 - y_0)e^{-u_1 t}. \tag{2.27}$$

If mutation occurs only from A to B but not the other way around, then A will die out and B will take over the whole population. We see that mutation can affect survival. Different mutation rates can introduce selection even in the absence of different reproductive rates.

2.3.1 Mutation Matrix

We can extend mutation dynamics to n different types. Let us introduce the mutation matrix, $Q = [q_{ij}]$. The probability that type i mutates to type j is given by q_{ij}. Since each type i has to produce itself or some other type, we have $\sum_{j=1}^{n} q_{ij} = 1$. Thus Q is a stochastic $n \times n$ matrix. A stochastic matrix is defined by the properties that (i) all entries are numbers from the interval $[0, 1]$ (so-called probabilities), (ii) there are as many rows as columns, and (iii) the sum of each row is 1. Stochastic matrices always have 1 as an eigenvalue, and no eigenvalue has an absolute value greater than 1.

Mutation dynamics can be written as

$$\dot{x}_i = \sum_{j=1}^{n} x_j q_{ji} - \phi x_i \qquad i = 1, \ldots, n \tag{2.28}$$

In vector notation we can write

$$\dot{\vec{x}} = \vec{x}Q - \phi\vec{x}. \tag{2.29}$$

Again the average fitness is just $\phi = 1$. The equilibrium is given by the left-hand eigenvector associated with eigenvalue 1:

$$\vec{x}^* Q = \vec{x}^*. \tag{2.30}$$

The point $\vec{x}^*$ denotes the unique globally stable equilibrium of the mutation dynamics.

2.4 MATING

One of the problems that Charles Darwin could not solve was the following: under random mating and blending inheritance, the variability in a population should rapidly decline. Yet it was clear that variability was needed for natural selection. If variability disappears, then natural selection has nothing upon which to act. Suppose there is a distribution of body size in a population. If children inherit the average body size of their parents, then after some time everybody is the same size. Under these circumstances, how can natural selection affect changes in body size?

The first part of the solution is that inheritance (on the level of genes) is not blending but particulate, as had been discovered by Gregor Mendel and published in 1866. That is, individuals have discrete genotypes that get reshuffled, not blended, during mating. Mendel's work was unknown to Darwin. The second step was a simple mathematical analysis, which was performed by the British mathematician G. H. Hardy, who was proud never to have done anything useful (= applied) in his life, only to have his name forever associated with a highly useful and very applied concept in population genetics. Moreover, Hardy's brief calculation was generalized by the German physician Wilhelm Weinberg.

Consider an infinitely large population of a diploid organism with two sexes and random mating (a diploid organism has two copies of its genome; humans and many other animals are diploid). Let us look at one particular gene locus and assume there are two alleles, A_1 and A_2. The alleles are variants of the same gene and might differ in one or a few point mutations. (Point mutation means that only one single base of the DNA sequence is changed.)

There are 3 different genotypes: A_1A_1, A_1A_2, A_2A_2. Let us denote their frequencies in the population by x, y, and z, respectively. Denote by p and q the frequencies of alleles A_1 and A_2. We have $x + y + z = 1$ and $p + q = 1$. Moreover,

$$
\begin{aligned}
p &= x + \frac{1}{2}y \\
q &= z + \frac{1}{2}y
\end{aligned}
\tag{2.31}
$$

Let us now assume random mating. In the next generation, the genotype frequencies are given by

$$
\begin{aligned}
x' &= p^2 \\
y' &= 2pq \\
z' &= q^2
\end{aligned}
\tag{2.32}
$$

For the allele frequencies in the next generation we have again

$$
\begin{aligned}
p' &= x' + \frac{1}{2}y' \\
q' &= z' + \frac{1}{2}y'
\end{aligned}
\tag{2.33}
$$

Combining (2.32) and (2.33), we observe that

$$
p' = p \qquad q' = q \tag{2.34}
$$

Therefore the allele frequencies remain unchanged from one generation to the next. Moreover, combining (2.32) and (2.34), we observe

$$
\begin{aligned}
x' &= p'^2 \\
y' &= 2p'q' \\
z' &= q'^2
\end{aligned}
\tag{2.35}
$$

From the first generation on, the genotype frequencies can be directly derived from the allele frequencies. Note that equation (2.35) need not hold for the initial genotype and allele frequencies. The Hardy-Weinberg law (expressed by equations 2.34 and 2.35) can be generalized to n alleles.

In summary, the Hardy-Weinberg law states that particulate inheritence preserves variation within a population under random mating.

SUMMARY

- Evolution requires populations of reproducing individuals.
- Asexual reproduction leads to exponential population growth (which will eventually be checked by resource limitation).
- Simple models of population growth in discrete time can give rise to very complicated dynamics.
- Selection arises when different types of individuals reproduce at different rates.
- Normally, the faster-reproducing (fitter) individual outcompetes the slower reproducing (less fit) individual.
- If there are many different types, then selection dynamics can lead to "survival of the fittest." All others become extinct.
- Sublinear growth rates lead to coexistence, "survival of all."
- Superlinear growth rates prevent invasion of a new type and thereby lead to "survival of the first."
- Mutation arises when reproduction is not perfectly accurate.
- Mutation promotes coexistence of different types.
- Asymmetric mutation can lead to selection even if all individuals have the same reproduction rate.
- The Hardy-Weinberg law states that random mating preserves genetic variation within a population.

3 FITNESS LANDSCAPES AND SEQUENCE SPACES

GENOMES ARE SEQUENCES of the four-letter alphabet A, T, C, G, denoting the nucleotides adenine, thymine, cytosine, and guanine. All living cells use double-stranded DNA to carry their genomic information. Many viruses also use DNA, but some viruses encode their genome in form of RNA. The genome length of organisms varies greatly, ranging from about 10^4 nucleotides for small viruses, to 10^6 for bacteria to 3×10^9 for humans. Curiously, newts and lungfish "need" an even larger genome than do humans (19×10^9 and 140×10^9, respectively). The evolutionary dynamics of genome size and genome organization is a fascinating topic.

If a cell wants to produce a particular protein, then the DNA of the corresponding gene is "transcribed" into messenger RNA (mRNA), which is in turn "translated" into protein. The transcription is done by particular enzymes called DNA-dependent RNA polymerases. The translation is performed by a complicated arrangement of RNA and proteins called ribosomes. The words "transcription" and "translation" were invented by the mathematician John von Neumann when he calculated how to build a self-reproducing machine. He came up with an architecture equivalent to the organization of cells some decades before molecular biology had been invented.

RNA also uses a four-letter alphabet, A, U, C, G. Thymine is replaced by uracil. Furthermore, the sugar backbone of RNA has an additional -OH (hydroxy) group, which makes the molecule less stable and more dynamic. DNA is a stable carrier of information. RNA also carries information, but in addition some RNAs have enzymatic activity.

Proteins consist of 20 amino acids. Each amino acid is encoded by a sequence of three letters of the RNA alphabet. This genetic code is essentially the same for all living cells, ranging from bacteria to humans to newts. Hence the genetic code is believed to have originated only once: in the first cell that is ancestor to all existing cells. A 4-letter alphabet generates 64 possible sequences of length 3. Since there are only 20 amino acids, the genetic code is redundant: some amino acids are encoded by more than one sequence. Some sequences are used to signal the end of the transcription process. We see that molecular biology adds a precise information-theoretic perspective to evolutionary dynamics.

3.1 SEQUENCE SPACE

In the green hills of Sussex lived an imaginative theoretical biologist, John Maynard Smith, who once pictured all proteins (of a certain length) arranged in such a way that nearest neighbors differed by a single amino acid. This was the origin of what we call "sequence space" as a concept in the human mind.

Let us consider all proteins of the modest length 100. Each position of the protein sequence is filled by one of 20 amino acids. Hence, this space has 100 dimensions and in total 20^{100} points. This number corresponds to 10^{130} proteins. In contrast, there are only some 10^{80} estimated protons in our universe. Nor is there any reason for us to consider only proteins of length 100; some proteins are much longer. We conclude there are many more possible proteins than available protons and hence evolution so far has and, for the remaining 10^{30} years that constitute the lifetime of our protons, will only explore a vanishingly small subset of all possible proteins.

What is true of proteins is also true of genes and genomes. We can imagine all nucleotide sequences of a certain length arranged in a way that nearest neighbors differ in one position. For sequence length L this generates a lattice

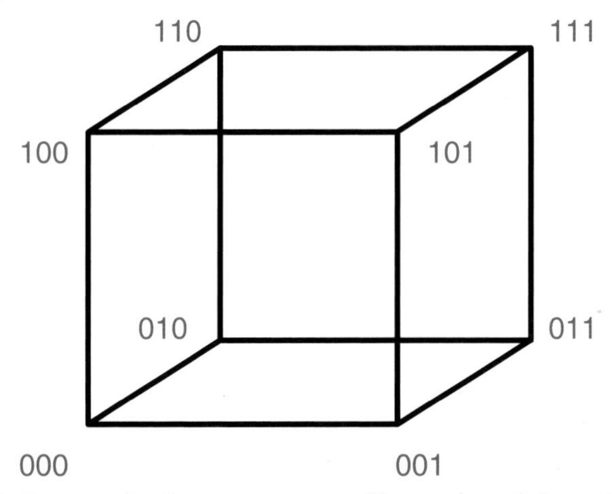

Figure 3.1 Genomes live in sequence space. The number of dimensions is given by the length of the genome. Small viruses live in 10,000 dimensions. Humans live in about 3 billion dimensions.

in an L-dimensional space. In each dimension there are 4 discrete possibilities. Hence there are 4^L possible sequences.

For writing computer programs, it is often convenient to use binary sequences, the fundamental strings of silicon thoughts. Moreover, everything from Shakespeare to *E. coli* can be encoded in binary sequences. For length L there are 2^L possibilities. In Figure 3.1, the binary sequence space for $L = 3$ is shown. The distance between 000 and 010 is one. The distance between 000 and 011 is 2 (and not $\sqrt{2}$). Hence sequence space is characterized not by a Euclidean metric but by a so-called Hamming metric or Manhattan metric. In Manhattan, if you are on 5th Avenue and 51st Street it takes 2 blocks to go to 6th Avenue and 52nd Street, not $\sqrt{2}$ blocks. This metric was introduced by Richard Hamming in information theory.

Let us compare the binary sequence space of length $L = 300$ with a three-dimensional cubic lattice containing the same number of points. There are

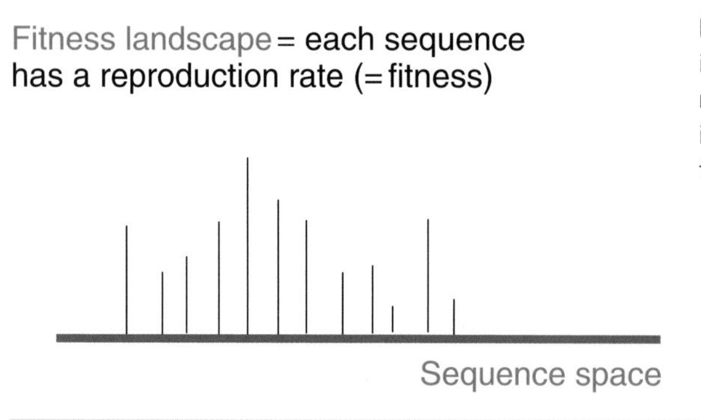
Fitness landscape = each sequence has a reproduction rate (= fitness)
Sequence space

Figure 3.2 The fitness landscape is a high-dimensional mountain range. Each genome (= each point in sequence space) gets assigned a fitness value.

$2^{300} \approx 10^{90}$ points. Imagine nearest neighbors are placed at a distance of 1 meter. The diagonal of the three dimensional cubic lattice has a length of about 10^{30} meters, which corresponds to about 10^{14} light years. In contrast, the longest distance in the L-dimensional hypercube is only 300 meters. Thus sequence space is characterized by short distances, but many dimensions. It is not far to move from one sequence to another, but there are many possible steps that lead in wrong directions. Evolution is a trajectory through sequence space. This trajectory needs an efficient guide.

3.2 FITNESS LANDSCAPES

The American population geneticist Sewall Wright invented the concept of a "fitness landscape" in the 1930s, but Manfred Eigen and Peter Schuster, collaborating in the 1970s, combined fitness landscape with sequence space. Consider a function that assigns to each genomic sequence a fitness value. Hence we build a mountain range on the foundation of an L-dimensional sequence space (Figure 3.2). This mountain range has $L + 1$ dimensions. The evolutionary process of mutation and selection explores this hyper-alpine mountain range.

The genomic sequence represents the genotype of an organism. The phenotype of an organism is given by its shape, behavior, performance and any kind of ecological interaction. The phenotype determines the fitness (reproductive rate) of the organism. There is a mapping from genotype to phenotype.

A quasispecies is a population of
reproducing RNA or DNA molecules

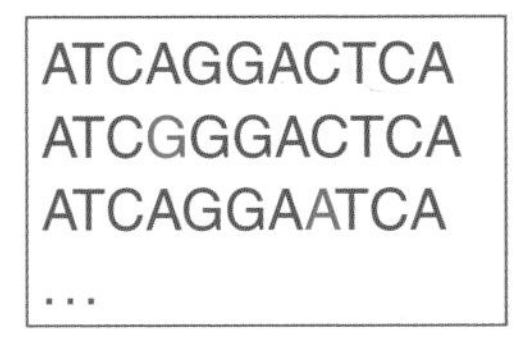

0000110011000110
0000110011100110
1000110011000010
...

4-nucleotide alphabet Binary alphabet

Figure 3.3 The ensemble of genomes of a natural population form a quasispecies: the genomes of different individuals are similar but not identical. Biology has chosen a four-letter alphabet consisting of the nucleotides A, T, C, and G for its genes. Most in silico evolution uses a binary alphabet for convenience. Sequence differences (mutations) are shown in red.

There is another mapping from phenotype to fitness. The fitness landscape is a convolution of these two mappings. It is a direct mapping from genotype to fitness.

The fitness landscape of certain problems can be determined experimentally. For example, HIV can generate point mutations that confer drug resistance. The relative growth rate of such mutants can be determined by in-vitro assays. In general, however, to understand the relationship between genotype, phenotype, and fitness is an extremely complicated problem. Much of biology, including developmental biology, molecular biology, post-genomics, and proteomics, is devoted to this very task.

3.3 THE QUASISPECIES EQUATION

A quasispecies is an ensemble of similar genomic sequences generated by a mutation-selection process (Figure 3.3). The term was introduced by the chemists Manfred Eigen and Peter Schuster. In chemistry the word "species" refers to an ensemble of identical molecules, for example, the species of all water molecules. But the species of all RNA molecules does not contain identical sequences, and therefore the term "quasispecies" was coined. Biologists are sometimes confused by this expression, because they relate it to the concept of a biological species.

We stay with binary sequences for convenience. We note that any genomic or other information can be encoded by binary sequences. Consider all binary sequences of length L. Enumerate all those sequences by $i = 0, 1, 2, \ldots, n$ where $n = 2^L - 1$. A natural enumeration is obtained if the sequence represents the binary description of the corresponding integer. For example, let $L = 4$. The sequence 0000 corresponds to $i = 0$, the sequence 0001 to $i = 1$, the sequence 0010 to $i = 2, \ldots$, the sequence 1111 to $i = 15$.

Imagine an infinitely large population of organisms, each carrying a genome of length L. Denote by x_i the relative abundance (= frequency) of those organisms that contain genome i. We have $\sum_{i=0}^{n} x_i = 1$. The genomic structure of the population is given by the vector $\vec{x} = (x_0, x_1, \ldots, x_n)$.

Denote by f_i the fitness of genome i. It is a non-negative real number. Thus genomes of type i are being reproduced at rate f_i. The fitness landscape is given by the vector $\vec{f} = (f_0, f_1, \ldots, f_n)$. The average fitness of the population, $\phi = \sum_{i=0}^{n} x_i f_i$, is the inner product of the vectors $\vec{x}$ and $\vec{f}$. We have $\phi = \vec{x}\vec{f}$.

During replication of a genome, mistakes can happen. The probability that replication of genome i results in genome j is given by q_{ij}. Here we again meet the mutation matrix $Q = [q_{ij}]$ of section 2.3. We remember that Q is a stochastic matrix: it has as many rows as columns; each entry is a probability, which means a number between 0 and 1; each row sums to one, $\sum_{j=0}^{n} q_{ij} = 1$.

The quasispecies equation (Figure 3.4) is given by

$$\dot{x}_i = \sum_{j=0}^{n} x_j f_j q_{ji} - \phi x_i \qquad i = 0, \ldots, n \tag{3.1}$$

Sequence i is obtained by replicating any sequence j at rate f_j times the probability that replication of sequence j generates sequence i. Each sequence is removed at rate ϕ to ensure that the total population size remains constant, $\sum_{i=0}^{n} x_i = 1$. Thus quasispecies dynamics are defined on the simplex, S_n.

In the limiting case of completely error-free replication, Q becomes the identity matrix: all diagonal entries are one, all off-diagonal entries are zero.

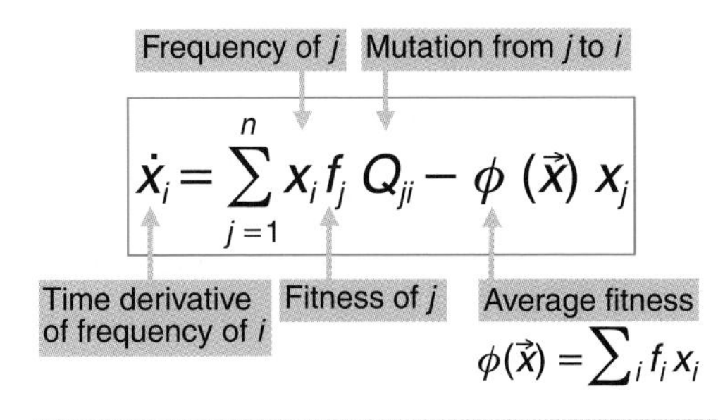

Figure 3.4 The quasispecies equation, formulated by Manfred Eigen and Peter Schuster, is one of the most important equations in theoretical biology. It describes the mutation and selection of an infinitely large population on a constant fitness landscape.

Consider an initial condition in the interior of the simplex, defined by $x_i > 0$ for all i. The quasispecies will converge to a homogeneous population that consists only of the fittest sequence. If $f_0 > f_i$ for all $i \neq 0$, then the stable equilibrium is given by $x_0 = 1$ and $x_i = 0$ for $i \neq 0$. If there are no errors, then the quasispecies equation (3.1) reduces to the selection equation (2.16) of section 2.2.1.

Let us now assume that errors occur. This means that (at least some) off-diagonal entries of Q are not zero. In many realistic contexts, the matrix Q is irreducible, which means it is possible to find a sequence of mutations from any one genome i to any other genome j. Furthermore, let $f_i > 0$ for at least some i. In this case, the quasispecies equation admits a single, globally stable equilibrium, $\vec{x}^*$, in the simplex S_n.

The equilibrium quasispecies, $\vec{x}^*$, does not necessarily maximize the average fitness ϕ. Consider again a fitness landscape with the property $f_0 > f_i$ for all $i \neq 0$. Then the population consisting only of sequence 0 will have a higher fitness than the equilibrium population $\vec{x}^*$. Thus, mutations reduce the average fitness at equilibrium.

Observe that (3.1) is a nonlinear differential equation. The term $-\phi x_i$ is of second order. Linear differential equations can always be solved, but nonlinear differential equations normally cannot be solved. This means for nonlinear differential equations the trajectories cannot always be written as explicit functions of time. The quadratically nonlinear quasispecies equation (3.1), however, can be solved as follows. First, define

$$\psi(t) = \int_0^t \phi(s) ds. \tag{3.2}$$

Note that

$$\dot{x}_i + \phi x_i = e^{-\psi} \frac{d(x_i e^{\psi})}{dt}. \tag{3.3}$$

Let us define

$$X_i(t) = x_i(t) e^{\psi(t)}. \tag{3.4}$$

Now $X_i(t)$ is given by the linear equation

$$\dot{X}_i = \sum_{j=0}^{n} X_j f_j q_{ji} \qquad i = 0, \ldots, n \tag{3.5}$$

This system of linear differential equations describes exponential growth of all the members of the quasispecies. The linear system (3.5) can be solved using standard techniques. Notice also that

$$X = \sum_{i=0}^{n} X_i = \left(\sum_{i=0}^{n} x_i \right) e^{\psi} = e^{\psi}. \tag{3.6}$$

This means, from equation (3.4), that we can write $x_i = X_i / X$, which in turn means that X_i can be interpreted as the absolute abundance of individuals with genome i. Also note that X, the total population size, grows as

$$\dot{X} = \dot{\psi} e^{\psi} = \phi X. \tag{3.7}$$

Therefore the total population size grows exponentially at a rate that is given by the average fitness, ϕ, of the population.

Let us combine the fitness landscape, $\vec{f}$, and the mutation matrix, Q, to obtain the mutation-selection matrix,

$$W = [w_{ji}] = [f_j q_{ji}]. \tag{3.8}$$

Quasispecies dynamics are determined by the properties of the matrix, W. In vector notation the quasispecies equation can be written as

$$\dot{\vec{x}} = \vec{x}W - \phi\vec{x}. \tag{3.9}$$

Hence the equilibrium of quasispecies dynamics is given by

$$\vec{x}W = \phi\vec{x}. \tag{3.10}$$

This is a standard eigenvalue problem. The average fitness, ϕ, is the largest eigenvalue of the matrix W. The left-hand eigenvector associated with this eigenvalue, with the proper normalization $\sum_{i=1}^{n} x_i = 1$, provides the equilibrium structure of the quasispecies. Generically, there is a unique and globally stable equilibrium.

3.4 A MUTATION MATRIX FOR POINT MUTATIONS

During the replication of a DNA or RNA genome, many types of mutational events can occur. "Point mutations" describe the change of one base for another. "Insertions" denote the addition of a string of bases to the existing sequence. "Deletions" characterize the reverse process, the loss of a string of bases. "Recombination" means that genetic material can be exchanged between two sequences. Here we will only deal with point mutations of binary sequences.

Let us consider the set of all sequences of a given length, L. The Hamming distance, h_{ij}, counts the number of positions that differ between sequences i and j. For example, the Hamming distance between the sequences 1010 and 1100 is two. Denote by u the probability that a mutation occurs in a specific position. Thus $1 - u$ is the probability that the mutation is copied correctly. We can write the probability that replication of sequence i results in sequence j as

$$q_{ij} = u^{h_{ij}}(1-u)^{L-h_{ij}}. \tag{3.11}$$

Hence a mutation has to occur in as many positions as differ between the sequences i and j, which is precisely the Hamming distance, h_{ij}. No mutation must occur in the remaining $L - h_{ij}$ positions.

Equation (3.11) is an elegant description of a mutation matrix that allows point mutations among binary sequences of constant length. It is assumed that the point mutation rate, u, is the same for all positions. It is further assumed that a mutation in one position is independent of a mutation in another position. Hence one error does not increase the probability of another error. There are no insertions and no deletions. All of these restrictions can be relaxed in principle, but doing so will lead to considerable complexity.

Let us use mutation matrix (3.11) to describe the human immunodeficiency virus as an example. The point mutation rate of HIV is approximately $u = 3 \times 10^{-5}$. The genome length of HIV is $L = 10^4$. Therefore the probability that the whole HIV genome is replicated without mutation is given by $(1 - u)^L \approx 0.74$. The probability that replication of the HIV genome results in a sequence that differs in one arbitrary position is given by $Lu(1 - u)^{L-1} = 0.22$. The probability that a particular one-error mutant, for example one that confers drug resistance or immune escape, is being produced is given by $u(1 - u)^{L-1} = 2.2 \times 10^{-5}$. If 10^9 newly infected cells are being produced each day, then any particular one-error mutant will arise 22,000 times each day. This number signifies the enormous potential of HIV (or other viruses or microbes) to escape from selection pressures that are meant to control them. We will revisit this topic in Chapter 10.

3.5 ADAPTATION IS LOCALIZATION IN SEQUENCE SPACE

The quasispecies equation (3.1) describes the movement of a population through sequence space. The quasispecies "feels" gradients in the mountain range of the fitness landscape. It attempts to climb uphill and reach local or global peaks (Figure 3.5). What are the conditions that this evolutionary walk will be successful? One such condition is the error threshold.

If the mutation rate u is too high, then the ability of the quasispecies to climb uphill and to remain on top of a mountain peak is impaired. In fact, we can show that for many natural fitness landscapes there is a maximum mutation rate, u_c, that is still compatible with adaptation. If the mutation rate exceeds this value, $u > u_c$, then adaptation is not possible.

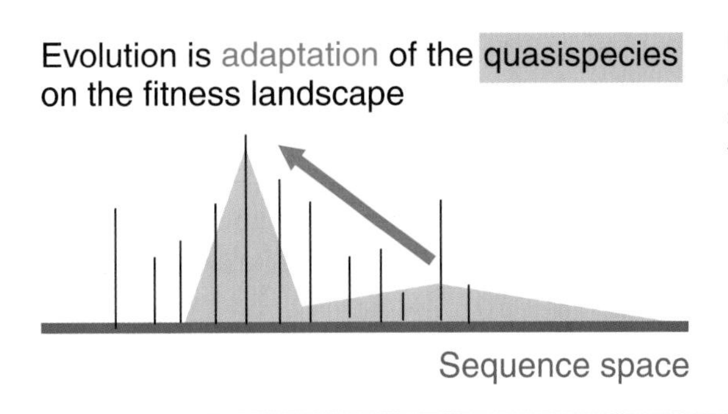

Evolution is adaptation of the quasispecies on the fitness landscape
Sequence space

Figure 3.5 Quasispecies love to climb mountains in high-dimensional spaces. The higher they get, the fitter they are. Adaptation means to go up.

Adaptation means that the quasispecies is able to find peaks in the fitness landscape and stay there. Suppose the fitness landscape contains only one peak. If the mutation rate is sufficiently low, then the equilibrium solution of equation (3.1) describes a quasispecies that is centered on this peak. Most sequences resemble the type with maximum fitness or nearby mutants. Sequences that are far away from the peak will have a very low frequency. (In population genetics, frequency means relative abundance.) We say the quasispecies is adapted to this peak. Similarly, we can say that the quasispecies distribution is localized at this peak. Adaptation means localization in sequence space. When the mutation rate of a quasispecies is zero, it contains only sequences with maximum fitness. When the mutation rate is very small, the quasispecies distribution is very narrow. As the mutation rate increases, the quasispecies distribution widens. There is a critical mutation rate, u_c, beyond which the equilibrium quasispecies no longer "feels" the peak. The quasispecies is no longer localized around the peak. Adaptation is lost. Strictly speaking, a well-defined "phase transition" from a localized to a delocalized state only occurs for infinite sequence length, but the phenomenon is striking already for binary sequences of length $L = 10$.

The maximum mutation rate, u_c, that is compatible with adaptation is called the "error threshold." Not all fitness landscapes have error thresholds. Narrow peaks of finite height have error thresholds. If a peak is so broad that most sequences in the sequence space are within the slopes of the peak, then an error threshold need not occur.

Quasispecies have a tendency to climb uphill. Starting from some random initial condition, $\vec{x}(0)$, the quasispecies equation (3.1) will tend to increase the average fitness, ϕ. But it is also easy to construct a counterexample. Suppose a certain sequence has maximum fitness, while all other sequences have lower fitness. If we start with a population that contains only the sequence with maximum fitness, then equation (3.1) will reduce the average fitness ϕ until an equilibrium between mutation and selection, a so-called mutation-selection balance has been reached.

Calculating the error threshold, u_c, for complex fitness landscapes is difficult, but the following simple fitness landscape provides the crucial insight. Consider all binary sequences of length L. The all-zero sequence, $00\ldots0$, has the highest fitness given by $f_0 > 1$. All other sequences have fitness 1. The all-zero sequence is sometimes called the "master sequence" or the wild type, while all other sequences are called "mutants."

The probability that the master sequence produces an exact copy of itself is given by $q = (1-u)^L$. The probability that the master sequence generates any mutant is given by $1-q$. The trick is to neglect the back mutation from the mutants to the master sequence. With this assumption the quasispecies equation (3.1) becomes

$$\begin{aligned} \dot{x}_0 &= x_0(f_0 q - \phi) \\ \dot{x}_1 &= x_0 f_0(1-q) + x_1 - \phi x_1 \end{aligned} \tag{3.12}$$

Here x_0 is the frequency of the master sequence, while x_1 is the sum of the frequencies of all the mutants. Clearly, $x_0 + x_1 = 1$. The average fitness is given by $\phi = f_0 x_0 + x_1$. System (3.8) collapses to a single equation

$$\dot{x}_0 = x_0[f_0 q - 1 - x_0(f_0 - 1)]. \tag{3.13}$$

If $f_0 q < 1$, then x_0 will converge to zero; the fittest sequence cannot be maintained in the population. If $f_0 q > 1$, then x_0 will converge to

$$x_0^* = \frac{f_0 q - 1}{f_0 - 1}. \tag{3.14}$$

Hence, the error threshold is given by

$$f_0 q > 1. \tag{3.15}$$

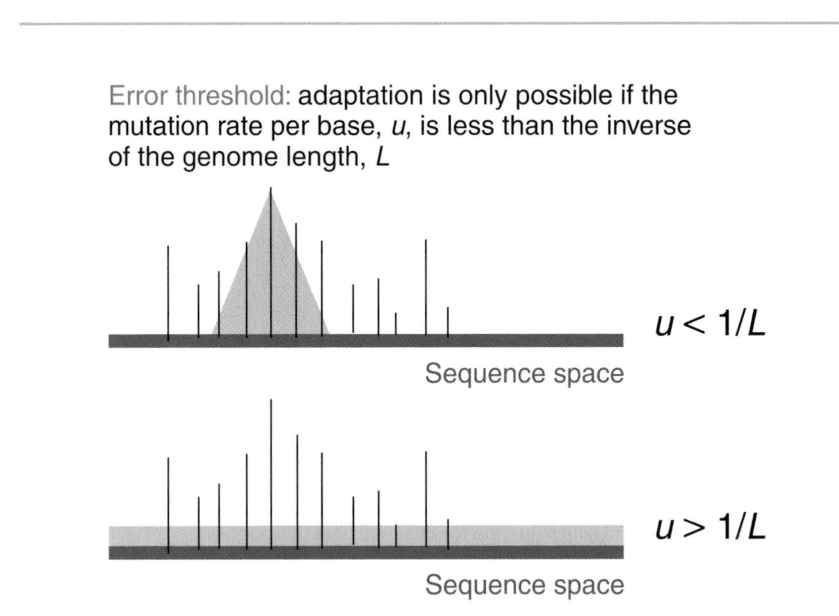
Error threshold: adaptation is only possible if the mutation rate per base, u, is less than the inverse of the genome length, L
u < 1/L
Sequence space
u > 1/L
Sequence space

Figure 3.6 Error threshold: a quasispecies can only maintain a peak in a fitness landscape if the mutation rate is less than the inverse of the genome length. This is a very general and beautiful result that must hold for any living organism. The beauty is not spoilt by two qualifying remarks that are necessary: (i) the genome length, L, has to be defined properly to include only those positions that affect fitness and (ii) there are some pathological landscapes where a peak can be maintained beyond the error threshold, for example if the peak is "infinitely" high or so wide that its presence can be felt by the majority of all possible sequences.

This inequality can be rewritten as $\log f_0 > -L \log(1-u)$. For small mutation rates, $u \ll 1$, we have $\log(1-u) \approx -u$. Therefore we obtain the condition

$$u < \frac{\log f_0}{L}. \tag{3.16}$$

If the fitness advantage of the master sequence is not too large and not too small, then $\log f_0$ is approximately 1. Now the error-threshold condition reduces to

$$u < 1/L. \tag{3.17}$$

Hence the maximum mutation rate that is still compatible with adaptation has to be less than the inverse of the genome length (Figure 3.6). In other

Table 3.1 Genome length (in bases), mutation rate per base, and mutation rate per genome for organisms ranging from DNA viruses to humans

Organism	Genome length in bases	Mutation rate per base	Mutation rate per genome
RNA viruses			
Lytic viruses			
$Q\beta$	4.2×10^3	1.5×10^{-3}	6.5
Polio	7.4×10^3	1.1×10^{-4}	0.84
VSV	1.1×10^4	3.2×10^{-4}	3.5
Flu A	1.4×10^4	7.3×10^{-6}	0.99
Retroviruses			
SNV	7.8×10^3	2.0×10^{-5}	0.16
MuLV	8.3×10^3	3.5×10^{-6}	0.029
RSV	9.3×10^3	4.6×10^{-5}	0.43
Bacteriophages			
M13	6.4×10^3	7.2×10^{-7}	0.0046
λ	4.9×10^4	7.7×10^{-8}	0.0038
T2 and T4	1.7×10^5	2.4×10^{-8}	0.0040
E. coli	4.6×10^6	5.4×10^{-10}	0.0025
Yeast *(S. cerevisiae)*	1.2×10^7	2.2×10^{-10}	0.0027
Drosophila	1.7×10^8	3.4×10^{-10}	0.058
Mouse	2.7×10^9	1.8×10^{-10}	0.49
Human *(H. sapiens)*	3.5×10^9	5.0×10^{-11}	0.16

Sources: Drake (1991, 1993) and Drake et al. (1998).

Note: Most organisms have a mutation rate per genome which is less than one, as predicted by the error threshold theory. Why $Q\beta$ and VSV have such a high mutation rate is at present unexplained.

words, the genomic mutation rate, uL, has to be less than one. In fact, this condition holds for most living organisms for which mutation rates have been measured (Table 3.1). For eukaryotes, the genome length L in this context should actually be defined as the total number of bases in the coding and regulatory regions of the DNA.

3.6 SELECTION OF THE QUASISPECIES

The following remarkable observation was first made by Peter Schuster and Jörg Swetina. Consider a fitness landscape that contains a high but narrow

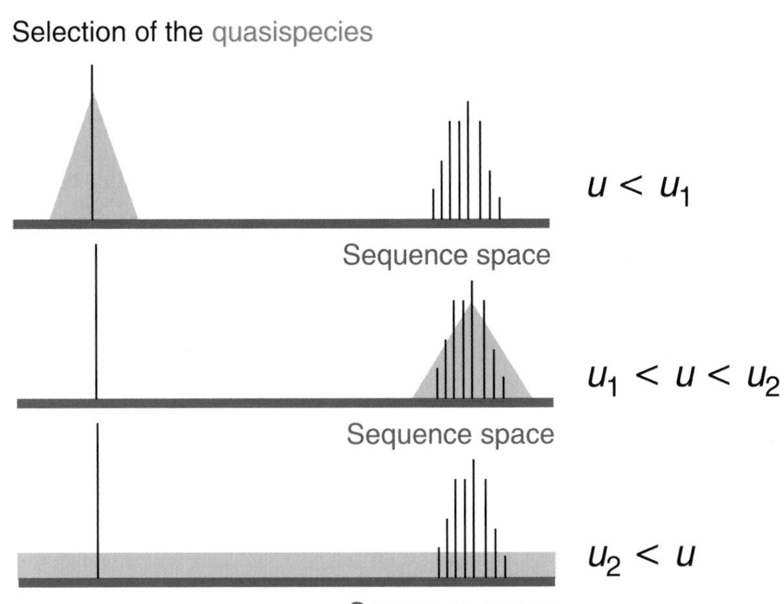

Figure 3.7 Consider a fitness landscape with two peaks. One is high but narrow, the other low but wide. If the mutation rate, u, is less than a critical value, u_1, then the higher peak is selected, indicated in blue. If the mutation rate, u, is greater than u_1, but less than the error threshold, u_2, then the lower peak is selected. If the mutation rate is greater than the error threshold, u_2, then neither peak can be maintained. For a given mutation rate, selection chooses the equilibrium quasispecies with maximum average fitness. "Survival of the fittest" is replaced by "survival of the quasispecies."

peak and in some distance a lower but broader peak (Figure 3.7). If the mutation rate is very small, the quasispecies at equilibrium will be centered around the higher peak. As the mutation rate increases, there is a sharp transition, and the quasispecies moves from the higher to the lower peak. The intuitive explanation is the following: for very small mutation rates only the maximum fitness matters, but for somewhat higher mutation rates the fitness of the neighboring sequences is also important. The second peak has a lower maximum fitness, but a broader ensemble of relatively good close-by neighbors. The first peak is like a brilliant person working alone, the second peak consists of a less brilliant person surrounded by a good team.

If mutation rates are sufficiently small, the quasispecies centered around the narrow peak has maximum fitness. But when mutation rates are higher, the quasispecies centered around the broader peak has maximum fitness. Beyond the error threshold neither peak can be maintained.

We conclude that selection does not always smile upon the fittest. For any given mutation rate, however, selection chooses the equilibrium distribution (the quasispecies) with maximum average fitness. "Selection of the fittest" is replaced by "selection of the quasispecies."

SUMMARY

- A quasispecies is a population of similar genomes.
- Quasispecies are formed by a mutation-selection process.
- In sequence space, all possible genomes of a certain length are arranged such that nearest neighbors differ by one point mutation. All sequences of length, L, can be arranged in a lattice that is embedded in an L-dimensional space.
- A fitness landscape is formed by assigning fitness values (reproductive rates) to all sequences. A fitness landscape is a high-dimensional mountain range over sequence space.
- Quasispecies live in sequence space and explore the fitness landscape.
- Quasispecies climb upward in the fitness landscape.
- The quasispecies equation describes deterministic evolutionary dynamics in terms of mutation and constant selection acting on an infinitely large population.
- Generically, the quasispecies equation has one globally stable equilibrium.
- At this equilibrium, the quasispecies consists not of solely the fittest genome but instead of a distribution of genomes in a mutation-selection balance.
- It is possible that this distribution does not contain the fittest genome at all. Hence "survival of the fittest" is replaced by "survival of the quasispecies."

- Adaptation is localization in sequence space. This is only possible if the mutation rate is below the error threshold.
- The error threshold states that the maximum possible mutation rate (per base) must be less than the inverse of the genome length (in bases).

EVOLUTIONARY GAMES 4

EVOLUTIONARY GAME THEORY means that the fitness of individuals is not constant, but depends on the relative proportions (frequencies) of the different phenotypes in the population: fitness is frequency dependent. Evolutionary game theory is the generic approach to evolutionary dynamics and contains as a special case constant selection.

Game theory was invented by John von Neumann and Oskar Morgenstern. They wanted to design a mathematical theory to study human behavior in strategic and economic decisions. Von Neumann was a Hungarian-born mathematician working at the Institute for Advanced Study, where he invented and revolutionized several fields of mathematics. We have already encountered him in Chapter 3 in connection with the terms "translation" and "transcription," which he invented when thinking about how to conceive of a machine that could reproduce itself. He built the first computer that held the program for the calculation in its memory rather than in its hardware. Incidentally, one of the first projects that this computer did in its spare time was a mathematical simulation of an evolutionary system.

John Nash came as a mathematics Ph.D. student to Princeton University with a remarkably short letter of reference: "This man is a genius." Nash invented a simple but important concept in game theory, which is now called the "Nash equilibrium." A Nash equilibrium is very similar to an evolutionarily stable strategy (ESS). Both concepts are important for evolutionary dynamics. Nash's Ph.D. thesis led to a single-page paper in the *Proceedings of the National Academy of Sciences USA* (1950), which earned him a Nobel Prize in economics in 1994.

William Hamilton and Robert Trivers were among the first to use game theoretical ideas in biology, but the field of evolutionary game theory was founded by the work of John Maynard Smith and others, including Peter Taylor, Josef Hofbauer, and Karl Sigmund.

John Maynard Smith and George Price introduced game theory to evolutionary biology and population thinking to game theory in a paper published in *Nature* in 1973. Traditional game theory typically analyzes an interaction between two players, for example, you and me. The question is how you can maximize your payoff in a game, given that you do not know what I will do. The concept of rationality comes into play. You may assume that I will act in order to maximize my payoff. Given this assumption, you will then behave to maximize your payoff. But nothing guarantees that I will behave rationally, and in fact many experimental games show that humans do not behave rationally.

Evolutionary game theory does not rely on rationality. Instead it considers a population of players interacting in a game. Individuals have fixed strategies. They interact randomly with other individuals. The payoffs of all these encounters are added up. Payoff is interpreted as fitness, and success in the game is translated into reproductive success. Strategies that do well reproduce faster. Strategies that do poorly are outcompeted. This is straightforward natural selection.

In Figure 4.1 we see two phenotypes. A can move while B cannot. A pays a certain cost for the ability to move, but also gains the associated advantage. Suppose the cost-benefit analysis leads to a fitness of 1.1 for A compared to a fitness of 1 for B. In this setting, fitness is constant, and A will certainly outcompete B. But imagine that the advantage of being able to move is large when few others are on the road, but diminishes as the highways get blocked up. In this case, the fitness of A is not constant, but is a declining function

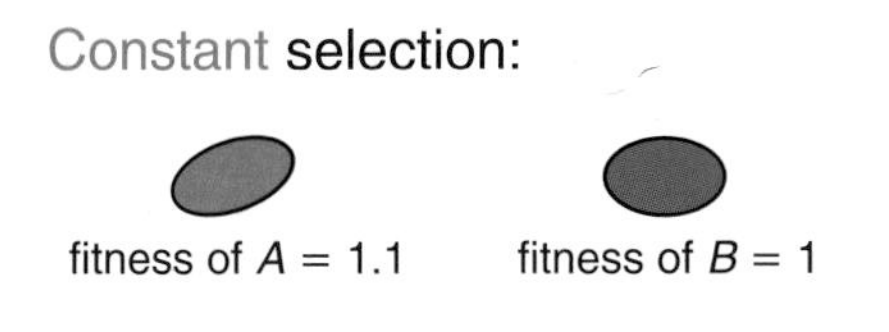

Frequency-dependent selection:

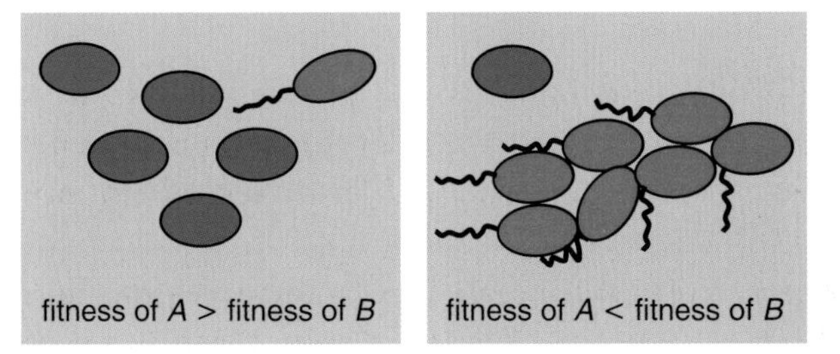

Figure 4.1 Constant selection means fitness neither depends on the composition of the population nor changes over time. For example, A has constant fitness 1.1, while B has constant fitness 1. In contrast, frequency-dependent selection means that fitness does depend on the relative abundance (= frequency) of individual types. A has the ability to move. If few other cells are moving, then A has a larger fitness than B. But if many other cells "are on the road," this fitness advantage is reversed (in this hypothetical example).

of the frequency of A. A has a higher fitness than B when A is rare, but has a lower fitness than B when A is common. What is the outcome of such a selection process?

Let us formalize the general case of frequency-dependent selection between two strategies A and B. Denote by x_A the frequency of A and by x_B the frequency of B. The vector $\vec{x} = (x_A, x_B)$ defines the composition of the population. Denote by $f_A(\vec{x})$ the fitness of A and by $f_B(\vec{x})$ the fitness of B. The selection dynamics can be written as

$$\begin{aligned} \dot{x}_A &= x_A[f_A(\vec{x}) - \phi] \\ \dot{x}_B &= x_B[f_B(\vec{x}) - \phi] \end{aligned} \tag{4.1}$$

The average fitness is given by $\phi = x_A f_A(\vec{x}) + x_B f_B(\vec{x})$.

Because $x_A + x_B = 1$ at all times, we can introduce the variable x with $x_A = x$ and $x_B = 1 - x$. We can write the fitness functions as $f_A(x)$ and $f_B(x)$.

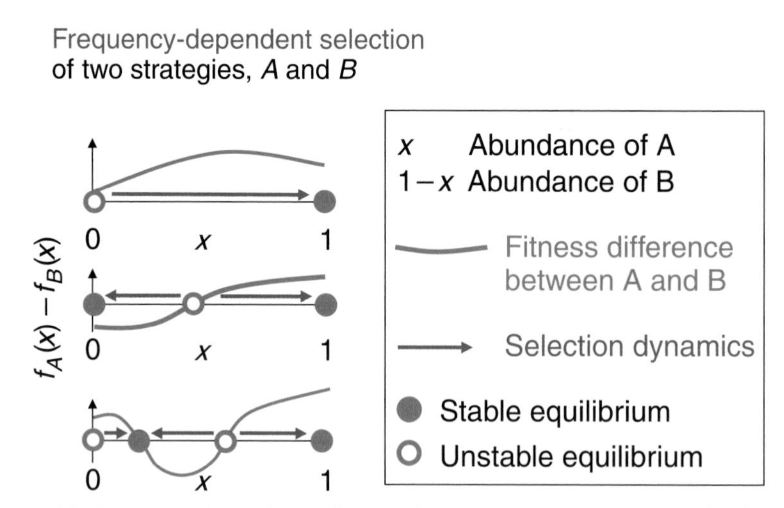

Figure 4.2 Frequency-dependent selection between two strategies can lead to multiple stable and unstable equilibria. The red curve, $f_A(x) - f_B(x)$, indicates the fitness difference between A and B as a function of the frequency, x, of A. If $f_A(x) - f_B(x) > 0$, then the frequency of A will increase, as shown by the blue arrows indicating the direction of selection dynamics. If $f_A(x) - f_B(x) < 0$, then the frequency of A will decline. Whenever $f_A(x) - f_B(x) = 0$, the frequency of A will not change. This condition characterizes an equilibrium of selection dynamics. If the slope of $f_A(x) - f_B(x)$ is positive at this point, then the equilibrium is unstable. If the slope is negative, then the equilibrium is stable. The points $x = 0$ and $x = 1$ are always equilibria. The equilibrium $x = 0$ is stable if $f_A(0) - f_B(0) < 0$. The equilibrium $x = 1$ is stable if $f_A(1) - f_B(1) > 0$.

System (4.1) leads to

$$\dot{x} = x(1 - x)[f_A(x) - f_B(x)]. \tag{4.2}$$

The equilibria of this differential equation are given by $x = 0$, $x = 1$, and all values $x \in (0, 1)$ that satisfy $f_A(x) = f_B(x)$. The equilibrium $x = 0$ is stable if $f_A(0) < f_B(0)$. Conversely, the equilibrium $x = 1$ is stable if $f_A(1) > f_B(1)$. An interior equilibrium, x^*, is stable if the derivatives of the functions f_A and f_B satisfy $f'_A(x^*) < f'_B(x^*)$. Figure 4.2 gives a graphical representation. There can be several stable and unstable equilibria in the interior of the interval $[0, 1]$.

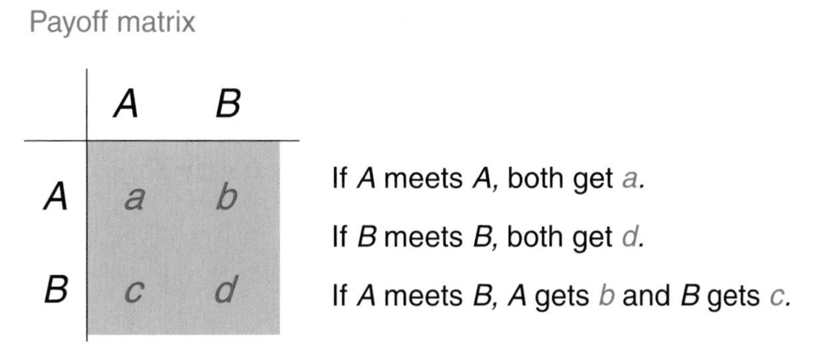

Payoff matrix
A
B
A
a
b
B
c
d
If A meets A, both get a.
If B meets B, both get d.
If A meets B, A gets b and B gets c.

Figure 4.3 A game between two strategies, A and B, is defined by a 2×2 payoff matrix.

4.1 TWO-PLAYER GAMES

Normally, a game with two strategies, A and B, is described by a payoff matrix

$$\begin{array}{c} \\ A \\ B \end{array} \begin{array}{c} \begin{array}{cc} A & B \end{array} \\ \begin{pmatrix} a & b \\ c & d \end{pmatrix} \end{array} \tag{4.3}$$

The payoff matrix is read in the following way: A gets payoff a when playing against A; A gets payoff b when playing against B; B gets payoff c when playing against A; B gets payoff d when playing against B (Figure 4.3).

The key idea of evolutionary game theory is to consider a population of A and B players and to equate payoff with fitness. If x_A is the frequency of A and x_B the frequency of B, then the expected payoff for A and B is respectively given by

$$\begin{aligned} f_A &= a x_A + b x_B \\ f_B &= c x_A + d x_B \end{aligned} \tag{4.4}$$

These equations assume that for each player the probability of interacting with an A player is x_A and the probability of interacting with a B player is x_B. Thus players meet each other randomly.

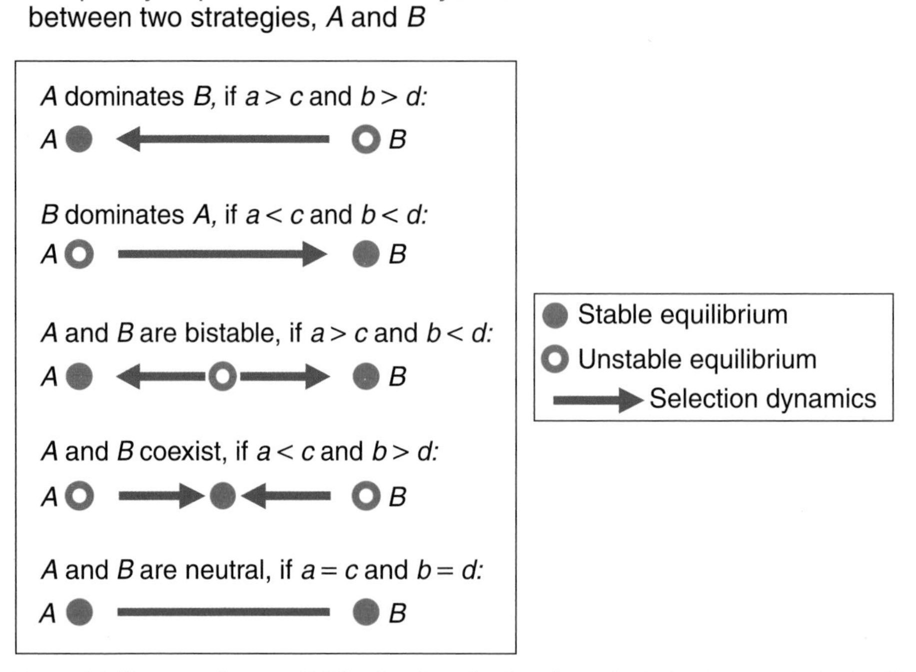

Figure 4.4 There are five possibilities for the selection dynamics between two strategies: (i) A dominates B, (ii) B dominates A, (iii) A and B are bistable, (iv) A and B coexist in a stable equilibrium, and (v) A and B are neutral variants of each other.

Let us now insert these linear fitness functions into equation (4.1). Let $x = x_A$ as before. We obtain

$$\dot{x} = x(1 - x)[(a - b - c + d)x + b - d]. \tag{4.5}$$

We want to classify the behavior of the nonlinear differential equation (4.5) depending on the ranking of the entries in the payoff matrix (4.1). There are five cases (Figure 4.4):

(i) A dominates B. This is the case if $a > c$ and $b > d$. If you were to play this game with another person, then no matter whether the other person chooses A or B it is best for you to play A. For a population of A and B players, this ranking of payoff values implies that the average fitness of A will always exceed that of B. Thus selection will favor A over B for any composition of the population. Selection will lead to the state where the whole population consists of A players, $x_A = 1$. More accurately, we say that A dominates B if $a \geq c$ and $b \geq d$, where at least one inequality must be strict.

(ii) B dominates A. This is the case if $a < c$ and $b < d$. This situation is the mirror image of case (i) with A and B exchanged. Again, more accurately, we say that A dominates B if $a \leq c$ and $b \leq d$, where at least one inequality must be strict.

(iii) A and B are bistable. This is the case if $a > c$ and $b < d$. If you were to play this game with another person, you should try to play the same choice as the other person. A is the best response for A. B is the best response for B. For the selection dynamics within a population, the outcome depends on the initial condition. There is an unstable equilibrium in the interior of the interval $[0, 1]$ given by $x^* = (d - b)/(a - b - c + d)$. If the initial condition, $x(0)$, is less than this value, $x(0) < x^*$, then the system will converge to all-B. If $x(0) > x^*$ then the system will converge to all-A.

(iv) A and B stably coexist. This is the case if $a < c$ and $b > d$. If you were to play this game with another person, you should always try to choose the opposite of what your opponent does. A is best response to B, and B is best response to A. A population of A and B players will converge to the interior, stable equilibrium

$$x^* = \frac{d - b}{a - b - c + d}. \tag{4.6}$$

(v) A and B are neutral. This is the case if $a = c$ and $b = d$. If you were to play this game with another person, then no matter what you chose you would always have exactly the same payoff as your opponent. Selection will not change the composition of the population. Any mixture of A and B is an equilibrium for selection dynamics.

4.2 THE NASH EQUILIBRIUM

The Nobel Prize–winning concept of a Nash equilibrium is defined in the following way. Imagine there is a game between two people. If both play a strategy that happens to be a Nash equilibrium, then neither person can deviate from this strategy and increase his payoff.

Consider the general payoff matrix between two strategies, A and B,

$$\begin{array}{c} \\ A \\ B \end{array} \begin{array}{c} \begin{array}{cc} A & B \end{array} \\ \begin{pmatrix} a & b \\ c & d \end{pmatrix} \end{array}$$

We have the following criteria:

(i) A is a strict Nash equilibrium if $a > c$.

(ii) A is a Nash equilibrium if $a \geq c$.

(iii) B is a strict Nash equilibrium if $d > b$.

(iv) B is a Nash equilibrium if $d \geq b$.

Let us explore the following game

$$\begin{array}{c} \\ A \\ B \end{array} \begin{array}{c} \begin{array}{cc} A & B \end{array} \\ \begin{pmatrix} 3 & 0 \\ 5 & 1 \end{pmatrix} \end{array} \tag{4.7}$$

If both players choose A, then one player can improve his payoff by switching to B. If both play B, then neither player can improve his payoff by switching to A. Hence B is a Nash equilibrium. Note that A is dominated by B. Note also that playing the Nash equilibrium, B, in this game leads to a lower payoff than playing the dominated strategy, A. The payoff matrix (4.7) is an example of the famous Prisoner's Dilemma game, which we will study in Chapter 5.

Consider the game

$$\begin{array}{c} \\ A \\ B \end{array} \begin{array}{c} \begin{array}{cc} A & B \end{array} \\ \begin{pmatrix} 3 & 1 \\ 5 & 0 \end{pmatrix} \end{array} \tag{4.8}$$

If both players choose A, then one player can improve his payoff by switching to B. If both play B, then again one player can improve his payoff by switching to A. Hence neither A nor B is a Nash equilibrium. This is an example of a Hawk-Dove game, which will be studied in section 4.6.

Finally, consider the game

$$\begin{array}{c} \\ A \\ B \end{array} \begin{array}{c} \begin{array}{cc} A & B \end{array} \\ \begin{pmatrix} 5 & 0 \\ 3 & 1 \end{pmatrix} \end{array} \tag{4.9}$$

If both players choose A, then neither player can improve his payoff by switching to B. If both play B, then again neither player can improve his payoff by switching to A. Hence both A and B are Nash equilibria.

4.3 EVOLUTIONARILY STABLE STRATEGY (ESS)

John Maynard Smith invented the important concept of an evolutionarily stable strategy when he was unaware of the Nash equilibrium in game theory. Imagine a large population of A players. A single mutant of type B is introduced. The game between A and B is given by the general payoff matrix (4.3), and the fitness functions are given by (4.4). What is the condition for selection to oppose the invasion of B into A?

Let us assume there is an infinitesimally small quantity of B invaders. Thus the frequency of B is ϵ. The frequency of A is $1-\epsilon$. For this population, the fitness of A is greater than the fitness of B if

$$a(1-\epsilon)+b\epsilon > c(1-\epsilon)+d\epsilon. \tag{4.10}$$

Canceling the ϵ terms, this inequality leads to

$$a > c. \tag{4.11}$$

If, however, it happens that $a=c$, then inequality (4.10) leads to

$$b > d. \tag{4.12}$$

Therefore we summarize: strategy A is ESS if either (i) $a>c$ or (ii) $a=c$ and $b>d$. This definition guarantees that selection will oppose the invasion of B into A. As we will see later, this concept holds only for infinitely large populations and for infinitesimally small quantities of the potential invader.

4.4 MORE THAN TWO STRATEGIES

Let us explore games with more than two strategies. The payoff for strategy S_i versus S_j is given by $E(S_i, S_j)$.

(i) Strategy S_k is a strict Nash equilibrium if

$$E(S_k, S_k) > E(S_i, S_k) \qquad \forall i \neq k \tag{4.13}$$

The symbol $\forall$ means "for all." Thus $\forall i \neq k$ reads "for all i not equal to k."

(ii) Strategy S_k is a Nash equilibrium if

$$E(S_k, S_k) \geq E(S_i, S_k) \qquad \forall i \tag{4.14}$$

(iii) Strategy S_k is ESS, if $\forall i \neq k$ we have either

$$E(S_k, S_k) > E(S_i, S_k) \tag{4.15}$$

or

$$E(S_k, S_k) = E(S_i, S_k) \quad \text{and} \quad E(S_k, S_i) > E(S_i, S_i). \tag{4.16}$$

Note that ESS guarantees that selection will oppose any potential invader. The same is true for a strict Nash equilibrium, but not for a Nash equilibrium. If $E(S_k, S_k) = E(S_j, S_k)$ and $E(S_k, S_j) < E(S_j, S_j)$ then S_k is still a Nash equilibrium, but selection will favor S_j invading S_k. Thus it makes sense to add a fourth definition.

(iv) Strategy S_k is stable against invasion by selection (let us call this "weak ESS") if $\forall i \neq k$ we have either

$$E(S_k, S_k) > E(S_i, S_k) \tag{4.17}$$

or

$$E(S_k, S_k) = E(S_i, S_k) \quad \text{and} \quad E(S_k, S_i) \geq E(S_i, S_i). \tag{4.18}$$

If a strategy is a strict Nash equilibrium then it is also an ESS. If a strategy is an ESS then it is also a weak ESS. If a strategy is a weak ESS then it is also a Nash equilibrium. Thus strict Nash implies ESS implies weak ESS implies Nash:

$$\text{strict Nash} \Rightarrow \text{ESS} \Rightarrow \text{weak ESS} \Rightarrow \text{Nash}. \tag{4.19}$$

All of these concepts play important roles when studying the evolutionary dynamics of frequency-dependent selection.

The concept of an "unbeatable strategy" was introduced by William Hamilton in his work on sex ratios preceding John Maynard Smith's work on evolutionary game theory. Strategy S_k is unbeatable if $\forall i \neq k$:

$$E(S_k, S_k) > E(S_i, S_k) \quad \text{and} \quad E(S_k, S_i) > E(S_i, S_i). \tag{4.20}$$

Therefore an unbeatable strategy dominates every other strategy. An unbeatable strategy is certainly a strict Nash equilibrium. An unbeatable strategy is the most you can ask for, but usually you are asking for too much. Unbeatable strategies are rare.

4.5 REPLICATOR DYNAMICS

Peter Taylor and Leo Jonker were the first to introduce a differential equation for evolutionary game dynamics. They were quickly followed by Christopher Zeeman (Warwick) as well as Peter Schuster, Josef Hofbauer, and Karl Sigmund (all three in Vienna).

Given what we have seen so far in this book, the equation that we will write down now is an obvious next step. Consider the interaction among n strategies. The payoff for strategy i when interacting with strategy j is given by a_{ij}. The $n \times n$-matrix $A = [a_{ij}]$ is called the "payoff matrix." Let x_i denote the frequency of strategy i. The expected payoff of strategy i is given by $f_i = \sum_{j=1}^{n} x_j a_{ij}$. The average payoff is given by $\phi = \sum_{i=1}^{n} x_i f_i$. Equating payoff with fitness, we obtain the replicator equation (Figure 4.5)

$$\dot{x}_i = x_i(f_i - \phi) \qquad i = 1, \ldots, n \tag{4.21}$$

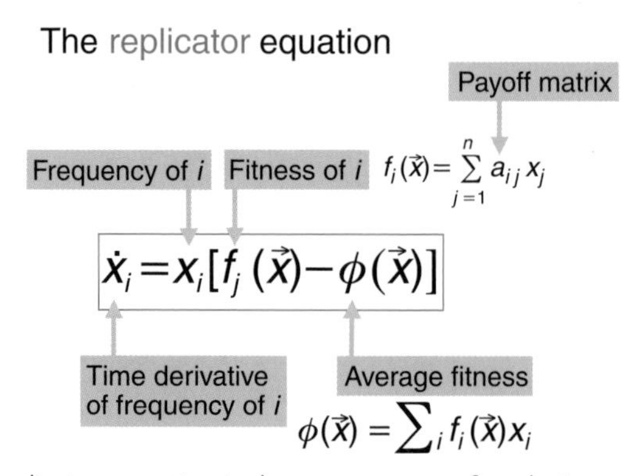

Figure 4.5 The replicator equation is the cornerstone of evolutionary game dynamics. It describes frequency-dependent selection among n different phenotypes (strategies) in an infinitely large population without mutation. In general, the fitness, f_i, of the phenotype, i, depends on the composition of the population, $\vec{x} = (x_1, \ldots, x_n)$. Usually f_i is a linear function of the frequencies x_j, and the coefficients of this function are the entries of the payoff matrix $A = [a_{ij}]$. The entry a_{ij} denotes the payoff for strategy i interacting with strategy j.

The difference between (4.21) and (2.16) is frequency-dependent as opposed to constant selection. The fitness values are linear functions of the frequencies.

Equation (4.21) is defined on the simplex S_n, which is given by $\sum_{i=1}^{n} x_i = 1$. We note that the interior of the simplex is invariant: if a trajectory starts in the interior, it will always remain in the interior; it might converge to the boundary, but it will never actually reach the boundary. Moreover, we note that each face of the simplex is invariant. A face is a subset of the simplex where at least one strategy has zero frequency. A strategy which is not there will not appear.

Replicator dynamics describe pure selection without mutation. Often, however, we ask the question whether a new strategy, which is not present in the population, could invade the population and increase in frequency. Thus, while not explicitly modeled, mutation is on the mind of the analyst.

The corners (vertices) of the simplex are fixed points of the replicator dynamics. Depending on the payoff matrix, A, there can be fixed points in the interior and in every face of the simplex.

4.5.1 Two Strategies

For $n = 2$, we have already derived a complete classification of all possible evolutionary dynamics in section 4.2. The simplex S_2 is the closed interval [0, 1]. The corner points, $x_1 = 0$ and $x_1 = 1$, are always fixed points of the replicator dynamics. In the interior, there can be at most one isolated fixed point. Such a fixed point exists if $(a_{11} - a_{21})(a_{12} - a_{22}) < 0$. This condition ensures that neither A dominates B nor B dominates A. Instead A and B can either be bistable or coexist. The interior fixed point is stable if A and B coexist which is the case for $a_{11} < a_{21}$ and $a_{12} > a_{22}$.

The neutral case, $a_{11} = a_{21}$ and $a_{12} = a_{22}$, implies that every point of the interval [0, 1] is an equilibrium point. In this case, the replicator dynamics do not move. The strategies A and B have identical fitness for any composition of the population. In such a case, we would say that replicator dynamics do not adequately describe the biological behavior. For any finite population size, the relative proportions of A and B will drift until eventually one strategy becomes extinct. We will discuss the evolutionary dynamics of finite populations in Chapter 6.

Constant selection between A and B is obtained as a special case of the replicator equation if $a_{11} = a_{12} \neq a_{21} = a_{22}$. Thus evolutionary game dynamics is the generic description of natural selection, with constant selection being a special case.

4.5.2 Three Strategies

For $n = 3$, interesting new dynamical features are possible. The phase space is the simplex S_3, which can be represented by a triangle with identical sides. There is a complete classification of all possible phase portraits.

Of special interest is the case where A is dominated by B, B is dominated by C, and C is dominated by A. This is the so-called Rock-Paper-Scissors game. In this well-known children's game, rock beats scissors which beat paper which beats rock.

Any 3×3 payoff matrix with cyclic domination of the three strategies characterizes a Rock-Paper-Scissors game. The analysis of the evolutionary game dynamics can be simplified by a trick that is applicable to any replicator equation: the dynamics of the replicator equation (4.21) remain unchanged if an arbitrary constant is added to each entry in a column of the payoff matrix. Thus by subtracting the diagonal element from each column, we can transform every payoff matrix into a payoff matrix that has only zero entries in the diagonal.

For example, the payoff matrix

$$A = \begin{array}{c} \\ R \\ S \\ P \end{array} \begin{array}{c} \begin{array}{ccc} R & S & P \end{array} \\ \begin{pmatrix} 4 & 2 & 1 \\ 3 & 1 & 3 \\ 5 & 0 & 2 \end{pmatrix} \end{array}$$

can be transformed to

$$A = \begin{array}{c} \\ R \\ S \\ P \end{array} \begin{array}{c} \begin{array}{ccc} R & S & P \end{array} \\ \begin{pmatrix} 0 & 1 & -1 \\ -1 & 0 & 1 \\ 1 & -1 & 0 \end{pmatrix} \end{array} \tag{4.22}$$

Both payoff matrices lead to identical replicator dynamics.

The payoff matrix (4.22) defines the symmetric Rock-Paper-Scissors game. The interior of the simplex S_3 contains a unique equilibrium point given by $(1/3, 1/3, 1/3)$. This point is stable, but not asymptotically stable. There are infinitely many periodic orbits surrounding this center. In fact all other points in the interior are on periodic orbits. The time average of each cycle is given by $(1/3, 1/3, 1/3)$. This situation is not generic: small deviations from the symmetry of A will change the phase portrait.

Note that payoff matrix (4.22) describes a zero-sum game. For a zero-sum game we have $a_{ij} = -a_{ji}$. The gain of one player is the loss of another; the average fitness of the population is always zero, $\phi = 0$. Thus the replicator equation becomes $\dot{x}_i = x_i f_i$ for all $i = 1, \ldots, n$.

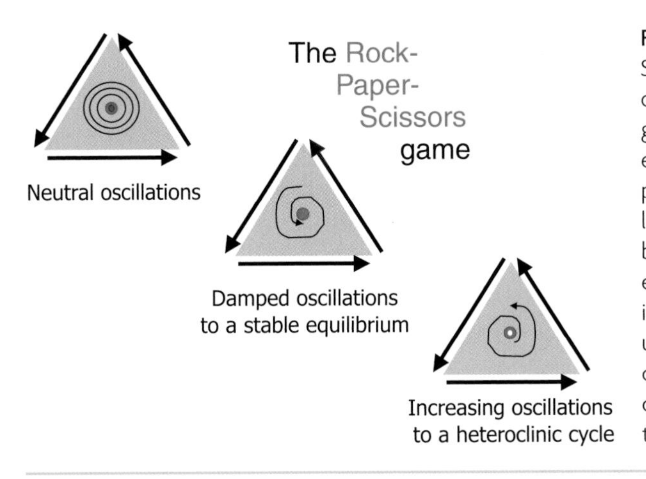

Figure 4.6 In the Rock-Paper-Scissors game, there is cyclic domination among three strategies. There is always one interior equilibrium. Depending on the payoff matrix, this interior equilibrium is (i) a center surrounded by neutral oscillations, (ii) a stable equilibrium that is approached in damped oscillations, or (iii) an unstable equilibrium, in which case there are oscillations of increasing amplitude converging to the boundary of the simplex.

The general Rock-Paper-Scissors game is given by the payoff matrix

$$A = \begin{pmatrix} 0 & -a_2 & b_3 \\ b_1 & 0 & -a_3 \\ -a_1 & b_2 & 0 \end{pmatrix} \tag{4.23}$$

There are two possibilities.

(i) if the determinant of A is positive (which is the case for $a_1a_2a_3 < b_1b_2b_3$), then there exists a unique interior equilibrium that is globally stable. The trajectories of the replicator dynamics starting from any interior initial condition will converge to this equilibrium in an oscillatory manner (Figure 4.6).

(ii) if the determinant of A is negative (which is the case for $a_1a_2a_3 > b_1b_2b_3$), then there exists a unique interior equilibrium that is unstable. The trajectories of the replicator dynamics starting from any interior initial condition will converge to the boundary of the simplex in oscillations with increasing amplitude. The boundary of the simplex is a heteroclinic cycle that is an attractor for all trajectories starting in the interior (a heteroclinic cycle is an invariant set containing saddle points on trajectories connecting them). For the differential equation, the oscillations will converge to the heteroclinic cycle without ever reaching it. For practical purposes, one of the three strategies will eventually become extinct, either by a rounding error of the computer program or by a random fluctuation in nature. Then only two strategies are left, and one will outcompete the other, leaving a single strategy in the end. It cannot be predicted which strategy will win. Thus the attracting heteroclinic cycle reveals a kind of deterministic unpredictability that is of a different type than chaos.

4.5.3 More than Three Strategies

For four species, the replicator dynamics are defined on the simplex S_4 given by a tetrahedron. This is the minimum number of dimensions that are needed for the replicator equation to allow limit cycles and chaotic attractors.

Finally, let us state some important results that hold in any dimension.

(i) An interior equilibrium of the replicator dynamics is given by the solution of the linear system of equations

$$f_1 = f_2 = \cdots = f_n \quad \text{and} \quad x_1 + x_2 + \cdots + x_n = 1. \tag{4.24}$$

Remember that $f_i = \sum_{j=1}^{n} a_{ij} x_j$. This system of n linear equations has either one or zero nondegenerate solutions. Thus there can be at most one isolated equilibrium in the interior of the simplex S_n.

(ii) If there is no equilibrium in the interior, then all trajectories converge to the boundary of the simplex. Thus there can be no chaotic attractor and no limit cycle in the interior if there is no equilibrium in the interior. This result is very helpful, because sometimes it is possible to show that a particular replicator equation admits no interior equilibrium. If this is the case, then we know that nothing more complicated can happen in the interior; coexistence of all strategies is impossible.

(iii) In degenerate cases, the replicator equation can admit a manifold of equilibria in the interior. These equilibria can be stable, but not asymptotically stable.

(iv) If a strategy is a strict Nash equilibrium or ESS, then the corner point of the simplex, which corresponds to a pure population of this strategy, is asymptotically stable.

Proofs of these assertions are given in the book by Hofbauer and Sigmund (1998).

4.6 HAWK OR DOVE?

Animals of the same species fight with each other. Conflicts arise over food, territory, or sex. Female lions fight to maintain their hunting grounds. Male chimpanzees fight for dominance of a group: the alpha male has to withstand challenges from other males, and in return gets the majority of matings. There are many examples of fierce and even deadly conflicts among animals. The human species, of course, is unsurpassed in its sad efficiency in inflicting murder, war, and genocide.

Often, however, conflicts between animals (including humans) do not escalate, but are fought out with those involved obeying certain limits. Ethology has a long-standing fascination with so-called conventional fights. A sequence of threatening signals and displays allow the contestants to assess each other's strength or determination before one of them simply walks away. Stags perform roaring matches followed by parallel walks and head pushing with interlocked antlers. Despite the lethal points of their antlers, only few contests lead to serious injury.

For a long time, biologists have accepted the following explanation: conventional fighting is frequently observed in nature because it is good for the species. Fighting that leads to serious injury is bad for the species. This argument, however, is problematic. There can certainly be selection between groups or whole species, but often stronger selection is put on the shoulders of individuals. If one individual in a population were to disobey the rules and escalate all fights by inflicting serious injury, then it might win many contests and thereby reproduce its genes more efficiently than others would.

Conventional fighting from the perspective of individual selection was first analyzed by John Maynard Smith. There are two strategies, hawks (H) and doves (D). While hawks escalate fights, doves retreat when the opponent escalates. The benefit of winning the fight is b. The cost of injury is c. If two hawks meet, then the expected payoff for each of them is $(b-c)/2$. The fight will escalate. One hawk wins, while the other is injured. Since both hawks are equally strong, the probability of winning or losing is 1/2. If a hawk meets a dove, the hawk wins and receives payoff b, while the dove retreats and receives payoff 0. If two doves meet, there will be no injury. One of them will eventually win. The expected payoff is $b/2$. Thus, the payoff matrix is given by

$$\begin{array}{c} \\ \text{H} \\ \text{D} \end{array} \begin{array}{c} \begin{array}{cc} \text{H} & \text{D} \end{array} \\ \begin{pmatrix} \frac{b-c}{2} & b \\ 0 & \frac{b}{2} \end{pmatrix} \end{array} \tag{4.25}$$

If $b < c$, then neither strategy is a Nash equilibrium. If everybody else plays "hawk," it is best to play "dove." If everybody else plays "dove," it is best to play "hawk." Thus hawks and doves can coexist. Selection dynamics will lead to a mixed population. At the stable equilibrium, the frequency of hawks is given by b/c. If the cost of injury is much larger than the benefit of winning the fight, $c \gg b$, then the equilibrium frequency of hawks will be small.

The name of the game is somewhat misleading, because normally we think of games between individuals of the same species. Moreover, real doves fight to death when confined in a cage.

4.6.1 Mixed Strategies

So far we have considered pure strategies that either play hawk all the time or play dove all the time. Let us now consider mixed strategies which play hawk with probability p and dove with probability $1-p$. There is no longer a discrete set of just two strategies but a continuous set of infinitely many strategies. The space of strategies is given by the closed interval $[0, 1]$. The payoff for strategy p_1 versus p_2 is given by

$$E(p_1, p_2) = \frac{b}{2}(1 + p_1 - p_2 - \frac{c}{b}p_1 p_2). \tag{4.26}$$

From this function we can infer that the strategy $p^* = b/c$ is evolutionarily stable. Note that

$$
\begin{aligned}
E(p^*, p^*) &= \frac{b}{2}(1 - \frac{b}{c}) \\
E(p, p^*) &= \frac{b}{2}(1 - \frac{b}{c}) \\
E(p^*, p) &= \frac{b}{2}(1 + \frac{b}{c} - 2p) \\
E(p, p) &= \frac{b}{2}(1 - \frac{c}{b}p^2)
\end{aligned}
\tag{4.27}
$$

We see that $E(p^*, p^*) = E(p, p^*)$ for all p. Therefore p^* is a Nash equilibrium, but not a strict Nash equilibrium. Since $E(p^*, p) > E(p, p)$ for all $p \neq p^*$, it follows that p^* is an evolutionarily stable strategy.

4.7 THERE IS ALWAYS A NASH EQUILIBRIUM

Consider a game given by an $n \times n$ payoff matrix A. There are n pure strategies, $S_1, \ldots, S_n$. Depending on the details of A, there may or may not exist a Nash equilibrium among those pure strategies. There is, however, always a Nash equilibrium if one considers the set of all mixed and all pure strategies. This is a deep result in game theory.

A strategy is given by the vector $\vec{p} = (p_1, \ldots, p_n)$. Here p_i is the probability of playing S_i. Clearly $\sum_{i=1}^{n} p_i = 1$. The payoff for strategy $\vec{q}$ versus strategy $\vec{p}$ is given by

$$E(\vec{q}, \vec{p}) = \sum_{i=1}^{n} \sum_{j=1}^{n} a_{ij} q_i p_j. \tag{4.28}$$

In vector notation we can write

$$E(\vec{q}, \vec{p}) = \vec{q} A \vec{p}. \tag{4.29}$$

It can be shown that for any payoff matrix A there exists at least one strategy $\vec{q}$ with the property that

$$\vec{q}A\vec{q} \geq \vec{p}A\vec{q} \qquad \forall \vec{p} \tag{4.30}$$

Thus $\vec{q}$ is a best reply to itself, a Nash equilibrium.

4.8 CHICKEN AND SNOWDRIFT

In the Chicken game, two cars head for each other at high speed. The loser is whoever chickens out first. The winner stays on the track. If neither driver chickens out, there is the substantial cost of a collision. Consider two strategies: A means you go for it. B means you chicken out after some time. The reward for winning is b, the cost of a collision is $-c$. If both players decide to chicken out, then the chance that you will win is $1/2$. Hence the payoff matrix is

$$\begin{array}{c} \\ A \\ B \end{array} \begin{array}{c} \begin{array}{cc} \mathrm{A} & \mathrm{B} \end{array} \\ \begin{pmatrix} -c & b \\ 0 & \frac{b}{2} \end{pmatrix} \end{array} \tag{4.31}$$

Comparing the entries in each column of the matrix leads to the same conclusion as in the Hawk-Dove game. It is always best to adopt the opposite strategy as your opponent. A mixed strategy between A and B is ESS.

In the Snowdrift game, two drivers are stuck on their way home, because the road is blocked by a snowdrift. You have a choice between cooperation and defection. Cooperation means that you get out of the car and shovel. Defection means that you remain in your car, relax, listen to music, and let the other person shovel. If both cooperate, the amount of work is only half as much. If neither cooperates, each is stuck in the snowdrift until the snowplow comes along. The benefit for getting home is b, the cost of shoveling in the cold is $-c$. The payoff matrix is

$$
\begin{array}{c} \\ \text{C} \\ \\ \text{D} \end{array}
\begin{array}{c}
\begin{array}{cc} \text{C} & \text{D} \end{array} \\
\begin{pmatrix} b-\frac{c}{2} & b-c \\ & \\ b & 0 \end{pmatrix}
\end{array}
\tag{4.32}
$$

If $b > c$, we have the same structure as in the Hawk-Dove game. It is best to defect when the other person cooperates. It is best to cooperate when the other person defects. If $b < c$, it is best to defect no matter what the other person does. This leads us to the Prisoner's Dilemma.

4.9 GAME THEORY AND ECOLOGY

Ecologists investigate the interaction of species and how their abundance changes over time. The logistic map, which we encountered in Chapter 2, is an important equation of ecology, demonstrating that complicated time series of species abundance can be generated by very simple rules. The fundamental mathematical approach to ecology is given by the Lotka-Volterra equation. The logistic map is a one-dimensional Lotka-Volterra equation in discrete time.

4.9.1 Predator and Prey

The hostilities between the Austrian and Italian navies during World War I disrupted fishing in the Adriatic Sea. After the war, it was observed that the number of predatory fish had increased. "Why does war favor sharks?" was the question that was posed to Italy's leading physicist of the time, Vito Volterra.

Volterra wrote down the following equations. Let x and y denote, respectively, the abundance of prey and predator fish. Prey reproduce at rate ax. They are eaten by predators at rate bxy. Predators die at rate cy and reproduce at rate dxy. Thus we have the following system of two nonlinear differential equations

$$
\begin{aligned}
\dot{x} &= x(a - by) \\
\dot{y} &= y(-c + dx)
\end{aligned}
\tag{4.33}
$$

If there are no predators, $y = 0$, then the population of prey fish grows exponentially as

$$x(t) = x(0)e^{at}. \tag{4.34}$$

If there are no prey, $x = 0$, then the population of predators declines exponentially as

$$y(t) = y(0)e^{-ct}. \tag{4.35}$$

The point $x = 0$ and $y = 0$ is a saddle point of the system. There is one interior equilibrium given by

$$x^* = c/d \quad \text{and} \quad y^* = a/b. \tag{4.36}$$

A linear stability analysis reveals that this equilibrium is neutrally stable. It is surrounded by infinitely many periodic orbits. The abundance of prey and the abundance of predators oscillate indefinitely. The oscillatory period is $2\pi/\sqrt{ac}$, but the amplitude depends on the initial condition. The time averages of the oscillations are given by the equilibrium values x^* and y^*. Neutral oscillations represent an ungeneric phenomenon, because small modifications of the differential equation (4.33) can destroy neutral stability.

Volterra could now answer the question why the reduced activity of fishing during the war had favored predators. Fishing decreases the reproductive rate of prey from a to $a - k$ and increases the death rate of predators from c to $c + m$. Hence, with fishing the time averages of the oscillations are given by

$$x_F^* = (c + m)/d \quad \text{and} \quad y_F^* = (a - k)/b. \tag{4.37}$$

The relative abundance of predators is greater in the absence of fishing:

$$y^*/x^* > y_F^*/x_F^*. \tag{4.38}$$

4.9.2 Kolmogorov's Predator-Prey Theorem

The neutral stability of equation (4.33) is a consequence of the particular assumptions for the interaction between prey and predator. There have been many extensions toward increased realism, such as including maximum carrying capacities for prey and predators and saturating functional responses of predator to prey. All such extensions tend to remove the unrealistic neutral stability. Instead, the abundance of predators and prey will usually converge to either a stable equilibrium or a stable limit cycle. In contrast to neutral oscillations, a stable limit cycle is a robust phenomenon. After small perturbations, the trajectory returns to the limit cycle. The period and amplitude do not depend on the initial condition, but are entirely determined by the parameters of the equation.

In 1936 the Russian mathematician A. N. Kolmogorov wrote a general predator-prey equation in the following form

$$\begin{aligned} \dot{x} &= xF(x, y) \\ \dot{y} &= yG(x, y) \end{aligned} \tag{4.39}$$

Here F and G are continuous functions with continuous first derivatives. Kolmogorov's theorem states that equation (4.39) has either a stable limit cycle or a stable equilibrium if the following conditions hold:

(i) $\partial F/\partial y < 0$

(ii) $x(\partial F/\partial x) + y(\partial F/\partial y) < 0$

(iii) $\partial G/\partial y < 0$

(iv) $x(\partial G/\partial x) + y(\partial G/\partial y) > 0$

(v) $F(0, 0) > 0$

Furthermore, there must exist constants, $A > 0$ and $B > C > 0$, such that

(vi) $F(0, A) = 0$

(vii) $F(B, 0) = 0$

(viii) $G(C, 0) = 0$

The biological interpretation of these conditions is illuminating: (i) the per capita rate of increase of prey is a decreasing function of predator abundance; (ii) for any given ratio of the two species, the rate of increase of prey is a decreasing function of population size; (iii) the per capita rate of increase of predators is a decreasing function of their abundance; (iv) for any given ratio of the two species, the rate of increase of predators is an increasing function of population size; (v) when both populations are small, the abundance of prey can increase; (vi) there exists a predator population size sufficiently large to prevent further increase of the prey population; (vii) there exists a critical prey population size, B, beyond which it can no longer grow even if there are no predators, which means the ecosystem has a maximum carrying capacity for the prey species; (viii) there exists a critical prey population size, C, below which predators cannot grow even if they are rare.

For a thorough discussion of such issues and many additional aspects of theoretical ecology, you must seek out Robert May's classic *Stability and Complexity of Model Ecosystems,* first published in 1973.

4.9.3 The Lotka-Volterra Equation

Equation (4.33) was also studied by the American biologist Alfred Lotka in the context of chemical kinetics and, therefore, became known as Lotka-Volterra equation. The general Lotka-Volterra equation describes the interaction of n species and is of the form

$$\dot{y}_i = y_i(r_i + \sum_{j=1}^{n} b_{ij} y_j) \qquad i = 1, \ldots, n \tag{4.40}$$

The abundance of species i is given by y_i, which is a non-negative number. Hence the equation is defined on the positive orthant R^n_+, which is given by the set of all points $(y_1, y_{21}, \ldots, y_n)$ with $y_i \geq 0$ for all i. The growth rate of species i is given by r_i. The interaction between species i and j is given by b_{ij}. The parameters r_i and b_{ij} can be positive, zero, or negative.

It turns out that the Lotka-Volterra equation (4.40) and the replicator equation (4.21) are equivalent. A replicator equation with n strategies can be transformed into a Lotka-Volterra equation with $n - 1$ species. The $n \times n$ matrix $A = [a_{ij}]$ defines the interaction of the n strategies in the replicator equation

$$\dot{x}_i = x_i \left(\sum_{j=1}^{n} a_{ij} x_j - \phi \right) \qquad i = 1, \ldots, n \tag{4.41}$$

The Lotka-Volterra equation

$$\dot{y}_i = y_i(r_i + \sum_{j=1}^{n-1} b_{ij} y_j) \qquad i = 1, \ldots, n-1 \tag{4.42}$$

with the parameters $r_i = a_{in} - a_{nn}$ and $b_{ij} = a_{ij} - a_{nj}$ is equivalent to (4.41).

Let $y = \sum_{i=1}^{n-1} y_i$. The equivalence can be shown with the transformation $x_i = y_i/(1+y)$ for $i = 1, \ldots, n-1$ and $x_n = 1/(1+y)$.

Therefore, whatever result holds for one equation will hold for the other. The equivalence of the Lotka-Volterra and replicator equations represents a beautiful bridge between aspects of theoretical ecology and evolutionary game theory.

We will see that theoretical ecology is the foundation for many studies in mathematical biology. For example, viruses and cells can behave as predator and prey. Immune cells, in turn, can "prey" on infected cells. Virus dynamics in infected hosts represent a "microecology."

SUMMARY

- Evolutionary game theory is the study of frequency-dependent selection.
- Games can be formulated in terms of a payoff matrix, which specifies the payoff for one strategy when interacting with another. Evolutionary game theory interprets payoff as fitness: successful strategies reproduce faster.
- A Nash equilibrium is a strategy with the following property: if two players adopt a Nash equilibrium, then neither player can improve her payoff by switching to another strategy.
- If an entire population (of infinite size) adopts an evolutionarily stable strategy (ESS), then no other strategy can invade.
- The replicator equation describes deterministic evolutionary game dynamics. For $n = 2$ strategies, there can be dominance, coexistence, bistability, or neutrality. For $n \geq 3$ strategies, there can be heteroclinic cycles. For $n \geq 4$ strategies, there can be limit cycles and chaos.

- In the Rock-Paper-Scissors game there can be either damped oscillations to a stable equilibrium or oscillations with increasing amplitude that eventually lead to the random elimination of two strategies.
- The Hawk-Dove game explains why many animals can resolve conflict without full-blown escalation.
- Chicken and Snowdrift are games that are similar to Hawk-Dove.
- The replicator equation is equivalent to the Lotka-Volterra equation of ecology. Evolutionary game theory and ecology have the same mathematical foundations.
- Kolmogorov's theorem specifies when a two-dimensional predator-prey system has a stable equilibrium or a stable limit cycle.

PRISONERS OF THE DILEMMA 5

TWO PEOPLE are suspected of having committed a joint crime. The suspects are confined to different rooms and cannot talk to each other. The police do not have sufficient evidence to convince a jury. The state attorney offers each of the suspects a deal: confess your crime, become a witness for the prosecution, and thereby avoid a prison sentence. If one prisoner confesses while the other does not, then the first will go free immediately; the second will receive a prison sentence of ten years. If both confess, they will each receive seven years. If neither of them confesses, then both will go free after one year because nothing can be proved. Here is the payoff matrix

$$\begin{array}{c} \\ \text{Remain silent} \\ \text{Confess} \end{array} \begin{array}{c} \begin{array}{cc} \text{Remain silent} & \text{Confess} \end{array} \\ \left(\begin{array}{cc} -1 & -10 \\ 0 & -7 \end{array} \right) \end{array} \tag{5.1}$$

What should they do? What does this have to do with evolutionary biology, anyway?

In fact, for biology the problem is as old as evolution itself. Evolutionary progress, the construction of new features, often requires the cooperation of simpler parts that are already available. For example, replicating molecules had to cooperate to form the first cells. Single cells had to cooperate to form the first multicellular organisms. The soma cells of the body cooperate and help the cells of the germ line to reproduce. Animals cooperate to form social structures, groups, and societies. Worker bees risk their lives to defend the beehive. Moreover they forgo their own reproductive potential and instead raise the offspring of another individual, the queen. In some bird species, helpers assist the parents in feeding their young. Humans cooperate on a large scale, giving rise to cities, states, and countries. Cooperation allows specialization. Nobody needs to know everything. But cooperation is always vulnerable to exploitation by defectors.

Imagine two individuals who can either cooperate, C, or defect, D. If both cooperate, then both get 3 points. If one cooperates, while the other defects, the cooperator gets 0 points while the defector gets 5. If both defect, they get 1 point each. The payoff matrix is

$$\begin{array}{c} \\ C \\ D \end{array} \begin{array}{c} \begin{array}{cc} C & D \end{array} \\ \begin{pmatrix} 3 & 0 \\ 5 & 1 \end{pmatrix} \end{array} \tag{5.2}$$

This payoff matrix has the same structure as the situation that was presented to the prisoners. What would you do? Cooperate or defect?

You should analyze the game as follows. Assuming the other person will cooperate, you have a choice between receiving 3 points for your cooperation or 5 points for your defection. Thus if the other person cooperates, it is best for you to defect. Assuming the other person will defect, you have a choice between receiving 0 points for your cooperation or 1 point for your defection. Thus if the other person defects, it is also best for you to defect. Therefore, no matter what the other person does, it is best for you to defect.

If the other person analyzes the game in the same rational way, then both of you will come to the conclusion to defect. Both of you will end up with only 1 point. This is less than the 3 points you could have received for mutual cooperation.

In the Prisoner's Dilemma you can either cooperate, *C*, or defect, *D*

The payoff matrix is:

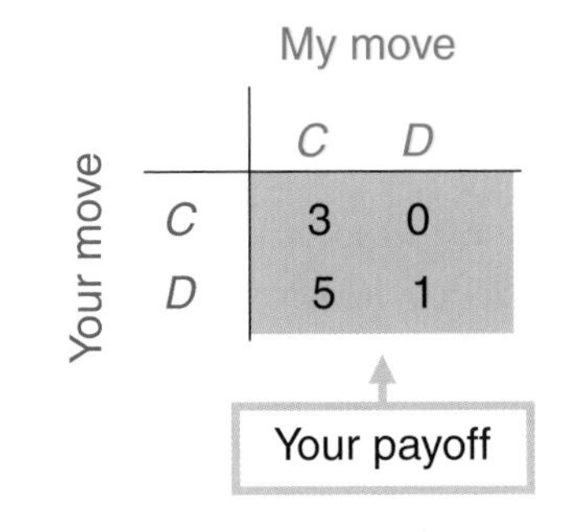

My move
C D
Your move
C 3 0
D 5 1
Your payoff

Figure 5.1 The Prisoner's Dilemma captures the essential problem of cooperation. Assuming that I will cooperate, you can get 3 points if you cooperate but 5 points if you defect. Assuming that I will defect, you can get 0 points if you cooperate but 1 point if you defect. Hence, no matter what I will do, it is better for you to defect. Defection is the "rational" (= payoff maximizing) strategy. If, however, I analyze the game in the same way that you do, then we both choose defection and both get 1 point. We could have received 3 points each, had we both chosen cooperation. But cooperation is "irrational." This is the dilemma.

This is the dilemma: rational players who act in order to maximize their payoff defect in the Prisoner's Dilemma (PD). Mutual cooperation leads to a higher payoff than mutual defection, but cooperation is irrational (Figure 5.1).

Do not take offense at the terms "rational" and "irrational" in this context. Note that the payoff describes exactly what the players want. There is no hidden agenda. The rewards (material or immaterial) are completely specified by the payoff matrix. Under this assumption, a rational player is defined as one who acts in a way so as to maximize her payoff. This is a simple and straightforward concept.

Experimental game theory, however, more often than not shows that humans do not behave rationally. They are guided by instincts that might have evolved via different situations. In the Prisoner's Dilemma, humans often try to cooperate. Only when they learn that it does not work will they switch to defection.

Returning to our prisoners, cooperation means not to cooperate with the state attorney but to cooperate with your partner and remain silent. If both of you remain silent, nothing can be proved. Defection means confession. If both of you defect, both will get a long prison sentence. No matter what your partner does, it is better for you to defect. The rational analysis suggests that both prisoners will confess and spend seven years in jail.

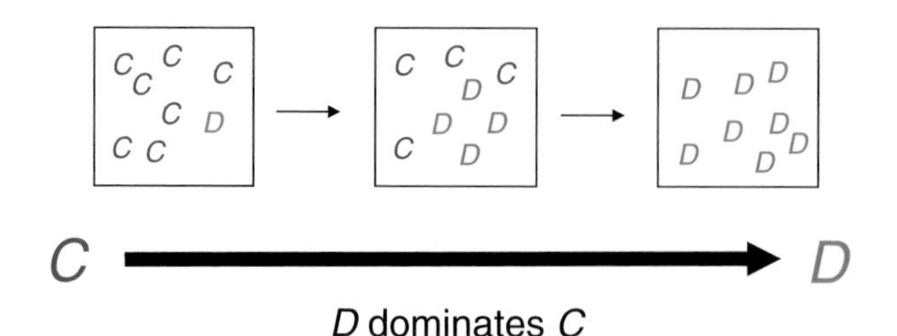

Figure 5.2 Natural selection chooses defection. In a mixed population of cooperators, C, and defectors, D, the latter always have a higher payoff. In the framework of evolutionary game theory, therefore, defectors reproduce faster and outcompete cooperators. Under natural selection, the average fitness of the population continuously declines. When the population consists only of defectors, the average fitness has reached its minimum. Specific mechanisms are needed for natural selection to favor cooperation.

If rationality leads to defection, which is a suboptimum outcome for both, then what does natural selection suggest? Imagine a population of cooperators and defectors. The frequency of cooperators is given by x, and the frequency of defectors by $1-x$. The average payoff for cooperators is $f_C = 3x$. The average payoff for defectors is $f_D = 5x + 1 - x = 4x + 1$. Thus defectors always have a higher fitness than cooperators. Defectors dominate cooperators. Natural selection steadily increases the frequency of defectors until cooperators have become extinct. Natural selection too favors defection (Figure 5.2).

5.1 DIRECT RECIPROCITY

Consider the payoff matrix

$$\begin{array}{c} \\ C \\ D \end{array} \begin{array}{c} \begin{array}{cc} C & D \end{array} \\ \begin{pmatrix} R & S \\ T & P \end{pmatrix} \end{array} \tag{5.3}$$

The game is a Prisoner's Dilemma if $T > R > P > S$. The temptation to defect, T, exceeds the reward for mutual cooperation, R, which is greater than the punishment, P, for mutual defection, which trumps the sucker's payoff, S. In addition, it is usually required that $R > (T + P)/2$. If this is not the case, then an agreement to alternate cooperation and defection leads to a higher payoff than pure cooperation in a repeated game.

Direct reciprocity refers to the concept that the game is not played just once but is repeated several times between the same two players. In this case, cooperation becomes a promising option. Imagine the game is repeated m times. Consider two strategies. "GRIM" cooperates on the first move and then cooperates as long as the opponent does not defect. If the opponent defects once it will switch permanently to defection. It will never forgive. The other strategy is "Always defect" (ALLD).

The payoff matrix is given as follows

$$\begin{array}{c} \\ \text{GRIM} \\ \text{ALLD} \end{array} \begin{array}{c} \begin{array}{cc} \text{GRIM} & \text{ALLD} \end{array} \\ \begin{pmatrix} mR & S + (m-1)P \\ T + (m-1)P & mP \end{pmatrix} \end{array} \tag{5.4}$$

If $mR > T + (m - 1)R$, then GRIM is a strict Nash equilibrium when competing with ALLD. In terms of a game between two players, if both follow this rule, then neither player can improve by switching to ALLD. In terms of evolutionary dynamics, if the whole population uses GRIM, then ALLD cannot invade: selection opposes ALLD at low frequency. GRIM is stable against invasion by ALLD if the number of rounds, m, exceeds a critical value,

$$m > \frac{T - P}{R - P}. \tag{5.5}$$

For the payoff values used in the matrix (5.2), we have $m > 2$.

Therefore, we seem to have found a mechanism to stabilize cooperation once it has been established. Note, however, that ALLD is also a strict Nash equilibrium, because mP is always greater than $S + (m - 1)P$. Thus we have no evolutionary mechanism to explain the emergence of cooperation. We will address this problem later.

There is an even more acute problem to deal with. Imagine we both know that we are playing the game for m rounds. There is no incentive for you to cooperate in the last round. For the last round, the same analysis applies as for the nonrepeated Prisoner's Dilemma. Realizing this point, we might choose to play GRIM with one small modification: we will certainly defect in the last round. Let us denote this strategy by GRIM*. The payoff matrix for GRIM versus GRIM* is given by

$$\begin{array}{c} \\ \text{GRIM} \\ \text{GRIM}^* \end{array} \begin{array}{c} \begin{array}{cc} \text{GRIM} & \text{GRIM}^* \end{array} \\ \begin{pmatrix} mR & (m-1)R+S \\ (m-1)R+T & (m-1)R+P \end{pmatrix} \end{array} \tag{5.6}$$

We notice that GRIM is dominated by GRIM*. A population of GRIM players can be invaded by a small fraction of GRIM*. Once everybody plays GRIM*, the same argument as before applies to the second-to-last round. It is not rational to cooperate in the penultimate round if the ultimate round is mutual defection. The argument can be repeated back to the first round. We can thus write down a sequence of strategies starting with GRIM that is dominated by the variant strategy that defects in the last round, which is dominated by the strategy that defects in the last two rounds and so on until we come to ALLD, which is a strict Nash equilibrium. In this strategy space, the only strict Nash equilibrium and the only evolutionarily stable strategy is ALLD.

We note again that humans do not use this backward induction in experimental situations. People might realize that it is best to defect in the end, but they do not carry the argument rigorously to its logical conclusion. Why not? One explanation is that the human strategic instincts are not formed by playing games with a well-defined number of rounds. Imagine that it is not quite known when the interaction will end. There might always be another round. There is always a tomorrow in our plans. History has a beginning, but it has no end.

5.1.1 Tomorrow Never Dies

Let us therefore study a repeated Prisoner's Dilemma with a variable, rather than a fixed number of rounds. Suppose that after each round there is a probability w that another round will be played. Hence the expected number of rounds is $\bar{m} = 1/(1-w)$. The payoff matrix between GRIM and ALLD becomes

$$\begin{array}{c} \\ \text{GRIM} \\ \text{ALLD} \end{array} \begin{array}{c} \begin{array}{cc} \text{GRIM} & \text{ALLD} \end{array} \\ \left(\begin{array}{cc} \bar{m}R & S+(\bar{m}-1)P \\ T+(\bar{m}-1)P & \bar{m}P \end{array} \right) \end{array} \tag{5.7}$$

GRIM is evolutionary stable if $\bar{m} > (T-P)/(R-P)$. Nothing changes except now there is no strategy that can simply defect on the last move.

But is GRIM an ideal strategy for the repeated Prisoner's Dilemma? What if the opponent defects only once, perhaps to see if he can get away with it? Is it best never to cooperate with such an individual again? A strategy with some mechanism of reconciliation might achieve a higher payoff than GRIM.

What is the set of all strategies for the repeated Prisoner's Dilemma? A deterministic strategy is a rule that assigns to any history of the game the decision to cooperate or to defect in the next move. A stochastic strategy is a rule that assigns to any history of the game the probability of cooperating (or defecting) in the next move. Each round of the game has four possible outcomes: CC, CD, DC, DD. There are $2^4 = 16$ deterministic strategies that consider only the last round of the game; for each of the four possible outcomes the strategy prescribes the decision to cooperate or to defect. Hence a deterministic strategy with a memory of one move can be encoded by a binary string of length 4. For example, 0000 means "always defect," while 1000 means cooperate only after a CC and defect otherwise. There are 2^{16} deterministic strategies that consider the last two moves of the game. There are 2^{4^m}-deterministic strategies that consider the last m moves. Stochastic strategies that consider the last m moves of the game span a 4^m dimensional strategy space. Each dimension is given by the interval $[0, 1]$, denoting a probability. For a game of arbitrary length, the space of all possible strategies has infinitely many dimensions. It is impossible to set up a computer simulation that considers every possible strategy in the repeated PD. How can we find successful strategies for direct reciprocity?

5.2 AXELROD'S TOURNAMENTS

The search for an optimum strategy led Robert Axelrod, a political scientist at the University of Michigan at Ann Arbor, to a brilliant idea. In 1978 he decided to conduct a Prisoner's Dilemma championship. He invited people from all over the world to submit strategies for the repeated game. The strategies had to be formulated in terms of computer programs. Every strategy played every other strategy. The payoffs from each encounter were added up. In the end, Axelrod analyzed which strategy had the highest total payoff.

There were fourteen contestants. Some programs were clever devices to deceive the opponent or to predict the opponent's behavior. The winning strategy was the simplest of all. It was Tit-for-tat (TFT). TFT starts with a cooperation and then does whatever the opponent did in the previous round. TFT will answer C for C and D for D. Playing against TFT is like playing the mirror image of yourself shifted by one round. TFT was submitted by the well-known game theorist Anatol Rapoport.

Axelrod published the results together with a comprehensive analysis. Then he invited people to submit strategies for a second world championship. This time sixty-three strategies were sent in. Many people submitted strategies that would have won the first tournament. Among these was Tit-for-two-tats submitted by John Maynard Smith. Only Rapoport fielded TFT. And it won again.

While it is possible to calculate the optimum strategy against a known set of contestants, it is difficult to predict the best strategy for an unknown set of contestants. Axelrod published a book, *The Evolution of Cooperation*, which became a best-seller. TFT was the unquestioned world champion in the repeated PD.

Axelrod praises TFT in his book for having a collection of important properties. It is "nice," which means it is never the first to defect. TFT never tries to get more than the opponent in a direct pairwise comparison. In each single tournament match, it will at most have received the same number of points as its opponent, but never more. The payoff sum over all matches, however, was higher for TFT than for any other strategy. Hence, TFT is not triumphant by directly beating other strategies. Its success lies in the fact that on average it gets a higher payoff from strategy X than other strategies get from strategy X. TFT is apparently very successful at inducing cooperation from other strategies.

Moreover, TFT is stable against invasion by ALLD if the average number of rounds, $\bar{m}$, is not too small. TFT playing ALLD will cooperate in the first round, but defect in every subsequent round. The payoff matrix for TFT versus ALLD is the same as for GRIM versus ALLD

$$\begin{array}{c} \\ \text{TFT} \\ \text{ALLD} \end{array} \begin{array}{c} \begin{array}{cc} \text{TFT} \qquad\qquad & \qquad \text{ALLD} \end{array} \\ \begin{pmatrix} \bar{m}R & S + (\bar{m} - 1)P \\ T + (\bar{m} - 1)P & \bar{m}P \end{pmatrix} \end{array} \tag{5.8}$$

TFT can resist invasion by ALLD if $\bar{m} > (T - P)/(R - P)$.

TFT has the advantage over GRIM that it can resume cooperation if the opponent cooperates. Unlike GRIM, TFT is not locked into permanent defection.

5.2.1 Achilles' Heel

Axelrod's original tournaments were conducted in an error-free digital universe, but real-world situations are permeated by mistakes. A "trembling hand" can lead to a misimplementation of one's own action. A "fuzzy mind" can cause the misinterpretation of the opponent's move.

It turns out that TFT has an Achilles' heel (Figure 5.3). In the presence of mistakes, two TFT players achieve a very low payoff. A single mistake moves the game from mutual cooperation to alternating between cooperation and defection. A second mistake can lead to mutual defection. In the long run, two TFT players, with a small chance of making mistakes, obtain the same payoff as two random players tossing coins to decide between C and D in every move. The payoff for two TFT players in the presence of small amounts of behavioral noise is given by

$$A(TFT, TFT) = \frac{R + T + P + S}{4}. \tag{5.9}$$

This payoff is certainly less than R, because $R > (T + S)/2$ and $R > P$. Hence TFT is weak in the presence of mistakes. As we shall soon see, mistakes imply that TFT can be invaded and even dominated by many other strategies.

Tit-for-tat cannot correct mistakes

TFT: $C\ C\ C\ \overset{*}{D}\ C\ D\ C\ D\ D\ D \ldots$

TFT: $C\ C\ C\ C\ D\ C\ D\ \underset{*}{D}\ D\ D \ldots$

Figure 5.3 The Achilles' heel of a world champion: Tit-for-tat cannot correct mistakes. If an error (red asterisk) occurs, then the game switches from mutual cooperation to alternating cooperation and defection. Another error can bring the game to mutual defection. Further errors lead back to cooperation. But in the long run, the expected payoff between two TFT players is the same as for two random players who flip coins to determine whether to cooperate or to defect. Errors reduce the performance of TFT.

There is another weakness even in the absence of mistakes. Imagine TFT playing "Always cooperate" (ALLC). The payoff matrix is given by

$$\begin{array}{c} \\ \text{TFT} \\ \text{ALLC} \end{array} \begin{array}{c} \begin{array}{cc} \text{TFT} & \text{ALLC} \end{array} \\ \begin{pmatrix} \bar{m}R & \bar{m}R \\ \bar{m}R & \bar{m}R \end{pmatrix} \end{array} \tag{5.10}$$

Both players cooperate on every move. Therefore TFT is neither a strict Nash equilibrium nor an evolutionarily stable strategy. In a finite population, random drift can lead from TFT to ALLC (Figures 5.4 and 5.5).

Axelrod's world series had no noise in terms of strategical mistakes or random drift. Could these effects lead to the downfall of the world champion, TFT?

5.3 REACTIVE STRATEGIES

As we have seen, the strategy space of the repeated PD is enormous. Any particular calculation can only explore a limited region of this space. Let us now consider the set of "reactive strategies." These strategies are given by two parameters: p denotes the probability of cooperating if the opponent has cooperated in the previous move; q denotes the probability of cooperating if

The repeated Prisoner's Dilemma

ALLC: *C C C C C C C . . .*
ALLD: *D D D D D D D . . .*

ALLD: *D D D D D D D . . .*
TFT: *C D D D D D D . . .*

TFT: *C C C C C C C . . .*
TFT: *C C C C C C C . . .*

Figure 5.4 Three simple strategies in the repeated Prisoner's Dilemma: Always cooperate (ALLC), Always defect (ALLD), and Tit-for-tat (TFT). ALLC is exploited by ALLD. ALLD can exploit TFT only on the first move; then TFT switches to defection. Two TFT players, in contrast, can cooperate on every move (if no errors occur). Therefore, in a game between ALLD and TFT, ALLD receives a slightly higher payoff than TFT, but two TFT players receive a much higher payoff still. TFT is successful for this reason.

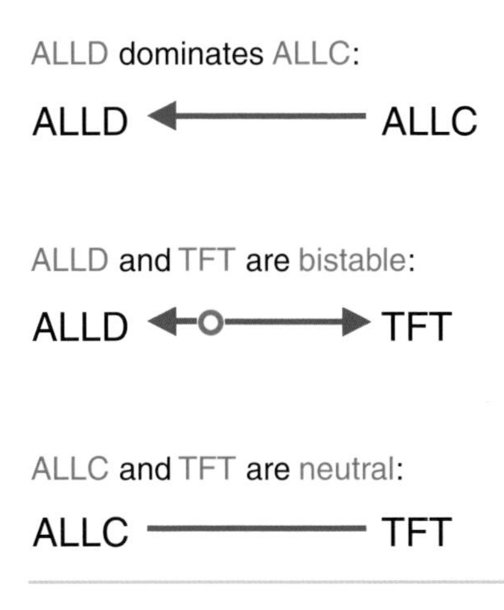
ALLD dominates ALLC:
ALLD
ALLC
ALLD and TFT are bistable:
ALLD
TFT
ALLC and TFT are neutral:
ALLC
TFT

Figure 5.5 Pairwise selection dynamics of the three basic strategies. In a mixed population of ALLD and ALLC, the former always has a higher fitness. In a mixed population of ALLD and TFT, ALLD has a higher fitness only when TFT is rare; otherwise TFT has a higher fitness. Selection dynamics are bistable. There is an unstable equilibrium (red circle) that acts as an invasion barrier: if the frequency of TFT is greater than this unstable equilibrium, then TFT can eliminate ALLD. In a mixed population of ALLC and TFT, all players have the same fitness. Therefore TFT is not evolutionarily stable.

the opponent has defected in the previous move. Note that reactive strategies have a very short memory: they only take into account what the opponent has done during the previous round. They do not even consider their own previous move.

A particular reactive strategy, $S(p, q)$, is a point in the unit square. Three corners of this square represent familiar strategies: (i) $S(0, 0)$ denotes ALLD; (ii) $S(1, 1)$ denotes ALLC; and (iii) $S(1, 0)$ denotes TFT. The fourth corner,

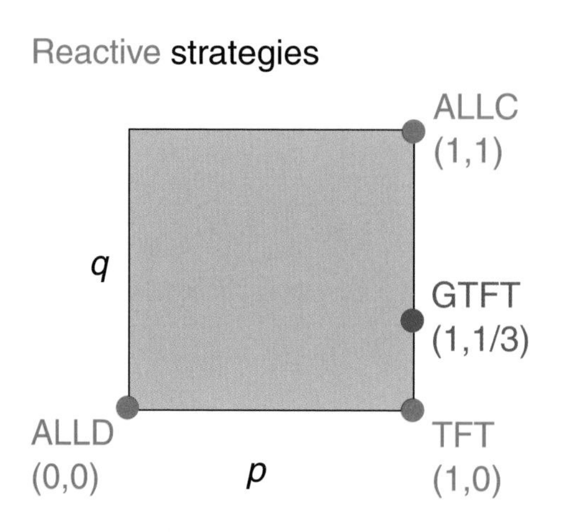

Figure 5.6 Reactive strategies are defined by two parameters: p denotes the probability of cooperating, if the opponent has cooperated in the previous round, while q denotes the probability of cooperating, if the opponent has defected in the previous round. A particular reactive strategy is given by a pair (p, q). Examples of reactive strategies include: ALLD $(0, 0)$, ALLC $(1, 1)$, and TFT $(1, 0)$. A TFT player with a 10% error rate is given by $(0.9, 0.1)$. It can be shown that TFT is the best catalyst for the emergence of cooperation, but once TFT has established cooperation it is rapidly replaced by Generous Tit-for-tat (GTFT). For the standard payoff values, $R = 3$, $T = 5$, $P = 1$, and $S = 0$, GTFT is given by $(1, 1/3)$. GTFT forgives on average every third defection. Natural selection among reactive strategies leads to the emergence of forgiveness.

$S(0, 1)$, represents the paradoxical strategy that cooperates when the opponent has defected and defects when the opponent has cooperated (Figure 5.6).

Can we derive a complete understanding of evolutionary dynamics within the set of reactive strategies? The repeated PD between two such strategies is a Markov chain on the state space (CC, CD, DC, DD). Let us label these states 1 to 4. State 1 means we both cooperate, CC. State 2 means I cooperate but you defect, CD. State 3 means I defect while you cooperate, DC. State 4 means we both defect, DD.

My reactive strategy is given by $S_1(p_1, q_1)$, while yours is $S_2(p_2, q_2)$. The Markov chain is defined by a 4×4 transition matrix, $M = [m_{ij}]$. The entry m_{ij} denotes the probability of moving from state i to state j. We have

$$M = \begin{array}{c} \\ CC \\ CD \\ DC \\ DD \end{array} \begin{array}{c} \begin{array}{cccc} CC & CD & DC & DD \end{array} \\ \begin{pmatrix} p_1p_2 & p_1(1-p_2) & (1-p_1)p_2 & (1-p_1)(1-p_2) \\ q_1p_2 & q_1(1-p_2) & (1-q_1)p_2 & (1-q_1)(1-p_2) \\ p_1q_2 & p_1(1-q_2) & (1-p_1)q_2 & (1-p_1)(1-q_2) \\ q_1q_2 & q_1(1-q_2) & (1-q_1)q_2 & (1-q_1)(1-q_2) \end{pmatrix} \end{array} \tag{5.11}$$

For example, the transition probability from state CD to state DC is given by $(1 - q_1)p_2$. This term is the product of two probabilities: $(1 - q_1)$ denotes the probability that I will defect given that you have defected in the previous round, while p_2 is the probability that you will cooperate given that I have cooperated in the previous round. All entries of the M matrix can be constructed in this way.

Let $\vec{x}_t$ denote the probability distribution of the game after t rounds. This means the vector $\vec{x}_t$ has four components that determine the probabilities that the game is in the four possible states, (CC, CD, DC, DD). We obtain the probability distribution after the next time step by multiplying the vector with the transition matrix

$$\vec{x}_{t+1} = \vec{x}_t M. \tag{5.12}$$

A stochastic matrix is regular if there exists a positive number k such that all elements of the matrix M^k are positive. If the matrix M is regular, then there exists a unique left-hand eigenvector, $\vec{x}$, associated with the eigenvalue one,

$$\vec{x} = \vec{x} M. \tag{5.13}$$

The components of this normalized eigenvector represent the stationary distribution of the Markov chain.

Let us introduce the quantities $r_1 = p_1 - q_1$ and $r_2 = p_2 - q_2$. The matrix M is regular if and only if $|r_1 r_2| < 1$. Because of the special symmetry of the M

matrix, the stationary distribution is of the form

$$\vec{x} = [s_1 s_2, s_1(1 - s_2), (1 - s_1)s_2, (1 - s_1)(1 - s_2)] \tag{5.14}$$

where

$$s_1 = \frac{q_2 r_1 + q_1}{1 - r_1 r_2} \tag{5.15}$$

and

$$s_2 = \frac{q_1 r_2 + q_2}{1 - r_1 r_2}. \tag{5.16}$$

Note that s_1 and s_2 represent the probabilities that players 1 and 2 cooperate in the stationary distribution. The expected payoff for strategy S_1 versus S_2 is given by

$$\begin{aligned} &E(S_1, S_2) \\ &\quad = Rs_1 s_2 + Ss_1(1 - s_2) + T(1 - s_1)s_2 + P(1 - s_1)(1 - s_2). \end{aligned} \tag{5.17}$$

Equations (5.15–5.17) provide the payoff for the game between any two strategies as long as $|r_1 r_2| < 1$. All the cases with $|r_1 r_2| = 1$ include deterministic strategies, for which the payoff can be easily calculated.

5.3.1 "In Silico" Evolution

Let us perform an experiment to understand the evolutionary dynamics of reactive strategies. Use a random number generator that returns a uniform distribution on [0, 1] to produce $n = 100$ reactive strategies. Let us use the same payoff values that Axelrod chose for his two tournaments: $R = 3$, $T = 5$, $S = 0$, and $P = 1$. Calculate the $n \times n$ payoff matrix using equation (5.17). Insert the matrix into the replicator equation. Assume that all n strategies are equally abundant at time $t = 0$. Observe the evolutionary trajectory.

In most cases, you will make the following observation. Many strategies converge to extinction. The most cooperative strategies disappear first. After some time, only a single strategy is left. This strategy is the one that is closest to ALLD. You have witnessed yet again the evolution of defection. You have also seen the destructive potential of frequency-dependent selection: the average

Generous tit-for-tat can correct mistakes

GTFT: *C C C D C D C C C C . . .*

GTFT: *C C C C D C C C C C . . .*

Figure 5.7 In contrast to TFT, GTFT can correct mistakes. With a certain probability, a sequence of cooperation and defection leads back to mutual cooperation. The expected payoff for two GTFT players is higher than for two TFT players.

fitness of the population declined steadily from about $(R + T + S + P)/4 = 9/4$ to $P = 1$.

In some cases, however, something remarkable happens. Suppose the initial ensemble contains, by chance or by design, a strategy that is close to TFT. Initially there is again an evolution toward ALLD. But when almost all strategies have disappeared, the embattled minority of TFT-like individuals fight the mighty ALLD. The reciprocators' frequency rises suddenly, while that of the defectors declines. But the reign of TFT is short-lived. Its moment of glory is brief and fleeting. TFT is soon replaced by strategies that are close to $p = 1$ and $q = 1/3$. Here the selective dynamics end.

Who is the new champion that has appeared on the cooperative wave initiated by TFT? The strategy $p = 1$ and $q = 1/3$ cooperates whenever the opponent has cooperated, but also cooperates one out of three times when the opponent has defected. We call this strategy "Generous Tit-for-tat" (GTFT). It is more forgiving than strict TFT when faced with a defection. This has two consequences.

(i) When one GTFT individual plays another, each receives an average payoff per round that is very close to the full reward for mutual cooperation, R. In contrast, two TFT players only obtain $(R + P + T + S)/4$. GTFT can make up for mistakes (Figure 5.7).

(ii) When GTFT plays ALLD, it has a slightly lower payoff than when TFT plays ALLD. Thus GTFT is less able to swing the tide away from defection than TFT is. Hence we need TFT to initiate cooperation. But once a cooperative population has been established, TFT will be replaced by GTFT.

5.4 GENEROUS TIT-FOR-TAT

For the general parameter values of the Prisoner's Dilemma, GTFT is defined as follows

$$\text{GTFT:} \qquad p = 1 \qquad q = \min\left\{1 - \frac{T-R}{R-S}, \frac{R-P}{T-P}\right\} \tag{5.18}$$

This is the highest level of forgiveness, q, that is still resistant against invasion by ALLD. Increasing q further would allow ALLD to dominate GTFT.

We can use equation (5.17) to show that GTFT is optimum in the following sense: among all reactive strategies that can resist invasion by ALLD, it affords the highest payoff for a population adopting it.

5.5 WIN-STAY, LOSE-SHIFT

Let us now consider the set of all stochastic strategies that base their decision whether to cooperate or to defect on the previous move of the game (including both the opponent's action and one's own action). Each strategy is thus defined by the conditional probabilities (p_1, p_2, p_3, p_4) to cooperate, given that the outcome of the last round was CC, CD, DC, or DD. Again the game between two such strategies, $S(p_1, p_2, p_3, p_4)$ and $S'(p'_1, p'_2, p'_3, p'_4)$, can be formulated as a Markov chain. The transition matrix of this process is given by

$$M = \begin{array}{c} \\ CC \\ CD \\ DC \\ DD \end{array} \begin{array}{c} \begin{array}{cccc} CC & CD & DC & DD \end{array} \\ \begin{pmatrix} p_1p'_1 & p_1(1-p'_1) & (1-p_1)p'_1 & (1-p_1)(1-p'_1) \\ p_2p'_3 & p_2(1-p'_3) & (1-p_2)p'_3 & (1-p_2)(1-p'_3) \\ p_3p'_2 & p_3(1-p'_2) & (1-p_3)p'_2 & (1-p_3)(1-p'_2) \\ p_4p'_4 & p_4(1-p'_4) & (1-p_4)p'_4 & (1-p_4)(1-p'_4) \end{pmatrix} \end{array} \tag{5.19}$$

M is a stochastic matrix. If M is regular, then it has a unique left-hand eigenvector, $\vec{x} = (x_1, x_2, x_3, x_4)$, associated with eigenvalue one,

$$\vec{x} = \vec{x}M. \tag{5.20}$$

This eigenvector represents the stationary distribution of the Markov chain. Thus x_1, x_2, x_3, and x_4 are the probabilities to be in state CC, CD, DC, and DD, respectively, after a large number of rounds. Hence the expected payoff for strategy S playing against S' is given by

$$E(S, S') = Rx_1 + Sx_2 + Tx_3 + Px_4. \tag{5.21}$$

Strictly speaking, this is the expected payoff per round when playing an infinitely repeated Prisoner's Dilemma. All of this is exactly as in section 5.3, but the transition matrix M no longer has the nice symmetry properties and hence we cannot, in general, write down a simple formula for the eigenvector $\vec{x}$.

The new strategy space is given by the four-dimensional cube $[0, 1]^4$. Reactive strategies are a subset of memory-one strategies fulfilling the constraints $p_1 = p_3$ and $p_2 = p_4$. ALLD is given by $(0, 0, 0, 0)$. ALLC is given by $(1, 1, 1, 1)$. TFT is given by $(1, 0, 1, 0)$. For Axelrod's payoff values, $R = 3$, $S = 0$, $T = 5$, and $P = 1$, which we will use now, GTFT is $(1, 1/3, 1, 1/3)$.

We performed a slightly different in-silico evolution experiment, adjusted to the much larger strategy space. We start with a homogeneous population using the random strategy given by $(1/2, 1/2, 1/2, 1/2)$. Every 100 generations, on average, a small amount of a new strategy is introduced. Selective dynamics are calculated using the replicator equation. The new strategy could become extinct, take over the whole population, or establish an equilibrium with the resident strategy. After some time, another random mutant is added. The new mutant is taken from a fixed random distribution over the whole strategy space. It is beneficial to use a distribution which has some bias toward the boundary of the hypercube, because the most relevant strategies live at the boundary. The dynamics differ from the classical replicator in that new mutants are added over time.

Originally this experiment was performed to confirm the success of GTFT. Indeed for some instantiations of this evolutionary process, GTFT is the ultimate winner. But more often we discovered—unexpectedly—that evolution came up with another strategy. The deterministic form of this strategy is given by $(1, 0, 0, 1)$. The strategy thus cooperates if the previous move was CC or DD. The strategy defects if the previous move was CD or DC. What does this mean?

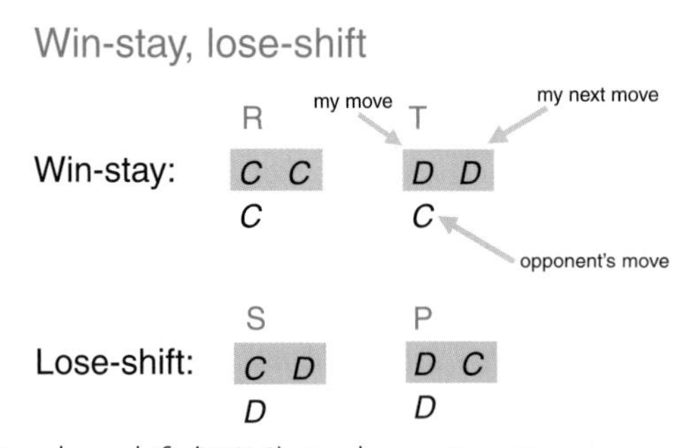

Figure 5.8 Win-stay, lose-shift (WSLS) simply monitors its own payoff: "if I am doing well, I will continue with the present action; if I am not doing well, I will try something else." The payoff R is obtained for mutual cooperation, CC. In this case WSLS will cooperate again. The even larger payoff T is obtained for DC. In this case WSLS will defect again. The payoffs R and T are considered a "win" and therefore WSLS "stays" with its current move. The payoff S is obtained for CD. In this case WSLS will shift from C to D. The payoff P is obtained for mutual defection, DD. In this case WSLS will shift from D to C. The payoffs S and P are considered a "loss" and therefore WSLS "shifts" to another move. It is remarkable that this simple and fundamental rule can outperform both TFT and GTFT in the repeated Prisoner's Dilemma.

Note that the strategy repeats its previous move whenever it has received a high payoff, T or R. The strategy shifts to the opposite move whenever it has received a low payoff, P or S. Thus the strategy follows the simple principle of Win-stay, lose-shift, WSLS (Figure 5.8).

Rapoport did consider WSLS, but noted that it attempts to cooperate with ALLD in every other round,

$$\begin{aligned} &\text{WSLS: } CDCDCDCD\ldots \\ &\text{ALLD: } DDDDDDDD \end{aligned} \tag{5.22}$$

Therefore he called it Simpleton.

Note however that WSLS is stable against invasion by ALLD if $R > (T + P)/2$. By coincidence, Axelrod's values have the ungeneric property that $R = (T + P)/2$. In this case, a variant of WSLS with $(1, 0, 0, 1 - \epsilon)$ is stable against invasion by ALLD. For $R < (T + P)/2$ we find that $(1, 0, 0, x)$ with $x < (R - P)/(T - R)$ is stable against invasion by ALLD.

Win-stay, lose-shift **can correct** mistakes

WSLS: $C\,C\,C\,\overset{*}{D}\,D\,C\,C\,C\ldots$

WSLS: $C\,C\,C\,C\,D\,C\,C\,C\ldots$

Win-stay, lose-shift **dominates** ALLC

WSLS: $C\,C\,C\,\overset{*}{D}\,D\,D\,D\,D\ldots$

ALLC: $C\,C\,C\,C\,C\,C\,C\,C\ldots$

Figure 5.9 Win-stay, lose-shift (WSLS) can correct mistakes. In a game between two WSLS players, a mistake (red asterisk) leads to one round of mutual defection, after which cooperation is resumed. WSLS dominates ALLC. After a mistake, WSLS switches to defection. Therefore WSLS leads to a more stable cooperation than TFT or GTFT. In contrast to WSLS, (i) TFT cannot correct mistakes and (ii) neither TFT nor GTFT is stable against random drift to ALLC.

WSLS has the nice property that it can correct occasional mistakes. If two WSLS strategists play each other, they will start with a sequence of cooperative moves and continue to cooperate until one of them defects by mistake. In the next round, both defect. The player who defected first received a very high payoff, was happy and defected again. The player who cooperated received a low payoff, was unhappy, and switched from cooperation to defection. Now both players defect, are unhappy, and switch back to cooperation. The sequence of moves is the following

$$\begin{aligned} &\text{WSLS: } CCCCC\overset{*}{D}DCCC\ldots \\ &\text{WSLS: } CCCCCCDCCC \end{aligned} \tag{5.23}$$

The asterisk indicates the erroneous move. The sequence of events seems compatible with human psychology: a mistake leads to a brief argument, after which friendship is resumed. WSLS has the advantage over TFT, because it can correct occasional mistakes. WSLS is a deterministic corrector, while GTFT is a stochastic one (Figure 5.9).

WSLS has an advantage over GTFT, too. Consider the game between WSLS and ALLC. Initially both cooperate, but after some time WSLS recognizes that

ALLC does not reciprocate defection. WSLS will then switch from mutual cooperation to exploitation. The sequence of moves is

$$\begin{aligned} &\text{WSLS: } CCCCC\overset{*}{D}DDDD\ldots \\ &\text{ALLC: } CCCCCCCCCC \end{aligned} \tag{5.24}$$

Again the asterisk indicates an erroneous move. Sadly, this behavior is also compatible with human psychology: unconditional (defenseless) cooperators tend to be exploited.

In summary, TFT is the best initiator of cooperation in a society that consists largely of defectors. Once cooperation has been established, however, TFT's unforgiving retaliation causes its downfall. Forgiving strategies like GTFT or WSLS take over. A GTFT population can drift to unconditional cooperation, which in turn invites defectors. A population using WSLS is stable against neutral drift to ALLC and can resist invasion by ALLD. WSLS is no simpleton (Figure 5.10).

SUMMARY

- Cooperation means that the donor pays a cost and the recipient gets a benefit.
- In evolutionary biology, cost and benefit are measured in terms of reproductive success.
- It is not obvious how natural selection can lead from competition to cooperation.
- Yet cooperation is abundant in biology.
- The Prisoner's Dilemma is a game that captures the essence of cooperation.
- In the Prisoner's Dilemma, cooperation is dominated by defection.
- The repeated Prisoner's Dilemma is a tool for studying direct reciprocity, which presents a mechanism for the evolution of cooperation.
- Tit-for-tat is a simple but successful strategy of direct reciprocity. Tit-for-tat cooperates on the first move and then does whatever the opponent did in the previous round.

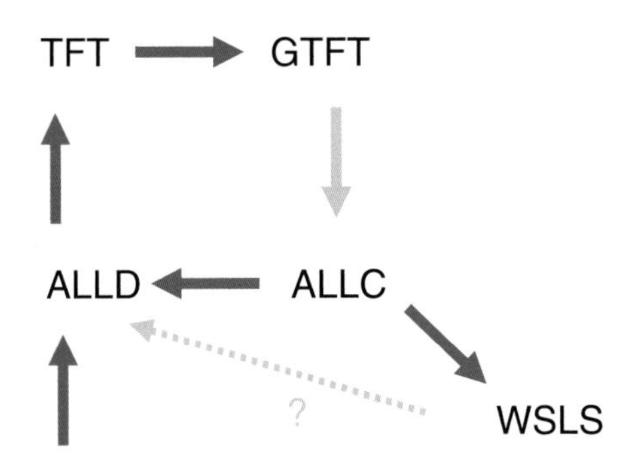

Figure 5.10 War and peace. Out of an initial "chaos" of random strategies, ALLD emerges as the first winner. Defection is born out of chaos. TFT is the best catalyst for initiating a cooperative society; only a small cluster of TFT players is needed to invade and replace ALLD. Once TFT is abundant, it is not here to stay. It is harmed by its own unforgiveness and replaced by GTFT. If everybody is nice and generous, the biological trait of retaliation is lost. ALLC emerges as a neutral mutant and takes over by random drift. A population of unconditional cooperators (ALLC) invites the rise of defectors (ALLD). There are ongoing oscillations of cooperative and defective societies in a struggle of war and peace. The cycles are interrupted by the emergence of WSLS, which can dominate ALLC and resist invasion by ALLD. Occasionally we observe evolutionary trajectories that dislodge WSLS (dotted arrow); the mechanism of this evolution is not yet fully understood.

- Tit-for-tat has two weaknesses: (i) it cannot correct mistakes; (ii) it cannot prevent neutral drift leading to "Always cooperate," ALLC.
- Generous Tit-for-tat cooperates whenever the opponent has cooperated, and sometimes even cooperates when the opponent has defected. Generous Tit-for-tat can correct mistakes.
- Evolutionary analysis of "reactive strategies" reveals that Tit-for-tat is a catalyst for the emergence of cooperation, but not the ultimate goal. It is replaced by Generous Tit-for-tat.
- Both strategies are outcompeted by Win-stay, lose-shift, which can correct mistakes and is stable against neutral drift to ALLC.

FINITE POPULATIONS 6

WE WILL NOW BEGIN to analyze evolutionary dynamics in finite populations. The abundance of individuals is given by integers rather than by continuous variables. The resulting evolutionary dynamics are no longer described by deterministic differential equations, but require a stochastic formulation.

The best approach for studying a biological problem is to try a deterministic description first and then move to a stochastic analysis only when the deterministic one misses relevant aspects. Usually differential equations are easier to analyze and interpret than stochastic processes, but many important biological effects only arise in a stochastic context. One such effect is neutral drift. In this chapter, we will study neutral drift and constant selection in populations of finite size.

6.1 NEUTRAL DRIFT

Consider a population of fixed size N. There are two types of individuals, A and B. They reproduce at the same rate. Therefore A and B are neutral variants with respect to selection. In any one time step, a random individual is chosen for reproduction and a random individual is chosen for elimination.

The Moran process

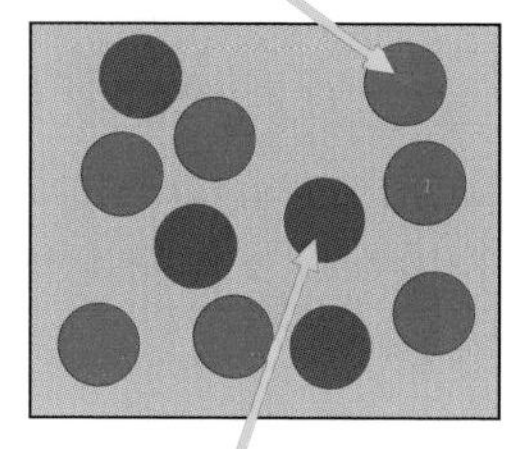
Choose one individual for reproduction
. . . and one for death

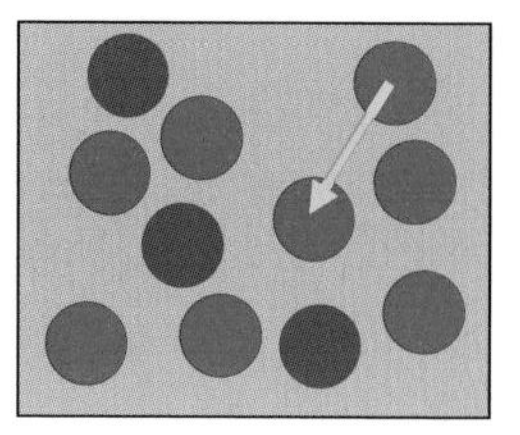
The offspring of the first individual replaces the second

Figure 6.1 The Moran process represents the simplest possible stochastic model to study selection in a finite population. In each time step, two individuals are chosen: one for reproduction and one for elimination. The offspring of the first individual will replace the second. Note that the two random choices could fall on the same individual; in this case, an individual will be replaced by its own offspring. The total population size is strictly constant.

We use sampling with replacement: the same individual could be chosen for reproduction and death. Reproduction occurs without mutation: A produces A and B produces B.

This stochastic process is named after the Australian population geneticist P. A. P. Moran, who invented it in 1958. The feature that in each time step there is always one birth and one death event ensures that the total population size is strictly constant (Figure 6.1). The only stochastic variable is the number of A individuals denoted by i. The number of B individuals is $N - i$. Stochastic processes with one variable are much easier to investigate than stochastic processes with two or more variables.

The Moran process is defined on the state space $i = 0, \ldots, N$. The probability of choosing an A individual (for birth or death) is given by i/N. The probability of choosing a B individual is given by $(N - i)/N$. There are four possibilities of what could happen in any one time step.

(i) An A individual could be chosen for reproduction and death. This event has probability $(i/N)^2$. After the event the number of A individuals is the same as before; the variable i has not changed.

(ii) A B individual could be chosen for reproduction and death. This event has probability $[(N - i)/N]^2$. After the event the number of B individuals is the same as before; the variable i has not changed.

(iii) An A individual could be chosen for reproduction and a B individual for death. This event has probability $i(N-i)/N^2$. After the event there is one more A individual than before; the variable i has changed to $i+1$.

(iv) A B individual could be chosen for reproduction and an A individual for death. This event also has probability $i(N-i)/N^2$. After the event there is one less A individual than before; the variable i has changed to $i-1$.

The transition matrix, $P = [p_{ij}]$, determines the probabilities of moving from any one state i to any other state j. P is an $(N+1) \times (N+1)$ stochastic matrix. All entries are probabilities. The sum over each row is one. For our stochastic process, the transition matrix is given by

$$\begin{aligned} p_{i,i-1} &= i(N-i)/N^2 \\ p_{i,i} &= 1 - p_{i,i+1} - p_{i,i+1} \\ p_{i,i+1} &= i(N-i)/N^2 \end{aligned} \tag{6.1}$$

All other entries are zero. Therefore the transition matrix is tri-diagonal. This is the defining property of "birth-death" processes. In any elementary stochastic step, the state variable i can only change by at most one (Figure 6.2).

For our particular birth-death process, we note that

$$p_{0,0} = 1 \qquad p_{0,i} = 0 \qquad \forall i > 0 \tag{6.2}$$

and

$$p_{N,N} = 1 \qquad p_{N,i} = 0 \qquad \forall i < N \tag{6.3}$$

The states $i = 0$ and $i = N$ are "absorbing states": once the process has reached such a state, it will stay there forever. The states $i = 1, \ldots, N-1$ are called transient. The process stays in the set of transient states only for some limited time. Eventually the population will consist of either all A or all B individuals. Although there is no selection, one of the two types will replace the other. Coexistence is not possible.

Since our stochastic process has two absorbing states, we can ask: starting in state i, what is the probability of reaching state N? In other words, given that

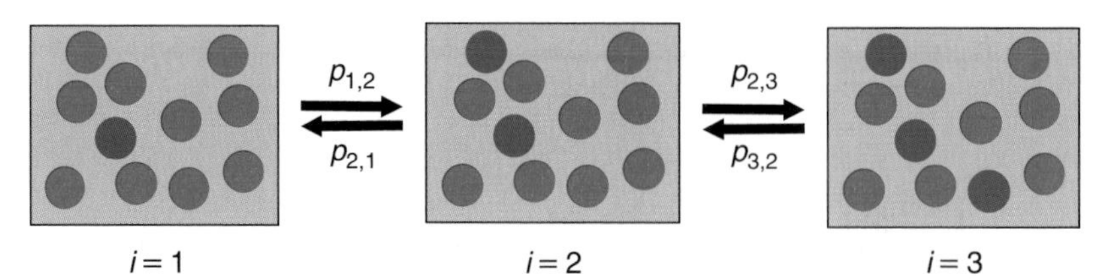
The Moran process is a birth-death process
$p_{1,2}$
$p_{2,1}$
$p_{2,3}$
$p_{3,2}$
$i=1$
$i=2$
$i=3$

There are two absorbing states: all-red and all-blue
$i=0$
$i=N$

Figure 6.2 The Moran process is a birth-death process. In each time step, the number, i, of blue individuals can only change by one at most. There are two absorbing states, $i = 0$ and $i = N$. In both cases, one type has taken over the entire population. No further change can occur (unless there is a new mutation).

we start with i many A individuals, what is the probability that eventually the whole population will consist of A individuals?

Let us do a formal calculation, which will be generalized in the next section. Denote by x_i the probability of ending up in state N when starting from state i. The probability of ending up in state 0 when starting from state i is given by $1 - x_i$, because there are no other absorbing states. We have

$$\begin{aligned} x_0 &= 0 \\ x_i &= p_{i,i-1}x_{i-1} + p_{i,i}x_i + p_{i,i+1}x_{i+1} \qquad \forall i = 1, \ldots, N-1 \qquad (6.4) \\ x_N &= 1 \end{aligned}$$

The probability of being absorbed in state N starting from i is given by the sum of the following three terms: (i) the probability of going from i to $i - 1$

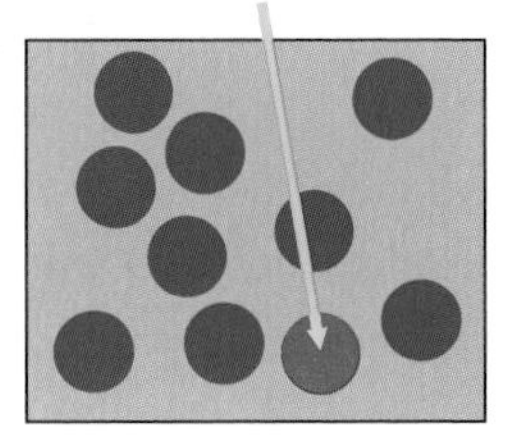

Neutral drift
The probability that a particular individual will become the ancestor of all individuals in the population is 1/N

Figure 6.3 In a finite population, if we wait long enough, the descendants of one particular individual will take over the entire population. If all individuals have the same fitness, then all individuals currently present in the population must have the same chance. Hence under neutral drift, the fixation probability (of the lineage) of any one individual is $1/N$.

multiplied by the probability of being absorbed from $i-1$; (ii) the probability of staying in i multiplied by the probability of being absorbed from i; (iii) the probability of going from i to $i+1$ multiplied by the probability of being absorbed from $i+1$. Thus we have a recursive equation for x_i. Note that $x_0 = 0$; from state 0 we can never reach state N. Moreover, we have $x_N = 1$; from state N, we will certainly reach state N, because we are already there.

Since $p_{i,i-1} = p_{i,i+1}$ and $p_{i,i} = 1 - 2p_{i,i+1}$, the solution of the linear system (6.4) is

$$x_i = i/N \qquad \forall i = 0, \ldots, N \tag{6.5}$$

The result is obvious. Since all individuals reproduce and die at the same rate, the chance that a particular individual will generate a lineage that will inherit the whole population must be $1/N$ (Figure 6.3). If there are i many A individuals, then the chance that one of them will make it (as opposed to one of the $N-i$ many B individuals) is simply i/N.

For each trajectory of our stochastic process, there are only two final possibilities: the trajectory reaches either state 0 or state N. The probability of being absorbed in 0 is one minus the probability of being absorbed in N. The probability of ending up in all-B when starting with $N-i$ many B individuals is given by $(N-i)/N$.

6.2 BIRTH-DEATH PROCESSES

Let us now perform the same calculation for a general birth-death process. A birth-death process is a one-dimensional stochastic process on a discrete state space, $i = 0, \ldots, N$. In each stochastic event, the state variable i can either remain unchanged or move to $i - 1$ or $i + 1$.

Denote by α_i the probability of a transition from i to $i + 1$. Denote by β_i the probability of a transition from i to $i - 1$. We have $\alpha_i + \beta_i \leq 1$. The probability of remaining in state i is given by $1 - \alpha_i - \beta_i$. Consider a birth-death process where $i = 0$ and $i = N$ are absorbing states. Therefore, we have $\alpha_0 = 0$ and $\beta_N = 0$. The transition matrix is of the form

$$P = \begin{pmatrix} 1 & 0 & 0 & \ldots & 0 & 0 & 0 \\ \beta_1 & 1-\alpha_1-\beta_1 & \alpha_1 & \ldots & 0 & 0 & 0 \\ \vdots & \vdots & \vdots & \ddots & \vdots & \vdots & \vdots \\ 0 & 0 & 0 & \ldots & \beta_{N-1} & 1-\alpha_{N-1}-\beta_{N-1} & \alpha_{N-1} \\ 0 & 0 & 0 & \ldots & 0 & 0 & 1 \end{pmatrix} \tag{6.6}$$

Denote by x_i the probability of reaching state N when starting from i. Clearly, $1 - x_i$ denotes the probability of reaching state 0 when starting from state i. In analogy to (6.4) we have

$$\begin{aligned} x_0 &= 0 \\ x_i &= \beta_i x_{i-1} + (1 - \alpha_i - \beta_i)x_i + \alpha_i x_{i+1} \qquad i = 1, \ldots, N-1 \\ x_N &= 1 \end{aligned} \tag{6.7}$$

In vector notation, we can write

$$\vec{x} = P\vec{x} \tag{6.8}$$

The absorption probabilities are given by the right-hand eigenvector associated with the largest eigenvalue, which is one, because P is a stochastic matrix.

Let us introduce the variables

$$y_i = x_i - x_{i-1} \qquad i = 1, \ldots, N \tag{6.9}$$

Note that $\sum_{i=1}^{N} y_i = x_1 - x_0 + x_2 - x_1 + \cdots x_N - x_{N-1} = x_N - x_0 = 1$. Let $\gamma_i = \beta_i/\alpha_i$. From equation (6.7) we find $y_{i+1} = \gamma_i y_i$. Therefore we have $y_1 = x_1$, $y_2 = \gamma_1 x_1$, $y_3 = \gamma_1\gamma_2 x_1$, and so on. If we sum all these expressions we get

$$x_1 = \frac{1}{1 + \sum_{j=1}^{N-1} \prod_{k=1}^{j} \gamma_k}. \tag{6.10}$$

From

$$x_i = x_1 \left(1 + \sum_{j=1}^{i-1} \prod_{k=1}^{j} \gamma_k \right), \tag{6.11}$$

we obtain

$$x_i = \frac{1 + \sum_{j=1}^{i-1} \prod_{k=1}^{j} \gamma_k}{1 + \sum_{j=1}^{N-1} \prod_{k=1}^{j} \gamma_k}. \tag{6.12}$$

Consider a population of one A individual and $N - 1$ B individuals. The probability that A takes over the whole population is called the fixation probability of A. We denote this probability by ρ_A. The idea is that a homogeneous population of B has produced a mutant of type A. We are interested in the probability of this mutant becoming fixed in the population, which means that it generates a lineage that takes over the whole population. Similarly, we denote by ρ_B the probability that a single B individual takes over a population that contains $N - 1$ A individuals. The fixation probabilities of A and B are respectively given by $\rho_A = x_1$ and $\rho_B = 1 - x_{N-1}$. Therefore, we have

$$\begin{aligned} \rho_A &= \frac{1}{1 + \sum_{j=1}^{N-1} \prod_{k=1}^{j} \gamma_k} \\ \rho_B &= \frac{\prod_{k=1}^{N-1} \gamma_k}{1 + \sum_{j=1}^{N-1} \prod_{k=1}^{j} \gamma_k} \end{aligned} \tag{6.13}$$

Note that the ratio of the fixation probabilities is simply given by the product over all γ_i

$$\frac{\rho_B}{\rho_A} = \prod_{k=1}^{N-1} \gamma_k. \tag{6.14}$$

If $\rho_B/\rho_A > 1$, then it is more likely that a single B mutant becomes fixed in an A population than the other way round.

The fixation probabilities that we have derived in this section hold for any selection scenario between A and B, including neutral drift, constant selection, and frequency-dependent selection.

6.3 RANDOM DRIFT WITH CONSTANT SELECTION

Let us now study the same process as before, but assume that A has fitness r while B has fitness 1. If $r > 1$, then selection favors A. If $r < 1$, then selection favors B. If $r = 1$, we are back to neutral drift. The fitness difference can be included in our process by modifying the probabilities of choosing A or B for reproduction.

The probability that A is chosen for reproduction is given by $ri/(ri + N - i)$. The probability that B is chosen for reproduction is given by $(N - i)/(ri + N - i)$. The probability that A is chosen for elimination is i/N. The probability that B is chosen for elimination is $(N - i)/N$. For the transition matrix, we obtain

$$\begin{aligned} p_{i,i-1} &= \frac{N-i}{ri+N-i}\frac{i}{N} \\ p_{i,i} &= 1 - p_{i,i+1} - p_{i,i+1} \\ p_{i,i+1} &= \frac{ri}{ri+N-i}\frac{N-i}{N} \end{aligned} \tag{6.15}$$

All other elements of the matrix are zero. Again we want to calculate the fixation probability, x_i, to reach state N starting from state i. Note that

$$\gamma_i = \frac{p_{i,i-1}}{p_{i,i+1}} = \frac{1}{r}. \tag{6.16}$$

The fixation probability of a new mutant with relative fitness r is:

$$\rho = \frac{1-1/r}{1-1/r^N}$$

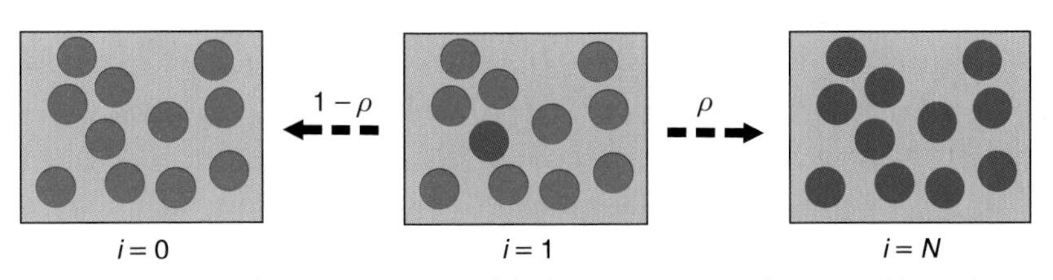

Figure 6.4 Suppose that a new mutant (blue) arises in a population and has relative fitness r. The lineage starting from this mutant can either become extinct or take over the whole population. The probability that the mutants will take over the population is given by the "fixation probability" $\rho = (1-1/r)/(1-1/r^N)$.

Therefore the probability of being absorbed in state N when starting in state i is given by

$$x_i = \frac{1-1/r^i}{1-1/r^N}. \tag{6.17}$$

The fixation probability of a single A individual in a population of $N-1$ B individuals (Figure 6.4) is

$$\rho_A = x_1 = \frac{1-1/r}{1-1/r^N}. \tag{6.18}$$

The fixation probability of a single B individual in a population of $N-1$ A individuals is

$$\rho_B = 1 - x_{N-1} = \frac{1-r}{1-r^N}. \tag{6.19}$$

The ratio of the two fixation probabilities is given by

$$\frac{\rho_B}{\rho_A} = r^{1-N}. \tag{6.20}$$

For the fixation probability of an advantageous A mutant, $r > 1$, in a large population, $N \gg 1$, we have the useful approximation

$$\rho_A = 1 - 1/r. \tag{6.21}$$

Even in the limit of an infinitely large population, $N \to \infty$, there is no guarantee that an advantageous mutant will take over. This is an important difference between deterministic and stochastic models of evolution. In a deterministic setting, an advantageous mutant is certain of victory regardless of how small r is as long as $r > 1$. In a stochastic setting, the chance of extinction always remains, no matter how large the population size N.

Let us consider some numerical examples for a population of size $N = 100$:

A 100% selective advantage, $r = 2$, leads to $\rho = 0.5$.

A 10% selective advantage, $r = 1.1$, leads to $\rho = 0.09$.

A 1% selective advantage, $r = 1.01$, leads to $\rho = 0.016$.

For a neutral mutant, $r = 1$, we have $\rho = 1/N = 0.01$.

A 1% selective disadvantage, $r = 0.99$, leads to $\rho = 0.0058$.

A 10% selective disadvantage, $r = 0.9$, leads to $\rho = 0.000003$.

We can also ask how often must a mutant with relative fitness r arise, before it has a probability 1/2 of taking over the population. The answer is $m = -\log 2/\log(1 - \rho)$. Again we consider a population of size 100. A mutant with $r = 2$ must arise once. A mutant with $r = 1.1$ must arise 7 times, while a mutant with $r = 1.01$ must arise 44 times. A neutral mutant, $r = 1$, must arise 69 times. Disadvantageous mutants with $r = 0.99$ or $r = 0.9$ must arise 119 times and about 234,861 times, respectively.

6.4 THE RATE OF EVOLUTION

Imagine a population of N reproducing individuals. All individuals are of the same type, A. Very rarely a mutation occurs which produces an individual of type B. Assume that mutation happens during reproduction. The mutation rate u represents the probability that the reproduction of A results in B. Thus $1 - u$ is the probability that reproduction of A occurs without mutation. For how long do we have to wait until a population of N A individuals will produce a B mutant? The rate at which a B mutant is being produced by the population is Nu. The time until the B mutant arises is exponentially distributed with mean $1/(Nu)$.

Suppose type B has a relative fitness r compared to fitness 1 of type A. Thus the probability that the new B mutant will take over the population is given by

$$\rho = \frac{1 - 1/r}{1 - 1/r^N}. \tag{6.22}$$

The rate of evolution from all-A to all-B is given by

$$R = Nu\rho. \tag{6.23}$$

The rate at which a B mutant is being produced is Nu. The probability that a B mutant reaches fixation is ρ. Hence the rate of transition from all-A to all-B is the product of these two terms.

If B is neutral, then $\rho = 1/N$ and the rate of neutral evolution is given by

$$R = u. \tag{6.24}$$

The rate of neutral evolution is independent of the population size and simply equals the mutation rate. This important result was derived by Motoo Kimura.

This insight is at the center of the so-called neutral theory of evolution. According to the neutral theory, the majority of mutations that can be observed—for example, when comparing genetic sequences of humans and chimpanzees—should be neutral. Advantageous mutations are extremely unlikely to occur in genes that have been optimized for millions of generations in the ancestors of these species. Deleterious mutations cannot be observed

Figure 6.5 The rate of producing neutral mutants is Nu, where N is the population size and u is the mutation rate. The fixation probability of a neutral mutant is $1/N$. Therefore the rate of neutral evolution is $R = Nu/N = u$: the rate of evolution equals the mutation rate. The population size cancels out. This relationship holds even if the total population size is changing over time. If the mutation rate is constant, then neutral mutants accumulate at a constant rate, giving rise to a "molecular clock." The figure shows three mutations that succeed in taking over the population. For each mutation that becomes fixed there are on average N unsuccessful attempts.

because they would be eliminated with high probability. Hence the majority of observed mutations in any phylogeny should be neutral (or nearly neutral).

The rate of accumulating neutral mutations is simply given by the mutation rate and is independent of the population size and fluctuations in the population size. If the mutation rate depends mostly on the accuracy of DNA replication, which in turn is performed by a well-optimized system of enzymes that has not changed much in all eukaryotes, then the rate of evolution is constant. The neutral theory provides a "molecular clock" (Figure 6.5).

There was once a heated controversy between the supporters and the opponents of the neutral theory of evolution. The extreme neutralist would say: all observable mutations, say between human and chimp, are neutral; hence neutral variation alone can explain the evolutionary divergence between these two species; adaptation is unimportant. The extreme adaptationist would say: neutral evolution is unimportant; it is not even evolution, because it represents random variation without adaptation; evolution always requires adaptation.

The resolution of the controversy is obvious for those who were never involved. Most molecular variation is neutral. Therefore neutrality is an excellent model for studying genetic variation. Neutrality is often a good assumption for building mathematical tools that help to calculate phylogenetic relationships among species. Certainly the overwhelming majority of mutations that have been fixed in populations since the origin of life have been neutral. Very occasionally, however, advantageous mutations come into play. These mutations are extremely important for determining the trajectory of evolution.

SUMMARY

- Evolutionary dynamics in populations of finite size require a stochastic theory.
- The Moran process is a birth-death process, which describes evolution in finite populations.
- If a finite population contains several different types, then without mutation eventually all but one type will be extinct. This is the case even if all types have the same fitness. This principle is called "neutral drift."
- In a population of size N, a neutral mutant will reach fixation with probability $1/N$.
- A mutant with relative fitness, r, will reach fixation with probability $\rho = (1 - 1/r)/(1 - 1/r^N)$.
- The rate of evolution is given by the product of the population size, N, the mutation rate, u, and the probability of fixation ρ.
- The rate of neutral evolution is given by the mutation rate, u, and is independent of the population size (because $\rho = 1/N$).
- If the mutation rate is constant, then neutral mutations accumulate in genomes at a constant rate. This effect is called the "molecular clock."
- The neutral theory of evolution recognizes the fact that the majority of mutations that become fixed in genomes are neutral.

GAMES IN FINITE POPULATIONS 7

IN CHAPTER 4, we discussed the traditional approach to evolutionary game dynamics via the replicator equation, which describes deterministic evolution in infinitely large populations. All of our understanding of frequency-dependent selection comes from this approach. We will now develop a framework for studying evolutionary game dynamics in finite populations. Finiteness requires stochasticity. The interplay of random drift and frequency-dependent selection will determine the outcome of evolutionary games. We will calculate fixation probabilities to decide whether selection favors one strategy over another. In a game between two strategies, A and B, the fixation probability of A is given by the probability that a single A player in a population of $N - 1$ B players generates a lineage of A that does not become extinct but instead takes over the whole population. If the fixation probability of A is greater than $1/N$, then selection favors A replacing B.

The intensity of selection plays an important role for game dynamics in finite populations. The game under consideration can have a strong or weak influence on the overall fitness of an individual. If the payoff makes a small contribution to fitness, then selection is weak. If the payoff makes a large contribution to fitness, then selection is strong. Some of our results only hold in the limit of weak selection. In the traditional replicator equation, in contrast, any parameter that describes the intensity of selection cancels out.

Biologists are interested in the concepts of a strict Nash equilibrium or an evolutionarily stable strategy, because natural selection protects populations of such strategies against invasion by mutants. We will see, however, that this implication only holds for deterministic dynamics of infinite populations. For stochastic dynamics of finite populations, we have to derive new conditions for evolutionary stability.

Risk dominance is an important concept in game theory, defined as follows: if two strategies, A and B, are best replies to themselves, then the risk-dominant strategy has the larger basin of attraction. We will see that in finite populations, however, the risk-dominant strategy need not have the larger fixation probability. Instead, we will encounter a $1/3$ law. If the basin of attraction of strategy B is less than $1/3$, then selection will favor the fixation of strategy A for sufficiently large N and weak selection.

7.1 ONE BASIC MODEL AND ONE-THIRD

Consider a game between two strategies, A and B, with payoff matrix

$$\begin{array}{c} \\ A \\ B \end{array} \begin{array}{c} \begin{array}{cc} A & B \end{array} \\ \begin{pmatrix} a & b \\ c & d \end{pmatrix} \end{array} \tag{7.1}$$

The total population size is N. The number of A individuals is i. The number of B individuals is $N - i$. For each individual, there are $N - 1$ other individuals. For each A individual, there are $i - 1$ other A individuals. For each B individual, there are $N - i - 1$ other B individuals. The probability that an A individual interacts (plays the game) with another A individual is given by $(i - 1)/(N - 1)$. The probability that an A individual interacts with a B individual is given by $(N - i)/(N - 1)$. The probability that a B individual interacts with another B individual is given by $(N - i - 1)/(N - 1)$. The probability that a B individual interacts with an A individual is given by $i/(N - 1)$. Hence, the expected payoff for A and B is, respectively, given by

$$F_i = \frac{a(i-1) + b(N-i)}{N-1}$$
$$G_i = \frac{ci + d(N-i-1)}{N-1} \tag{7.2}$$

The index i indicates that these quantities represent the expected payoff in a population that contains i many A individuals.

In the traditional framework of evolutionary game dynamics, the expected payoff is interpreted as fitness. Individuals reproduce, either genetically or culturally, with a rate that is proportional to their payoff. Let us introduce a parameter w that measures the intensity of selection. The fitness of A and B is given by

$$f_i = 1 - w + wF_i$$
$$g_i = 1 - w + wG_i \tag{7.3}$$

The intensity of selection, w, is a number between 0 and 1. If $w = 0$, the game does not contribute to fitness. Strategies A and B are neutral variants. If $w = 1$, selection is strong; the fitness is entirely determined by the expected payoff. The limit $w \to 0$ characterizes the case of weak selection, where the payoff provides only a small contribution to fitness. Figure 7.1 illustrates the basic model of evolutionary game dynamics in finite populations.

It is important to note that the parameter w, which quantifies the intensity of selection, cancels out in deterministic replicator dynamics of infinite populations, but plays a crucial role in the stochastic process describing finite populations. We will obtain elegant results in the limit of weak selection.

Consider a Moran process between A and B. The frequency-dependent fitness values are given by equation (7.3). The state variable, i, denotes the number of A individuals. The probability to move from i to $i+1$ is given by

$$p_{i,i+1} = \frac{if_i}{if_i + (N-i)g_i} \frac{N-i}{N}. \tag{7.4}$$

The probability to move from i to $i-1$ is given by

$$p_{i,i-1} = \frac{(N-i)g_i}{if_i + (N-i)g_i} \frac{i}{N}. \tag{7.5}$$

Games in finite populations

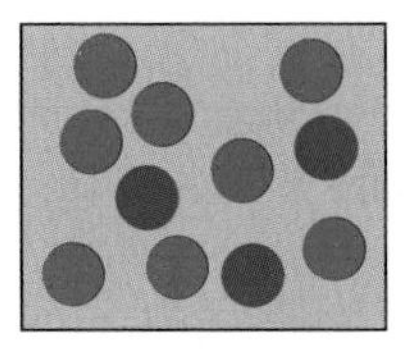

There are i players of type A and $N - i$ players of type B

The fitness of A is: $f_i = 1 - w + w\dfrac{a(i-1) + b(N-i)}{N-1}$

The fitness of B is: $g_i = 1 - w + w\dfrac{ci + d(N-i-1)}{N-1}$

The parameter w denotes the intensity of selection

Figure 7.1 We can study evolutionary game dynamics in finite populations of size N. Each individual can interact with $N - 1$ other individuals. The expected payoff for each individual is determined from these interactions. The parameters a, b, c, and d represent the entries of the payoff matrix. The parameter w, a number between 0 and 1, denotes the intensity of selection. If $w = 1$, then the fitness of an individual is identical to its payoff. If $w = 0$, all individuals have the same fitness. Small w denotes the case of weak selection: the game under consideration makes only a small contribution to the total fitness of an individual. In any one time step, one individual is chosen for reproduction proportional to fitness, while a second individual is chosen for elimination at random. The total population size is constant.

The probability that the process remains in state i is simply

$$p_{i,i} = 1 - p_{i,i+1} - p_{i,i-1}. \tag{7.6}$$

All other transitions have zero probability.

Note that $p_{0,0} = 1$ and $p_{N,N} = 1$. Therefore the process has two absorbing states, $i = 0$ and $i = N$. If the population has reached either one of these states, then it will stay there forever. Any mixed population of A and B will eventually end up in either all-A or all-B. We want to calculate the fixation probabilities of A and B (Figure 7.2).

Fixation probability of strategy A

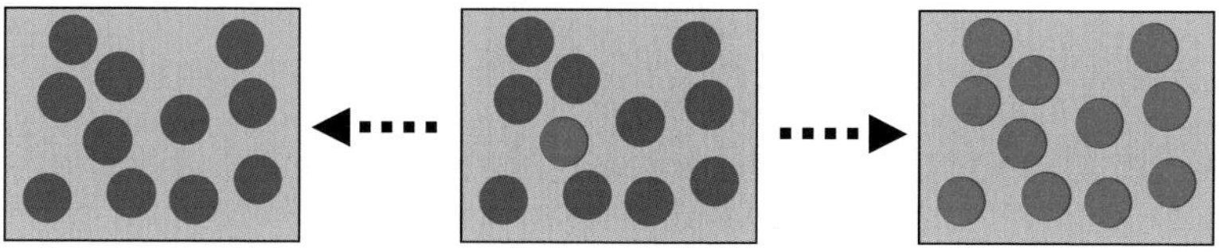

$$\rho = \frac{1}{1+\sum_{k=1}^{N-1}\prod_{i=1}^{k}\frac{g_i}{f_i}}$$

If $\rho > 1/N$, then selection favors A replacing B

If $\rho < 1/N$, then selection opposes A replacing B

Figure 7.2 The fixation probability, ρ, of a strategy under frequency-dependent selection can be calculated. For a neutral mutant, $\rho = 1/N$. Selection favors the fixation of the invading strategy if $\rho > 1/N$. Selection opposes the fixation of the invading strategy if $\rho < 1/N$.

For the backward to forward transition ratio, we obtain

$$\frac{p_{i,i-1}}{p_{i,i+1}} = \frac{g_i}{f_i}. \tag{7.7}$$

Using equation (6.13) of Chapter 6, the fixation probability of A is given by

$$\rho_A = 1/\left(1+\sum_{k=1}^{N-1}\prod_{i=1}^{k}\frac{g_i}{f_i}\right). \tag{7.8}$$

The ratio of the fixation probabilities is

$$\frac{\rho_B}{\rho_A} = \prod_{i=1}^{k}\frac{g_i}{f_i}. \tag{7.9}$$

For weak selection . . .

$\rho > 1/N$

is equivalent to

$a(N-2) + b(2N-1) > c(N+1) + d(2N-4)$

for $N=2$:	$b > c$
$N=3$:	$a+5b > 4c+2d$
$N=4$:	$2a+7b > 5c+4d$
$N=5$:	$3a+9b > 6c+6d$
. . .	
large N:	$a+2b > c+2d$

Figure 7.3 In the limit of weak selection, $w \to 0$, the condition $\rho > 1/N$ is equivalent to a simple inequality that is linear in N. For any given population size, therefore, a simple condition determines whether or not selection favors the fixation of a strategy.

Let us consider the limit of weak selection. A Taylor expansion of equation (7.8) for $w \to 0$ leads to

$$\rho_A \approx \frac{1}{N}\frac{1}{1-(\alpha N-\beta)w/6}. \tag{7.10}$$

Here $\alpha = a + 2b - c - 2d$ and $\beta = 2a + b + c - 4d$.

If $\rho_A > 1/N$, then selection favors the fixation of A. From equation (7.10), we see that $\rho_A > 1/N$ is equivalent to $\alpha N > \beta$. This condition can be written as

$$a(N-2) + b(2N-1) > c(N+1) + d(2N-4). \tag{7.11}$$

For a population of only two individuals, $N = 2$, we have

$$b > c. \tag{7.12}$$

This result makes sense: in a mixed population of one A and one B individual, the former has payoff b and the latter has payoff c; hence if $b > c$, then A is more likely to become fixed than B (Figure 7.3).

For large population size, inequality (7.11) leads to

$$a + 2b > c + 2d. \tag{7.13}$$

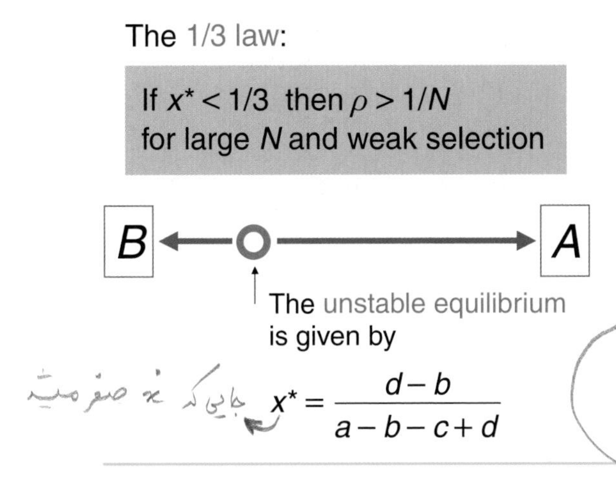

The 1/3 law:
If x* < 1/3 then ρ > 1/N
for large N and weak selection
B
A
The unstable equilibrium
is given by
x* = (d − b)/(a − b − c + d)

Figure 7.4 There is a surprising $1/3$ law for evolutionary games in finite populations. Consider two strategies A and B in a bistable relationship, $a > c$ and $b < d$. The unstable equilibrium (of the replicator equation) occurs at a frequency of A given by $x^* = (d - b)/(a - b - c + d)$. For finite population dynamics, we find that selection favors strategy A, which means $\rho_A > 1/N$, if $x^* < 1/3$. In other words, a strategy has a fixation probability greater than $1/N$, if it has a higher fitness at frequency 1/3. This simple condition holds for weak selection and large population size.

How can we interpret this condition?

Consider a game with $a > c$ and $b < d$. Both A and B are best replies to themselves. Consider the limit of large population size. If the frequency of A is high, then A has a larger fitness than B. If the frequency of B is high, then B has a larger fitness than A. There is a point where the two fitnesses are equal. This point can be calculated by setting $F_i = G_i$ in equation (7.2). For large N, this equilibrium point is reached at a frequency of A given by

$$x^* = \frac{d - b}{a - b - c + d}. \tag{7.14}$$

In the replicator equation, this expression denotes the unstable equilibrium between A and B.

Inequality (7.13) leads to

$$x^* < 1/3. \tag{7.15}$$

Therefore, if the unstable equilibrium occurs at a frequency of A which is less than $1/3$, then in a large finite population of size N, in the limit of weak selection, the probability that a single A mutant takes over the whole population is greater than $1/N$. In this case, selection favors the fixation of A in B. The condition $x^* < 1/3$ also means that the basin of attraction of B is less than $1/3$ (Figure 7.4).

If A dominates B, then $a > c$ and $b > d$. In this case, $x^* < 0$ and inequality (7.13) always holds. Therefore, if A dominates B, then selection will favor the fixation of A and oppose the fixation of B in a sufficiently large population. But the dominated strategy, B, can still be favored in a small population if $b < c$. In this case, there will be a critical population size, N_c. If $N < N_c$, then selection might favor the dominated strategy B. If $N > N_c$, selection will favor the dominant strategy A.

7.2 EVOLUTIONARY STABILITY IN FINITE POPULATIONS

These results have immediate consequences for the concept of evolutionary stability. The well-known definition of an evolutionarily stable strategy is motivated by selection dynamics in infinite populations. For payoff matrix 7.1, strategy B is ESS if either (i) $d > b$ or (ii) both $d = b$ and $a < c$. These conditions imply that selection opposes the spread of infinitesimally small fractions of A in infinitely large populations of B.

For finite population size N, we propose that B is an evolutionarily stable strategy, ESS_N, if two conditions hold: (i) selection opposes A invading B, which means that a single mutant A in a population of B has a lower fitness; and (ii) selection opposes A replacing B, which means $\rho_A < 1/N$, for any $w > 0$ (Figure 7.5).

The first condition is equivalent to

$$b(N-1) < c + d(N-2). \tag{7.16}$$

The second condition, for small w, is equivalent to

$$a(N-2) + b(2N-1) < c(N+1) + d(2N-4). \tag{7.17}$$

For $N = 2$, both conditions reduce to $b < c$. For large populations, the two conditions lead to $b < d$ and $x^* > 1/3$, respectively. Hence for small populations the traditional ESS concept is neither necessary nor sufficient; for large populations, it is necessary but not sufficient (Figure 7.6). If we consider a game with many different strategies, then the two conditions must hold in pairwise comparison with every other strategy.

Evolutionary stability in finite populations

	A	B
A	a	b
B	c	d

B is ESS_N if

1. Selection opposes A invading B:

$$b(N-1) < c + d(N-2)$$

2. Selection opposes A replacing B:

$$a(N-2) + b(2N-1) < c(N+1) + d(2N-4)$$

Figure 7.5 There are two logical requirements for evolutionary stability in finite populations. Selection has to protect an evolutionarily stable strategy (ESS) against the invasion and fixation of a mutant strategy. If the fitness of a single mutant is less than the fitness of the resident, then selection opposes invasion. If the fixation probability of the mutant is less than $1/N$, then selection opposes fixation. The first condition is always a simple linear inequality in N. The second condition is a simple linear inequality in N for weak selection.

For small N $(N=2)$

B is ESS_N if

1. $b < c$
2. $b < c$

Traditional ESS condition is neither necessary nor sufficient

For large N

B is ESS_N if

1. $b < d$
2. $x^* < 1/3$

Traditional ESS condition is necessary but not sufficient

Figure 7.6 The smallest possible population size for an evolutionary game is $N = 2$. In this case, both conditions for evolutionary stability reduce to $b < c$. For large N, the invasion condition is $b < d$ and the fixation condition is $x^* < 1/3$. Hence for small finite populations, the traditional ESS concept is neither necessary nor sufficient to confer protection by selection. For large finite populations, the traditional ESS concept is necessary but not sufficient.

Two examples

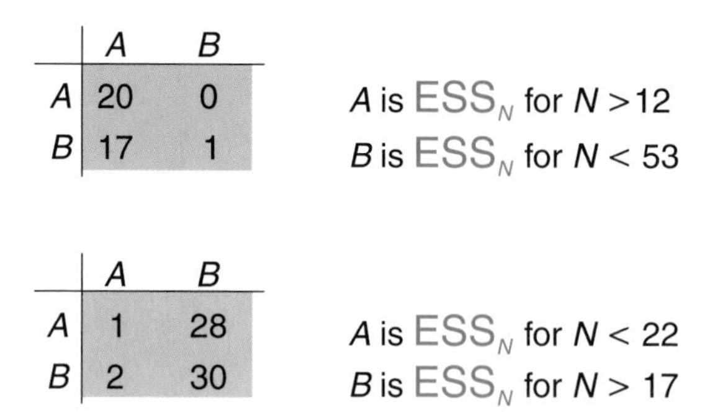

A B
A 20 0
B 17 1
A is ESS$_N$ for N >12
B is ESS$_N$ for N < 53
A B
A 1 28
B 2 30
A is ESS$_N$ for N < 22
B is ESS$_N$ for N > 17

Figure 7.7 Whether one or the other strategy is ESS$_N$ depends on the population size. Two interesting examples are shown. In the first case, both A and B are strict Nash equilibria. In finite populations, however, B is the only ESS$_N$ for $N = 2, 3, \ldots, 12$, both strategies are ESS$_N$ for $N = 13, \ldots, 52$, and A is the only ESS$_N$ for $N \geq 53$. In the second example, B dominates A. In finite populations, however, A is the only ESS$_N$ for $B \geq 2, 3, \ldots, 17$, both strategies are ESS$_N$ for $N = 18, \ldots, 21$, and only B is ESS$_N$ for $N \geq 22$.

The motivation of the ESS$_N$ concept is as follows. If a strategy is ESS$_N$, then a single mutant of any other strategy must have a lower fitness. Therefore selection opposes the initial spread of any other strategy. As we have seen, however, in a finite population it is possible that the fixation of a strategy is favored by selection although its initial increase is opposed by selection. Thus the second condition demands that a strategy is only ESS$_N$ if the fixation probability of every other strategy is less than the neutral threshold, $1/N$. In summary, we simply require that a homogeneous ESS$_N$ population be protected by selection against invasion and replacement. These requirements represent a natural extension of the original ESS concept formulated by John Maynard Smith for infinitely large populations and deterministic evolutionary dynamics. Two specific examples are discussed in Figure 7.7.

If $d > b$, then B is both a strict Nash equilibrium and an ESS in comparison with A. A strict Nash equilibrium implies protection by selection against replacement in the following sense: for a given payoff matrix (7.1), with $d > b$ and for any given intensity of selection, $0 < w \leq 1$, we have $\rho_A \to 0$ as

$N \to \infty$. For every finite population size, N, however, selection can favor the fixation of strategy A.

7.3 RISK DOMINANCE

Sometimes it is of interest to ask whether A is more likely to replace B than vice versa. Let ρ_A and ρ_B denote the respective fixation probabilities. In the case where both A and B are best replies to themselves and in the limit of weak selection and large population size, we find that $\rho_A > \rho_B$ is equivalent to

$$a + b > c + d. \qquad (7.18)$$

This condition means that A is risk dominant. If both A and B are best replies to themselves, $a > c$ and $b < d$, then the risk-dominant strategy has the larger basin of attraction. Inequality (7.18) can be written as $x^* < 1/2$.

Let ρ_A denote the probability that a single A player reaches fixation in a population of B. Let ρ_B denote the probability that a single B player reaches fixation in a population of A. We have

$$\frac{\rho_A}{\rho_B} = \prod_{i=1}^{N-1} \frac{f_i}{g_i}. \qquad (7.19)$$

For weak selection (small w) we find

$$\frac{\rho_A}{\rho_B} = 1 + w\left[\frac{N}{2}(a + b - c - d) + d - a\right]. \qquad (7.20)$$

This equation can also be obtained from equation (7.10) and its symmetric counterpart, $\rho_B = (1/N)/[1 - (\alpha' N - \beta')w/6]$, where $\alpha' = -2a - b + 2c + d$ and $\beta' = -4a + b = c + 2d$.

It follows that $\rho_A > \rho_B$ is equivalent to

$$(N - 2)(a - d) > N(c - b). \qquad (7.21)$$

For large N, this means $a - c > d - b$. Hence if both A and B are strict Nash equilibria, then the risk-dominant equilibrium has a higher fixation probability. For general N and w, however, risk dominance does not decide

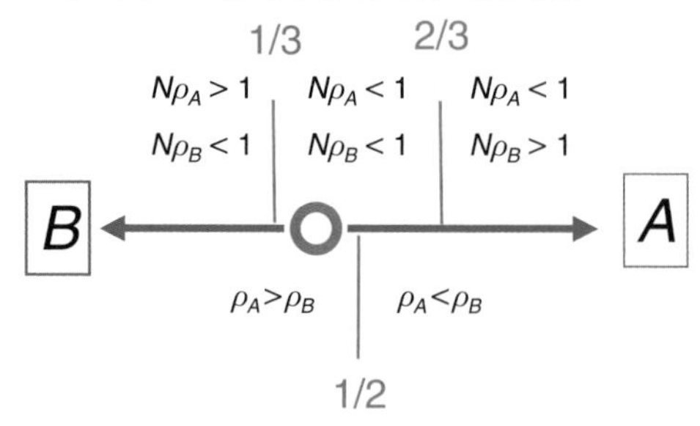

Figure 7.8 The figure illustrates the relationship between the 1/3 law and risk dominance. The fixation probabilities of strategies A and B are given by ρ_A and ρ_B. The unstable equilibrium, x^*, is illustrated by the red circle. If $x^* < 1/3$, then $N\rho_A > 1 > N\rho_B$; selection favors A and opposes B. If $x^* > 2/3$, then $N\rho_B > 1 > N\rho_A$; selection favors B and opposes A. If $1/3 < x^* < 2/3$, then both $N\rho_A$ and $N\rho_B$ are less than one; selection opposes the fixation of both strategies. Risk dominance is determined by x^* being greater or less than $1/2$. If $x^* < 1/2$, then strategy A is risk-dominant; it has the larger basin of attraction. If $x^* > 1/2$, then strategy B is risk-dominant. For evolutionary game dynamics in finite populations, we find that $x^* < 1/2$ is equivalent to $\rho_A > \rho_B$, while $x^* > 1/2$ is equivalent to $\rho_A < \rho_B$. All these relationships hold in the limit of large population size and weak selection. In general, however, risk dominance does not determine the ranking of the fixation probabilities.

whether ρ_A is greater than ρ_B. Figure 7.8 shows the relationship between risk dominance and the 1/3 law.

Note that both ρ_A and ρ_B can be less than $1/N$. In this case, selection opposes replacement in either direction. It is also possible to find conditions where both ρ_A and ρ_B are greater than $1/N$. In this case, selection favors replacement in either direction.

7.4 TIT-FOR-TAT CAN INVADE "ALWAYS DEFECT"

In the nonrepeated Prisoner's Dilemma, cooperators are dominated by defectors. In the repeated PD, the same two players meet more than once, and there are many conceivable strategies that allow cooperative behavior which cannot be invaded by defectors (see Chapter 5). One such strategy is Tit-for-tat, which cooperates in the first round and then does whatever the opponent did in the previous round. If the number of rounds is greater than a critical value, then neither "Always defect" (ALLD) nor TFT can be invaded by the other. If everybody plays ALLD, then TFT has a lower fitness. If everybody plays TFT, then ALLD has a lower fitness. Thus TFT can maintain cooperation, but likewise ALLD can maintain defection. The question is, How can cooperation get established?

The notion that ALLD resists invasion by TFT is derived from concepts of evolutionary stability and game dynamics of infinite populations. If everybody in an infinitely large population uses ALLD, then a small fraction of TFT players have a lower payoff. Therefore every invasion attempt by TFT is eliminated by natural selection.

The payoff matrix for TFT and ALLD in a Prisoner's Dilemma that is repeated for m rounds, on average, is given by

$$\begin{array}{c} \\ \text{TFT} \\ \text{ALLD} \end{array} \begin{array}{c} \begin{array}{cc} \text{TFT} & \text{ALLD} \end{array} \\ \begin{pmatrix} mR & S + (m-1)P \\ T + (m-1)P & mP \end{pmatrix} \end{array} \tag{7.22}$$

Recall that the Prisoner's Dilemma is defined by $T > R > P > S$. If the average number of rounds, m, exceeds a critical value,

$$m > \frac{T - P}{R - P}, \tag{7.23}$$

then ALLD does not dominate TFT. Each strategy is stable against invasion by the other strategy.

Let us now study the evolutionary game dynamics of TFT and ALLD in finite populations. We can use the payoff matrix (7.22) together with equation (7.8) to calculate the fixation probability, ρ, of a lineage starting from a single TFT player in a population of ALLD. Figure 7.9 shows that $N\rho$ is a one-humped function of N. For a wide choice of parameter values, there is an intermediate range of population sizes, N, where selection favors TFT. Thus

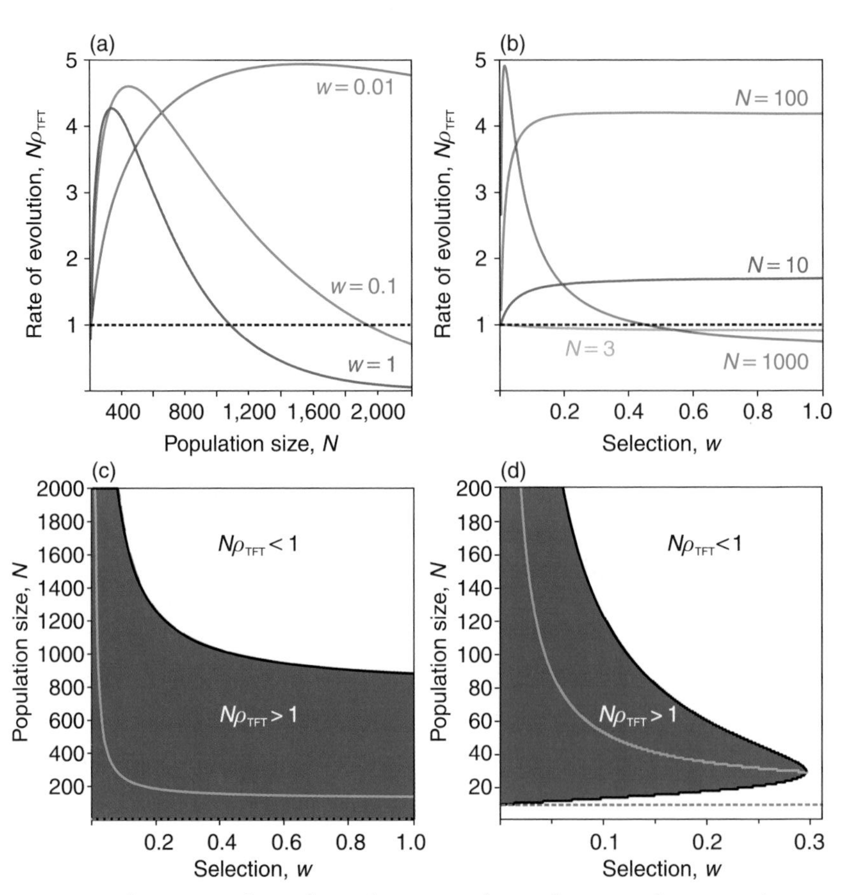

Figure 7.9 Selection can favor the replacement of ALLD by TFT in finite populations. (a) The rate of evolution, $N\rho_{TFT}$, is a one-humped function of population size N. There is an intermediate range of N that leads to positive selection of TFT, $N\rho_{TFT} > 1$. (b) $N\rho_{TFT}$ is shown as function of w, the intensity of selection. For small N, we have $N\rho_{TFT} < 1$ for all w. For larger N we have $N\rho_{TFT} > 1$ for all w. For even larger N we have $N\rho_{TFT} > 1$ as long as w is below a certain threshold. (c, d) The blue-shaded region indicates the parameter region where $N\rho_{TFT} > 1$. The light blue line shows the optimum value of N for given w maximizing $N\rho_{TFT}$. The broken red line indicates $N_{min} = (2a + b + c - 4d)/(a + 2b - c - 2d)$, which is the predicted minimum population size required for positive selection of TFT in the limit of weak selection. Parameter choices: $R = 3$, $T = 5$, $P = 1$, $S = 0$; $n = 10$ rounds for (a–c) and $n = 4$ rounds for (d).

Selection can favor TFT replacing ALLD

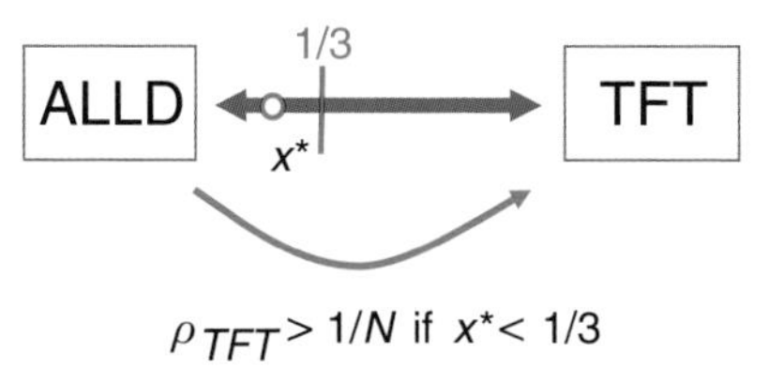
1/3
ALLD
TFT
x*
ρ_TFT > 1/N if x* < 1/3

Figure 7.10 In a finite population, the fixation probability of a single TFT mutant in a population of ALLD can be greater than $1/N$. This is the case if the unstable equilibrium (the invasion barrier) is less than 1/3, a condition that can be easily fulfilled in the repeated Prisoner's Dilemma.

the invasion and replacement of ALLD by TFT, starting from a single individual of TFT, can be favored by natural selection. Interestingly, there are critical minimum and maximum population sizes that allow positive selection of TFT. In very small populations, there is a strong effect of spite: helping another individual leads to a significant disadvantage; in a population of size $N = 2$, TFT always has a lower fitness than ALLD. In very large populations, it is too unlikely that TFT will reach the invasion barrier, x^*, when starting with a single player. Thus neither small nor large but intermediate population sizes are optimum for initiating cooperation.

Combining the payoff matrix (7.22) and condition (7.11) we obtain

$$m > \frac{T(N+1) + P(N-2) - S(2N-1)}{(R-P)(N-2)}. \tag{7.24}$$

This inequality determines the minimum number of rounds required for selection to favor TFT replacing ALLD for a given population size N. Note that we need at least a population size of $N = 3$. For a large population size, we obtain the condition

$$m > \frac{T + P - S}{R - P}. \tag{7.25}$$

This inequality ensures that the basin of attraction of ALLD is less than 1/3.

Let us consider the payoff values $R = 3$, $T = 5$, $P = 1$, and $S = 0$ as a numerical example. For $N = 3$, we need $m > 10.5$ rounds. For $N = 4$, we need $m > 6.75$ rounds. For large N, we only need $m > 3$ rounds.

SUMMARY

- The Moran process can be extended to study evolutionary game dynamics in populations of finite size.
- The intensity of selection is of crucial importance. The game's payoff can make a large contribution to fitness (strong selection) or a small contribution (weak selection).
- The fixation probabilities determine whether selection favors the replacement of an existing strategy by an incoming mutant.
- Natural selection favors A replacing B in a sufficiently small population provided $b > c$.
- Natural selection favors A replacing B in a sufficiently large population and for weak selection (small w) provided $a + 2b > c + 2d$. If A and B are best replies to themselves, then this inequality means that the basin of attraction of B is less than $1/3$.
- The analysis leads to natural conditions for evolutionary stability in finite populations. These conditions specify whether a given resident strategy is protected by selection against invasion and replacement by any mutant strategy.
- The traditional ESS and Nash conditions are neither necessary nor sufficient to imply protection by selection in finite populations.
- Even if A is risk dominant over B, then B can have a greater fixation probability than A. Only in the limit of weak selection and large population size does risk dominance determine the ranking of the fixation probabilities.
- In a finite population, natural selection can favor the replacement of "Always defect" by a cooperative strategy (such as Tit-for-tat), when starting from a single individual using that strategy.

EVOLUTIONARY GRAPH THEORY 8

UP TO NOW we have studied evolutionary dynamics in homogeneous populations, where all individuals are in equivalent positions. Let me now introduce a general framework to analyze the effect of population structure on evolutionary dynamics. We will do this by placing the individuals on the vertices of a graph. The edges of the graph determine competitive interaction. If there is an edge from vertex i to j, then in a genetic setting the offspring of i can replace j. In a cultural setting, some information (an idea) can spread from i to j.

The graph can represent spatial structure among plants or animals in an ecosystem. The graph can also describe the architecture of cells in a multicellular organism, including the cellular differentiation hierarchy. For example, stem cells divide into progenitors, which divide into differentiated cells. The organs of many multicellular animals have such a design, which can delay the onset of cancer (as we shall see in Chapter 12). The graph might also represent relationships in a social network of humans. In this context, the dynamics on the graph describe cultural evolution and the spread of new inventions and ideas. Obviously, human societies are never homogeneous. Individuals in central positions may be more influential than others.

We will ask whether, on the one hand, particular graphs can accelerate the rate of evolution by increasing the fixation probabilities of advantageous mutants. On the other hand, can we find graphs that reduce the fixation probabilities of such mutants? Can certain graphs completely eliminate the effect of selection? Is it possible to characterize all those graphs that have the same evolutionary dynamics (in terms of fixation probability) as unstructured populations? We will assume that the graph does not change on the time scale under consideration. The extension to graphs that change over time is an important task that lies ahead.

This chapter describes first steps into a largely unexplored territory. I have included it in this book because I think that many investigations will follow based on those first steps. The general question of how population structure affects evolutionary dynamics is hugely important and has been a long-standing topic in population genetics. A deeper, mathematical understanding of cultural evolution in human society requires the study of evolutionary dynamics on social networks. Although the main part of this chapter deals with constant selection, the final section looks ahead to games on graphs and states a fascinating result for the evolution of cooperation on graphs.

8.1 THE BASIC IDEA

Label all individuals in the population with $i = 1, 2, \ldots, N$. At each time step, a random individual is chosen for reproduction. The probability that the offspring of i replaces j is given by w_{ij}. Hence the process is determined by an $N \times N$ matrix, $W = [w_{ij}]$. Note that all entries of W are probabilities, which means they are numbers between 0 and 1. Moreover, the offspring of any one individual has to go somewhere. Therefore the sum $\sum_{j=1}^{N} w_{ij}$ must be equal to one. The matrix W is stochastic.

We can imagine all individuals occupying the vertices of a graph. If $w_{ij} > 0$, there is an edge from vertex i to j. If $w_{ij} = 0$, there is no edge leading from vertex i to j. The matrix W defines a weighted digraph. Digraphs can have two edges between vertices i and j: one going from i to j; the other one going from j to i (Figure 8.1).

The idea that the offspring of one individual replaces another individual is taken from the Moran process. This process is recovered as the special case

Evolutionary graph theory

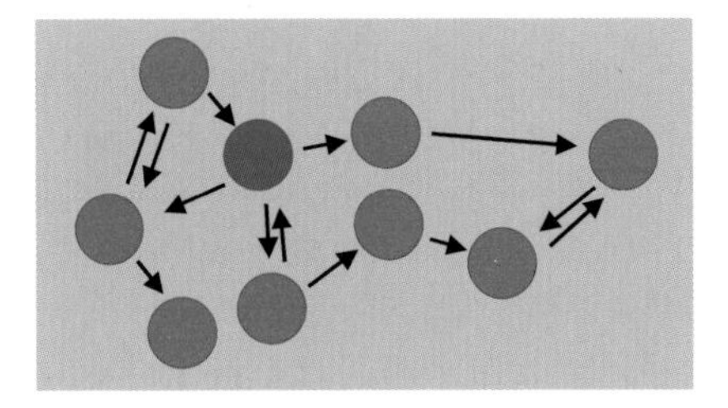

1. The individuals occupy the vertices of a graph
2. The edges determine where to place the offspring

Figure 8.1 Evolutionary graph theory is a powerful approach to study the effect of population structure on evolution. The individuals occupy the vertices of the graph. The edges denote reproduction. In each time step, an individual is chosen for reproduction at random but proportional to its fitness. The offspring replaces an adjacent individual with a probability that is proportional to the weight of the edge. We can interpret evolutionary graph theory as describing either genetic reproduction or cultural imitation.

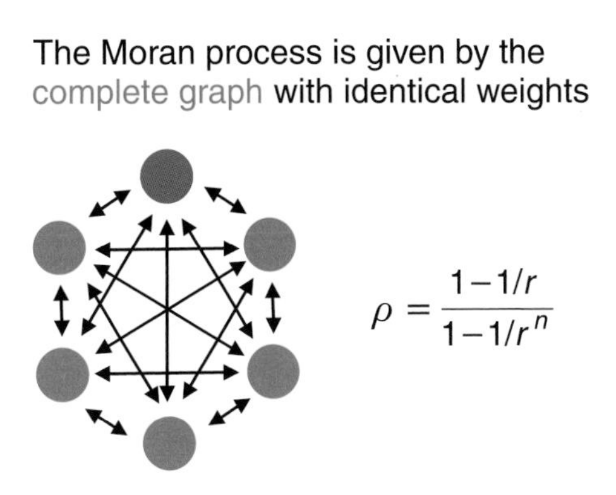
The Moran process is given by the
complete graph with identical weights
$\rho = \frac{1-1/r}{1-1/r^n}$

Figure 8.2 An unstructured population is given by a complete graph: there is an edge between any two vertices. All edges have the same weight. The evolutionary process is equivalent to the Moran process with its well-known fixation probability.

of the complete graph with identical weights, $w_{ij} = 1/N$ for all i and j. The complete graph is defined by the property that all possible edges exist: the offspring of any one individual can replace any other individual (Figure 8.2).

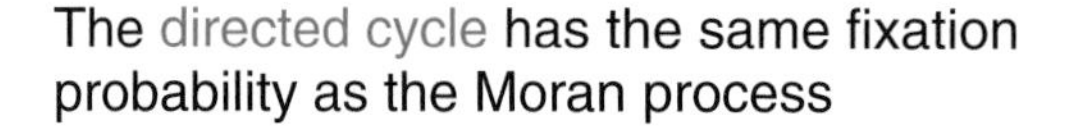

The directed cycle has the same fixation probability as the Moran process

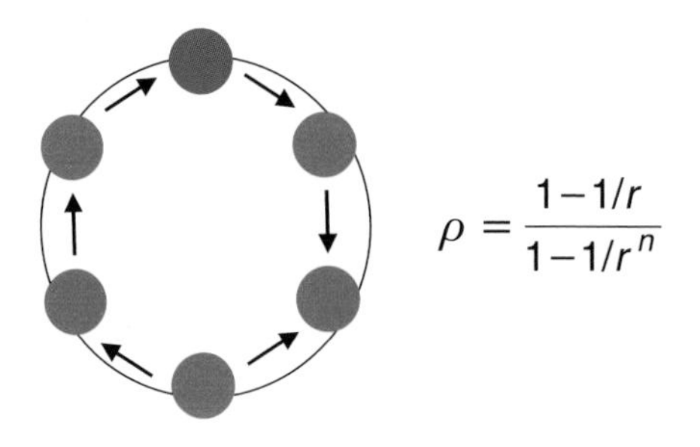

Figure 8.3 If the graph is a directed cycle, then each individual can place its offspring into the one adjacent place. It is easy to show that the fixation probability of a single mutant with relative fitness r is the same as in the Moran process.

8.2 FIRST OBSERVATIONS

The first question we ask is the following: what is the fixation probability of a new mutant that arises at a random position on a graph?

8.2.1 The Directed Cycle

As a first example, we consider a directed cycle of size N (Figure 8.3). The W matrix is given by

$$W = \begin{pmatrix} 0 & 1 & 0 & \dots & 0 & 0 \\ 0 & 0 & 1 & \dots & 0 & 0 \\ 0 & 0 & 0 & \dots & 0 & 0 \\ \vdots & \vdots & \vdots & \ddots & \vdots & \vdots \\ 0 & 0 & 0 & \dots & 0 & 1 \\ 1 & 0 & 0 & \dots & 0 & 0 \end{pmatrix} \tag{8.1}$$

Initially all individuals are of type A. After some time, a mutant B is generated that has relative fitness r. This B individual gives rise to a lineage, which will eventually die out or take over the whole population. Starting from one B mutant, only one cluster of B individuals can emerge. It is not possible for this cluster to break into two or more fragments. This fact makes the calculation of the fixation probability straightforward.

Let m denote the number of B individuals. In order to reduce m by one, the A individual immediately preceding the B cluster in the directed cycle must be chosen for reproduction. Thus the probability to go from m to $m-1$ is given by

$$p_{m,m-1} = \frac{1}{N-m+rm}. \tag{8.2}$$

In order to increase m by one, the B individual at the end of the cluster has to be chosen for reproduction. Therefore the probability of going from m to $m+1$ is given by

$$p_{m,m+1} = \frac{r}{N-m+rm}. \tag{8.3}$$

The ratio of these two probabilities is

$$\gamma_m = \frac{p_{m,m-1}}{p_{m,m+1}} = \frac{1}{r}. \tag{8.4}$$

This quantity is independent of m and identical to what is obtained in the Moran process with constant selection.

From equation (6.13), the fixation probability of a birth-death process is given by

$$\rho = \frac{1}{1+\sum_{k=1}^{N-1}\prod_{m=1}^{k}\gamma_m}. \tag{8.5}$$

Therefore we obtain here

$$\rho = \frac{1-1/r}{1-1/r^N}. \tag{8.6}$$

The fixation probability on a directed cycle is identical to the fixation probability in the Moran process.

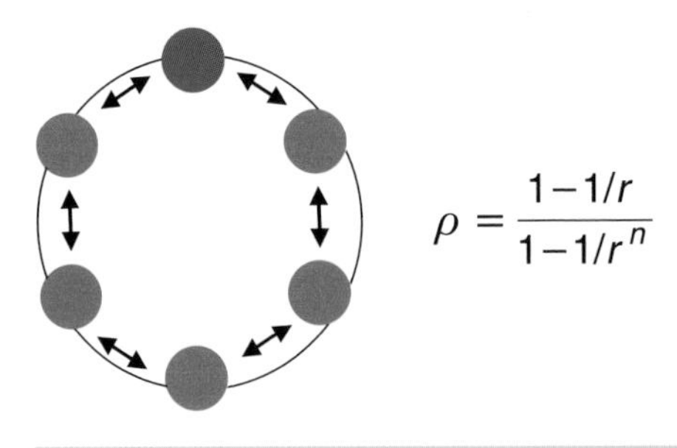

The cycle has the same fixation probability as the Moran process
$\rho = \frac{1-1/r}{1-1/r^n}$

Figure 8.4 If the graph is a cycle, then each individual can place its offspring into any one of the two adjacent places. Again the fixation probability of a single mutant with relative fitness r is the same as in the Moran process.

8.2.2 The Cycle

As a second example, we consider the (bidirected) cycle shown in Figure 8.4. Any two neighbors are connected by two edges: one going in one direction, the other going in the opposite direction. All edges have the same weight. We have

$$W = \begin{pmatrix} 0 & 1/2 & 0 & \dots & 0 & 1/2 \\ 1/2 & 0 & 1/2 & \dots & 0 & 0 \\ 0 & 1/2 & 0 & \dots & 0 & 0 \\ \vdots & \vdots & \vdots & \ddots & \vdots & \vdots \\ 0 & 0 & 0 & \dots & 0 & 1/2 \\ 1/2 & 0 & 0 & \dots & 1/2 & 0 \end{pmatrix} \tag{8.7}$$

As before, starting from a single B mutant, there can only be one cluster of B. Again it is easy to confirm that

$$p_{m,m-1} = \frac{1}{N-m+rm} \quad \text{and} \quad p_{m,m+1} = \frac{r}{N-m+rm} \tag{8.8}$$

Thus the birth-death process on the cycle is described by the same transition matrix as the process on a directed cycle. Hence we obtain the same fixation probability as before.

8.2.3 The Line and the Burst

Let us now consider a linear array as shown in Figure 8.5. From vertex i the offspring can be placed into vertex $i+1$. Vertex N places its offspring onto itself. No edge leads to vertex 1. We have

$$W = \begin{pmatrix} 0 & 1 & 0 & \dots & 0 & 0 \\ 0 & 0 & 1 & \dots & 0 & 0 \\ 0 & 0 & 0 & \dots & 0 & 0 \\ \vdots & \vdots & \vdots & \ddots & \vdots & \vdots \\ 0 & 0 & 0 & \dots & 0 & 1 \\ 0 & 0 & 0 & \dots & 0 & 1 \end{pmatrix} \tag{8.9}$$

What is the fixation probability of a randomly placed mutant? The answer is very simple. It is

$$\rho = 1/N. \tag{8.10}$$

With probability $(N-1)/N$ the mutant arises in positions $i = 2, \dots, N$, and its lineage will be eliminated eventually. With probability $1/N$, however, the mutant arises in position $i = 1$, and its offspring lineage will take over the population. The fixation probability is totally independent of the relative fitness, r, of the mutant. Thus the line has a fixation probability that differs from the Moran process.

Another graph that has a different fixation behavior as the Moran process is the "burst," which is also shown in Figure 8.5. There is one central vertex and $N-1$ peripheral vertices. Edges lead from the center to the periphery. We have

$$W = \begin{pmatrix} 0 & 1/(N-1) & 1/(N-1) & \dots & 1/(N-1) & 1/(N-1) \\ 0 & 1 & 0 & \dots & 0 & 0 \\ 0 & 0 & 1 & \dots & 0 & 0 \\ \vdots & \vdots & \vdots & \ddots & \vdots & \vdots \\ 0 & 0 & 0 & \dots & 1 & 0 \\ 0 & 0 & 0 & \dots & 0 & 1 \end{pmatrix} \tag{8.11}$$

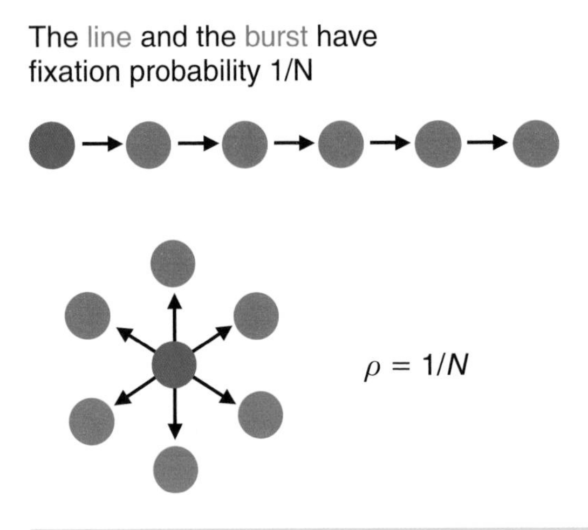
The line and the burst have
fixation probability 1/N
ρ = 1/N

Figure 8.5 Here are two graphs that do not have the same fixation probability as in the Moran process. Both for the "line" and for the "burst," the fixation probability of a randomly placed mutant is $\rho = 1/N$, independent of the fitness of this mutant. In the line, a mutant can only take over the population if it arises in the left-most position. In the burst, the mutant must arise in the central position. These two graphs are suppressors of selection, because all mutants—irrespective of their fitness—have the same fixation probability as a neutral mutant in the Moran process.

A new mutant can only reach fixation if it arises in the center. The chance that a randomly placed mutant originates in the center is $1/N$. Hence the fixation probability is again independent of the relative fitness, r, of the new mutant. The burst and the line have the same fixation probability.

8.2.4 Balancing Drift and Selection

The fixation probability of the Moran process,

$$\rho_M = \frac{1 - 1/r}{1 - 1/r^N}, \tag{8.12}$$

defines a particular balance between natural selection and random drift. If a graph, G, has the same fixation probability as the Moran process, then we say that this graph is ρ-equivalent to the Moran process; it has the same balance of selection and drift.

If, for an advantageous mutant, $r > 1$, the fixation probability on G is greater than the fixation probability in the Moran process, $\rho_G > \rho_M$, then the graph G favors selection over drift. It increases the fixation probability of an advantageous mutant. Therefore graph G is an amplifier of selection.

If, for an advantageous mutant, the fixation probability on G is less than the fixation probability in the Moran process, $\rho_G < \rho_M$, then the graph G favors drift over selection. It reduces the fixation probability of an advantageous mutant. Therefore graph G is a suppressor of selection.

Similarly if, for a disadvantageous mutant, $r < 1$, the fixation probability on G is greater (less) than the fixation probability in the Moran process, then the graph G is a suppressor (amplifier) of selection.

If $\rho_G = 1/N$ for any r, then the graph G is the strongest possible suppressor of selection; it completely eliminates the effect of selection.

We have seen that the cycle and the directed cycle are both ρ-equivalent to the Moran process, whereas the line and the burst completely eliminate selection.

8.3 THE ISOTHERMAL THEOREM

The temperature of a vertex is defined as the sum of all weights that lead into that vertex. The temperature of vertex j is given by

$$T_j = \sum_{i=1}^{N} w_{ij}. \tag{8.13}$$

A vertex with a high temperature will change more often than a vertex with a low temperature. If all the vertices have the same temperature, then a graph is isothermal. We have the following "isothermal theorem": a graph is ρ-equivalent to the Moran process if and only if it is isothermal (Figure 8.6).

For an isothermal graph we have $\sum_{i=1}^{N} w_{ij} = \text{constant}$. Since $\sum_{j=1}^{N} w_{ij} = 1$, it follows that $\sum_{i=1}^{N} w_{ij} = 1$. Therefore a graph is ρ-equivalent to the Moran process if and only if W is a doubly stochastic matrix, which means that all rows and all columns sum to one.

Let us prove the isothermal theorem. The configuration of a population on a graph can be described by a binary vector, $\vec{v} = (v_1, \ldots, v_N)$. If the vertex i is occupied by type A, then $v_i = 0$. If the vertex i is occupied by type B, then $v_i = 1$. Therefore the vector $\vec{v}$ describes a two-coloring of the graph. Denote by m the total number of B individuals. Thus $m = \sum_i v_i$. The probability that

The isothermal theorem

The temperature of a vertex is the sum of all weights leading into that vertex

$$T_j = \sum_i w_{ij}$$

If all vertices have the same temperature, then the fixation probability is equivalent to the Moran process

Figure 8.6 The isothermal theorem characterizes all those graphs that have the same fixation probability as the unstructured population (described by the Moran process). The temperature of a vertex determines how often the individual in this vertex will be replaced. A hot vertex changes more often than a cold vertex. If all vertices have the same temperature, then the matrix $W = [w_{ij}]$ is doubly stochastic and the graph is isothermal.

m increases by one is given by

$$p_{m,m+1} = \frac{r \sum_i \sum_j w_{ij} v_i (1 - v_j)}{rm + N - m}. \tag{8.14}$$

The probability that m decreases by one is given by

$$p_{m,m-1} = \frac{\sum_i \sum_j w_{ij} (1 - v_i) v_j}{rm + N - m}. \tag{8.15}$$

The fixation probability is the same as in the Moran process if for any coloring $\vec{v}$ we have

$$\frac{p_{m,m-1}}{p_{m,m+1}} = \frac{1}{r}. \tag{8.16}$$

This is the case if

$$\sum_i \sum_j w_{ij} (1 - v_i) v_j = \sum_i \sum_j w_{ij} v_i (1 - v_j). \tag{8.17}$$

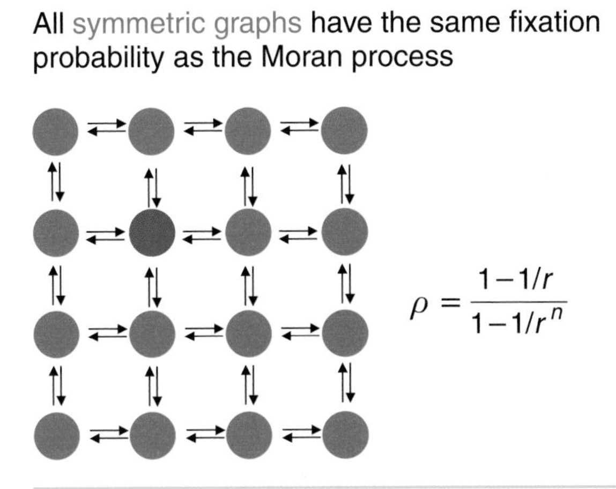
All symmetric graphs have the same fixation probability as the Moran process
$\rho = \frac{1-1/r}{1-1/r^n}$

Figure 8.7 A symmetric graph is defined by the property $w_{ij} = w_{ji}$ for all i and j. This means the weight of the connection from vertex i to j is the same as from vertex j to i. It turns out that all symmetric graphs have the same fixation probability as the Moran process. All spatial lattices (square, hexagonal, triangular) are symmetric graphs.

This equality must hold for any vector $\vec{v}$. In particular, it must hold for all vectors of the form $v_k = 1$ and $v_i = 0$ for all $i \neq k$. In this case equation (8.17) reduces to

$$\sum_j w_{kj} = \sum_j w_{jk} \qquad \forall k \tag{8.18}$$

Since $\sum_j w_{kj} = 1$, we have

$$\sum_j w_{jk} = 1 \tag{8.19}$$

and therefore equation (8.18) means that the matrix W is doubly stochastic and the corresponding graph is isothermal.

The cycle and the directed cycle are both isothermal. All symmetric graphs, $w_{ij} = w_{ji}$, are isothermal (Figure 8.7). The cycle is symmetric. Most spatial lattices that have been investigated in evolutionary dynamics are symmetric. But many asymmetric graphs are also isothermal. The directed cycle, for example, is asymmetric but isothermal.

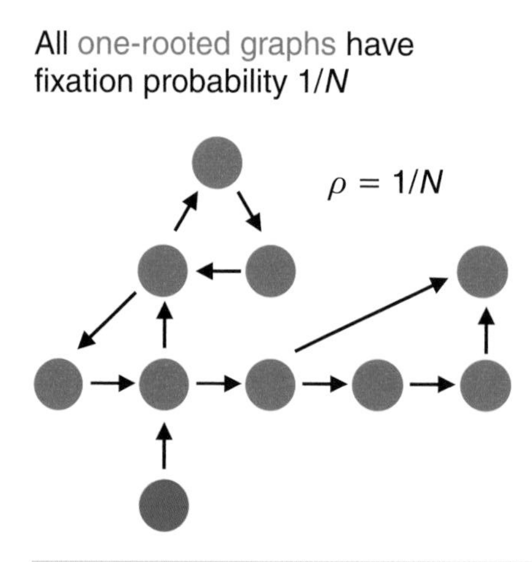
All one-rooted graphs have fixation probability 1/N
ρ = 1/N

Figure 8.8 It is easy to see that all one-rooted graphs have fixation probability $1/N$ regardless of the fitness of the mutant. Only a mutant that arises in the root generates a lineage that will take over the population. The probability that a randomly placed mutant arises in the root is $1/N$.

The line, however, is not isothermal. The $i = 1$ vertex has temperature 0. The vertices $i = 2, \ldots, N - 1$ have temperature 1. The vertex N has temperature 2. Therefore the line is not ρ-equivalent to the Moran process. The burst is also not isothermal; the central vertex has temperature 0 while all other vertices have temperature 2.

8.4 SUPPRESSING SELECTION

A root is a vertex that has no edge leading into it. A root has zero temperature. If a graph is one-rooted, then it has fixation probability $1/N$. The new mutant must arise at the root, otherwise it cannot take over the whole population. A randomly placed mutant arises at the root with probability $1/N$. Therefore every one-rooted graph completely eliminates selection (Figure 8.8).

If a graph has multiple roots, then any lineage arising from a single mutant can never take over the whole population. If a mutant arises in one of the roots, then it will give rise to a lineage that will never become extinct. Thus, graphs with multiple roots allow the coexistence of different lineages (Figure 8.9).

It is easy to construct suppressors of selection that have a fixation probability of advantageous mutants that is somewhere between $1/N$ and ρ_M. Subdi-

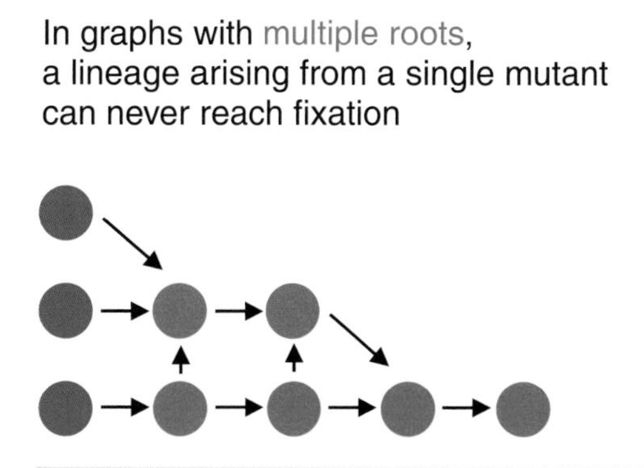
In graphs with multiple roots,
a lineage arising from a single mutant
can never reach fixation

Figure 8.9 If a graph has multiple roots, then the lineage arising from a single mutant can never take over the entire graph. If a mutant arises in a vertex that is not a root, then it can only generate a transient lineage. If a mutant arises in a root, then it will generate a lineage that cannot become extinct. Graphs with multiple roots promote diversity.

vide the population into two compartments with population sizes N_1 and N_2. The total population size is $N = N_1 + N_2$. The first compartment is placed on a complete graph. Edges lead from the first compartment into the second compartment, but not the other way around. The second compartment is on an arbitrary graph with the only constraint that all vertices of the second compartment must be reachable from the first compartment. Thus the first compartment is a source; the second compartment is a sink. The fixation probability of this graph is

$$\rho_G = \frac{1 - 1/r}{1 - 1/r^{N_1}}. \tag{8.20}$$

For advantageous mutants, $r > 1$, we have

$$1/N < \rho_G < \rho_M(N). \tag{8.21}$$

In general, graphs that have small upstream and large downstream populations tend to be suppressors of selection.

8.5 AMPLIFYING SELECTION

The balance between drift and selection, as determined by the fixation probability of the complete graph, can also be tilted toward selection. Consider the

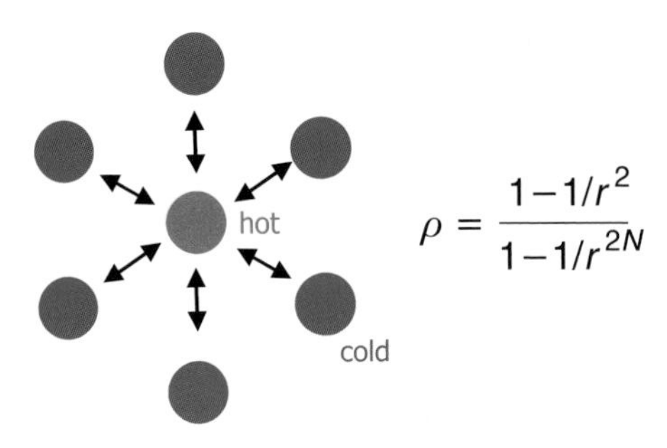

Figure 8.10 It is possible to find graphs that amplify selection over drift. The "star" is a good example. For large N, a mutant with relative fitness r has a fixation probability $\rho = (1 - 1/r^2)/(1 - 1/r^{2N})$. Therefore a relative fitness r on a star is equivalent to a relative fitness r^2 in the Moran process. The star is an amplifier of selection.

star structure shown in Figure 8.10. As the population size, N, becomes large, the fixation probability of a randomly placed mutant approaches

$$\rho_M = \frac{1 - 1/r^2}{1 - 1/r^{2N}}. \tag{8.22}$$

Hence the star is an amplifier of selection. An advantageous mutant with relative fitness $r > 1$ behaves like an advantageous mutant with fitness r^2 in a standard Moran process. A disadvantageous mutant, $r < 1$, has a probability of fixation that is equivalent to an even greater fitness disadvantage, r^2, in the Moran process.

Can we construct even more powerful amplifiers? The superstar shown in Figure 8.11 amplifies a selective difference r to r^k, where k is the length of each loop in the graph. As the number of leaves and vertices within each leaf grows, the fixation probability becomes

$$\rho_M = \frac{1 - 1/r^k}{1 - 1/r^{2k}}. \tag{8.23}$$

By increasing k, we can guarantee the fixation of any advantageous mutant, $\rho \to 1$ if $r > 1$, and guarantee the extinction of any disadvantageous mutant, $\rho \to 0$ if $r < 1$.

The funnel, shown in Figure 8.12, is another potent amplifier. There are $k + 1$ layers, labeled $j = 0, \ldots, k$. Layer 0 contains only a single vertex. Layer

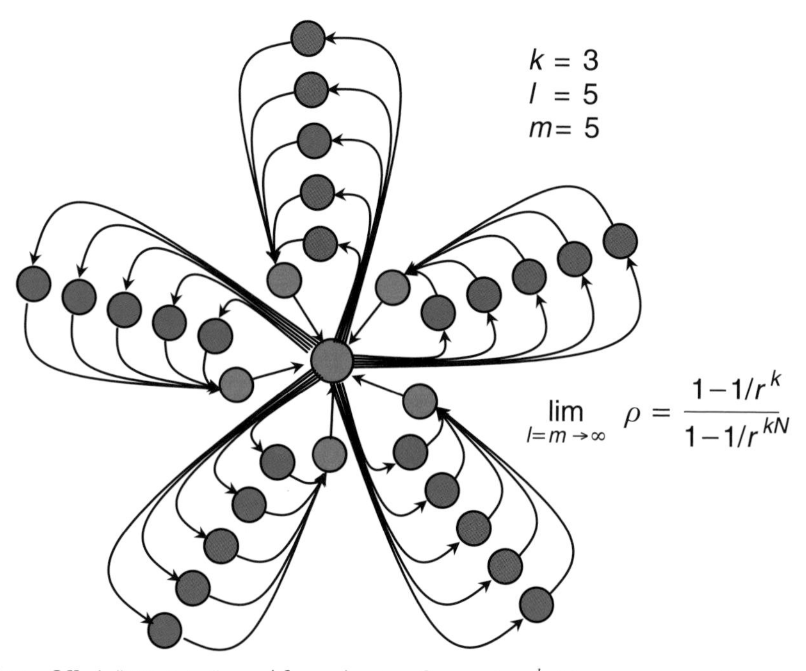

Figure 8.11 A "superstar" amplifies selection from r to r^k. The parameters l and m denote the number of leaves and the number of loops in a leaf, respectively. The parameter k denotes the length of each loop. The amplification from r to r^k holds in the limit of large l and m. In the limit of large k, the superstar guarantees fixation of any advantageous mutant and elimination of any disadvantageous mutant. The coloring indicates hot vertices (red) and cold vertices (blue).

j contains m^j vertices. All edges that originate from vertices in layer j lead into $j - 1$. All edges that originate from the single vertex in layer 0 lead into layer k. As k increases, the fixation probability of any advantageous mutant converges to 1.

Computer simulations show that scale-free networks are mild amplifiers. This is of particular interest because scale-free networks, including small-world networks, have been observed in various circumstances. Scale-free net-

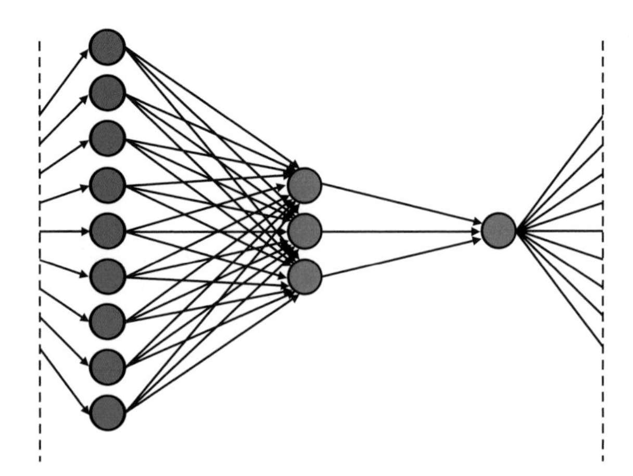
The funnel is a strong amplifier of selection

Figure 8.12 The "funnel" is another potent amplifier of selection. There is a single vertex in one layer. All edges leading into that vertex come from a preceding layer with m vertices. All edges leading into that layer come from a preceding layer with m^2 vertices, and so on. All outgoing edges of the single vertex wrap around to lead into the largest layer. In the limit of large m and many layers, the fixation probability converges to one for any advantageous mutant and to zero for any disadvantageous mutant. Again the coloring indicates hot vertices (red) and cold vertices (blue). The superstar and the funnel were invented by Erez Lieberman.

works are defined by the property that they have a degree distribution which is a straight line in a log-log plot. The degree of a vertex is the number of edges connected to this vertex.

8.6 CIRCULATIONS

We can also design a more elegant version of evolutionary dynamics on graphs. Instead of first choosing a vertex for reproduction and then choosing again where to place its offspring, we can simply choose an edge. In this case w_{ij} can be any non-negative number, and W need not be a stochastic matrix. Edge ij is chosen with a probability proportional to w_{ij} multiplied by the fitness of its tail, which is the fitness of the individual at vertex i.

The circulation theorem
The fixation probability is the same as in the Moran process, if and only if the graph is a circulation

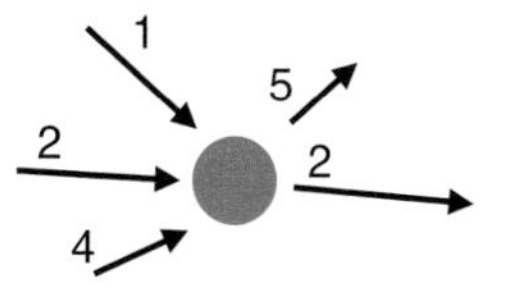

Sum over incoming weights = Sum over outgoing weights

Figure 8.13 In an extended approach to evolutionary graph theory, the matrix W is no longer stochastic; instead, the weights w_{ij} are arbitrary non-negative numbers. In each time step, an edge is chosen with a probability proportional to its weight multiplied by the fitness of the individual at its tail. If edge ij is chosen, then the offspring of i will replace j. In this framework, a graph has the same fixation behavior as the Moran process if and only if it is a circulation. A graph is a circulation, if for each vertex the sum of incoming weights equals the sum of outgoing weights. Circulations constitute an important set of graphs that arise in many different contexts.

In this framework, a graph G is ρ-equivalent to the Moran process if and only if it is a circulation (Figure 8.13). A circulation is defined by the property

$$\sum_{j=1}^{N} w_{kj} = \sum_{j=1}^{N} w_{jk} \qquad \forall k = 1, \ldots, N \tag{8.24}$$

This means that for each vertex k, the sum over all weights entering it must equal the sum over all weights leaving it. The proof of this "circulation theorem" is equivalent to that of the "isothermal theorem." Note that every isothermal graph is a circulation, but not every circulation is isothermal.

8.7 GAMES ON GRAPHS

The next step is to study evolutionary game dynamics on graphs. The general task would be to calculate the fixation probability of a certain strategy, A, competing with another strategy, B. In principle, there can be two different graphs: the interaction graph, H, determines who plays with whom, while the replacement graph, G, specifies the reproductive events (who learns from whom or who is replaced by whose offspring). Classifying all games on all combinations of graphs is a vast (and perhaps even impossible) undertaking. Here I can mention only one specific, though most illuminating, example.

Let us suppose that the replacement graph and the interaction graph are the same, $H = G$, and study the interaction between cooperators, C, and defectors, D. A cooperator helps all of its neighbors. For each neighbor, the cooperator pays a cost, c, and the neighbor receives a benefit, b. Defectors do not provide any help. They have no costs, but they can benefit by receiving help from adjacent cooperators. Each individual occupies the vertex of a graph. The payoffs from all interactions are summed. At first, let us consider regular graphs of degree k: each individual has exactly k neighbors. We consider the case of weak selection: the fitness of an individual is a constant plus w times the payoff. Weak selection means that w is small.

Consider three different update rules for the game dynamics.

1. "Birth-death" process: In each time step, an individual is selected for reproduction proportional to its fitness. The offspring replaces a random neighbor. It turns out that for any choice of the parameter values, b and c, the fixation probability of cooperators, ρ_C, is always less than $1/N$, while the fixation probability of defectors, ρ_D, is always greater than $1/N$:

$$\rho_C < 1/N < \rho_D. \tag{8.25}$$

   In this "birth-death" process, selection always favors defectors.

2. "Death-birth" process: In each time step, a random individual is chosen to die. The neighbors compete for the empty site proportional to their fitness. In this case, we find that cooperators are advantageous and defectors are disadvantageous, $\rho_C > 1/N > \rho_D$, if

$$b/c > k. \tag{8.26}$$

   This surprisingly simple rule is the crucial condition for the evolution of cooperation on regular graphs given the "death-birth" update rule.

3. Imitation process: In each time step, a random individual is chosen to update its strategy. It will either stay with its own strategy or imitate a neighbor's strategy proportional to fitness. Therefore the focal individual's own payoff also affects the update dynamics. In this case we find that cooperators are advantageous and defectors are disadvantageous, $\rho_C > 1/N > \rho_D$, if

$$b/c > k + 2. \tag{8.27}$$

For $k = 2$, a regular graph is a cycle. In this case the three results can be obtained from direct calculations. All that is required is to check whether the boundary between a cluster of cooperators moves in favor of the cooperators or the defectors (Figure 8.14). For the "birth-death" process, only the payoff of the two individuals right at the boundary matters. Clearly the defector has a higher payoff than the cooperator. The boundary always moves in favor of the defectors. Selection promotes defection, in this case. For the other two update rules, the payoff of the four individuals that are closest to the boundary determines the outcome. There are always two cooperators and two defectors. Again the defector at the boundary has a higher payoff than the cooperator at the boundary. But the second cooperator has a higher payoff than the second defector. Therefore cooperation could be favored. A simple calculation shows that this is precisely the case, if $b/c > 2$ for the "birth-death" rule and $b/c > 4$ for the imitation rule.

For $k > 2$, the three findings can be obtained via a complicated calculation that uses "pair-approximation." In this technique, one keeps track of the average frequency of cooperators and defectors as well as the average frequency of all pairs, CC, CD, DC, and DD. Strictly speaking, pair approximation is formulated for Bethe lattices (or Cailey trees), where every individual has exactly k neighbors and there are no loops.

The findings were confirmed by computer simulations for lattices and random regular graphs. There is excellent agreement between the simulation results and the calculation that uses pair approximation. Moreover, the simple rules $b/c > k$ and $b/c > k + 2$ also hold for random graphs and scale-free networks.

Games on cycles

1. "Birth-death" process: defectors always win

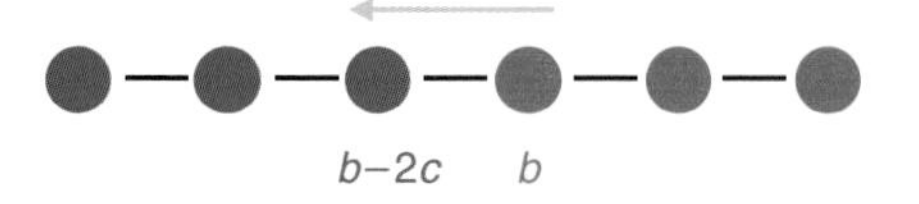

2. "Death-birth" process: cooperators win if $b/c > 2$

3. "Imitation" process: cooperators win if $b/c > 4$

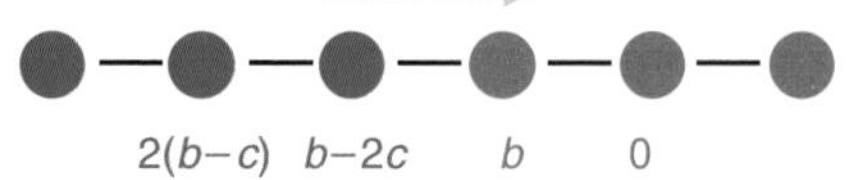

Figure 8.14 Games on graphs can be studied by assuming that individuals interact with their nearest neighbors and thereby accumulate payoff. This figure illustrates the evolution of cooperation on a one-dimensional graph, a cycle. Cooperators pay a cost c for each neighbor. Each neighbor of a cooperator receives benefit b. In the "birth-death" update rule, selection always favors defectors, because only the payoffs of the two individuals at the boundary matter. For the "death-birth" process, the payoff of the next two individuals is also important; in this case, selection favors cooperators provided $b/c > 2$. For the "imitation" process, cooperators win if $b/c > 4$. All calculations are performed in the limit of weak selection and large population size. To calculate the fixation probability of either cooperators or defectors, we only have to analyze whether the boundary between a large cluster of cooperators and a large cluster of defectors moves in one direction or the other, because the lineage arising from one individual always forms a single cluster. A cluster of cooperators (or defectors) cannot break into pieces.

SUMMARY

- Evolutionary graph theory is a powerful approach to study the effect of population structure on evolution.
- The graph can represent the spatial configuration of a population, the differentiation hierarchy of cells in a multicellular organism, or a social network.
- The vertices of the graph are individuals. The (weighted) edges denote reproduction.

- Reproduction can be genetic or cultural. In the first case, the offspring of one individual replaces another individual in an adjacent vertex. In the second case, cultural information spreads from one vertex to the next.
- We study the fixation probability of a randomly placed mutant with relative fitness r.
- If a graph is isothermal, then it has the same fixation behavior as the unstructured population.
- Graphs that are not isothermal can change the balance between drift and selection.
- Amplifiers of selection increase the probability of fixation of advantageous mutants and reduce the probability of fixation of disadvantageous mutants. Suppressors of selection work in the opposite direction.
- The star, superstar, and funnel are amplifiers of selection.
- Scale-free graphs are amplifiers of selection.
- In an extended framework, all graphs that are circulations have the same fixation behavior as the unstructured population.
- We can also study games on graphs. A simple rule for the evolution of cooperation on graphs is $b/c > k$: selection favors cooperation if the benefit-to-cost ratio exceeds the number of neighbors.

SPATIAL GAMES 9

LET US NOW STUDY the deterministic evolutionary dynamics of spatial games. The members of a population are arranged on a two (or higher)-dimensional array. In each round, every individual plays the game with its immediate neighbors. After this, each site is occupied by its original owner or by one of the neighbors, depending on who scored the highest payoff in that round. These rules specify a deterministic cellular automaton. John von Neumann stood at the beginning of both game theory and cellular automata. In the theory of spatial games, these two approaches meet for the first time.

We will see that spatial effects can dramatically change the outcome of frequency-dependent selection. In space, strategies can coexist that exclude each other in a homogeneous setting. Moreover, spatial games have fascinating mathematical properties and a rich dynamical behavior. We will encounter spatial chaos, dynamic fractals, and evolutionary kaleidoscopes. Our goal is to formulate the simplest possible theory for deterministic spatial evolutionary game dynamics.

9.1 SPACED OUT

Consider an evolutionary game between two (or more) strategies. Each player occupies a position on a spatial grid and interacts with all of its neighbors. The payoffs from these interactions are added up. In the next generation, depending on the payoff, each player retains its current strategy or adopts the strategy of a neighbor.

We want to design a completely deterministic spatial game. This can be achieved with the following two rules: (i) each player adopts the strategy with the highest payoff in its neighborhood and (ii) all players are updated in synchrony.

Figure 9.1 illustrates the rules of the game for a square lattice and the Moore neighborhood; each cell has 8 nearest neighbors defined by a king's move on a chessboard. A player will retain its current strategy if it has a higher payoff than all of its neighbors. Otherwise the player will adopt the strategy of that neighbor that has the highest payoff. Note that the fate of a cell depends on its own strategy, the strategies of the 8 neighbors, and the strategies of their neighbors. Thus 25 cells in total determine what will happen to a cell. In the terminology of cellular automata, the transition rules are complex, but in terms of an evolutionary game they can be stated simply and naturally.

We are studying deterministic evolutionary game dynamics (without mutation) in a population with spatial structure. The transition rules are entirely deterministic. The outcome of the game depends only on the initial configuration of the population and the payoff matrix.

9.2 SPATIAL COOPERATION

As a specific example, we will explore the most interesting evolutionary game, the struggle between cooperators, C, and defectors, D. We will find that spatial games lead to a fascinating new mechanism for the evolution of cooperation, called "spatial reciprocity."

Consider the following Prisoner's Dilemma payoff matrix

Spatial games

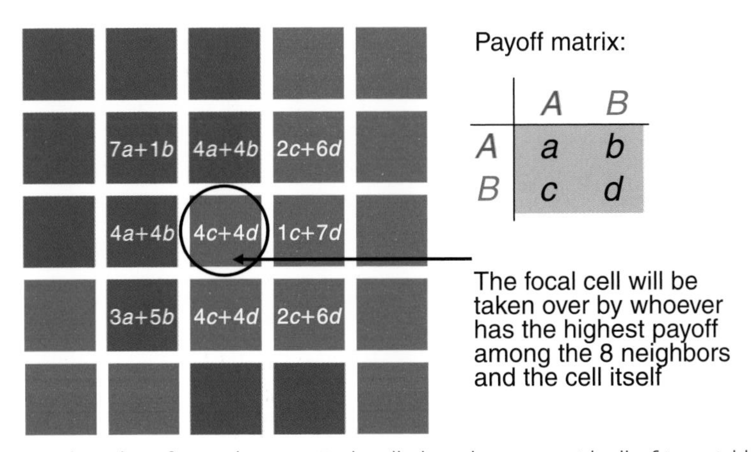

Figure 9.1 The rules of spatial games. Each cell plays the game with all of its neighbors. In this example, we use a square lattice and the Moore neighborhood, where each cell has 8 neighbors. The payoff for each player is evaluated. Subsequently each player compares its own payoff with that of its neighbors and adopts the strategy of whoever has the highest score. The fate of each cell depends on the state of all 25 cells in the 5×5 square that is centered around the cell.

$$\begin{array}{c} \\ C \\ D \end{array} \begin{array}{c} \begin{array}{cc} C & D \end{array} \\ \begin{pmatrix} 1 & 0 \\ b & \epsilon \end{pmatrix} \end{array} \tag{9.1}$$

If two cooperators interact, both receive one point. If a defector meets a co-operator, the defector gets payoff $b > 1$, while the cooperator gets payoff zero. The interaction between two defectors leads to the very small positive payoff ϵ. This payoff matrix is designed to keep things as simple as possible. For exploring different evolutionary dynamics, we vary the single parameter, b, and we choose to set $\epsilon \to 0$.

On the square lattice with the Moore neighborhood, each individual has 8 neighbors. Therefore, the possible payoffs for a cooperator are given by the set $\{1, 2, 3, \ldots, 8\}$. The possible payoffs for a defector are given by the set $\{b, 2b, 3b, \ldots, 8b\}$. The discrete nature of the possible payoff values means that there are only discrete transition points for b that can influence the dynamics. For $1 < b < 2$, these transitions occur at

$$8/7 = 1.1428\ldots$$

$$7/6 = 1.166\ldots$$

$$6/5 = 1.2$$

$$5/4 = 1.25$$

$$8/6 = 1.333\ldots$$

$$7/5 = 1.4$$

$$3/2 = 1.5$$

$$8/5 = 1.6$$

$$5/3 = 1.666\ldots$$

$$7/4 = 1.75$$

$$9/5 = 1.8$$

Figure 9.2 shows typical distributions of cooperators and defectors for different values of the parameter b. All simulations are performed on a 100×100 square lattice. There are periodic boundaries, which means that the edges of the square are wrapped around to generate a torus. This geometry has the advantage that all positions on the grid are equivalent. There are no boundary effects. The initial configuration is obtained at random with half of the cells being cooperators, the other half defectors.

The color code is as follows:

Blue represents a C that was a C in the previous generation.

Red represents a D that was a D in the previous generation.

Green represents a C that was a D in the previous generation.

Yellow represents a D that was a C in the previous generation.

Therefore blue and red indicate static cells, while green and yellow show changing cells. If a picture contains only red and blue, then it is a fixed point of the evolutionary dynamics: nothing has changed from the last generation, and nothing will change anymore. The more green and yellow cells, the more changes are occurring.

For $b = 1.10$, we observe a rather static pattern. Most cells are cooperators. There are isolated lines of defectors, which do not change. There are a few isolated single defectors, which generate squares of 9 defectors only to oscillate back to a single defector in the next generation. For $b = 1.15$, the lines of defectors oscillate at the end. There are many oscillating positions including isolated defectors. For $b = 1.24$, the lines of defectors start to be connected. There are a few oscillating positions. There are single defectors that oscillate to squares of 9 defectors, then to crosses of 5 defectors and back to single defectors. For $b = 1.35$, there is a pulsating network of defectors. Lines oscillate between thickness one and three. For $b = 1.55$, there is an irregular but static network of defectors permeating a world which is still dominated by cooperators.

For $b = 1.65$, the tide has turned. Defectors have won the majority. Cooperators survive in clusters. The picture is neither static, nor oscillatory, but highly dynamic. The clusters of cooperators always try to expand. They collide, break into pieces, and disappear. New clusters are being formed all the time. The system will certainly run into a cycle eventually (there are only finitely many states), but the transient can be longer than the lifetime of the universe. For $b = 1.70$, the pattern is again very static. There are mostly defectors. Cooperators survive in a few clusters.

Is it possible to understand these observations?

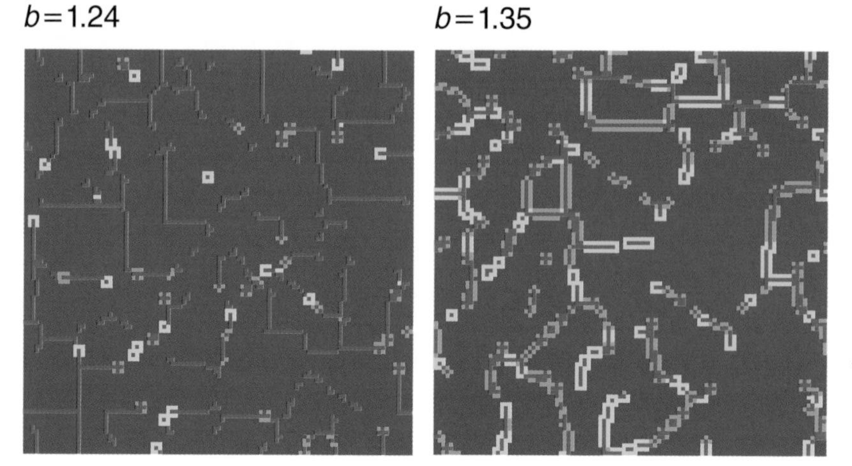

Figure 9.2 The spatial Prisoner's Dilemma displays an amazing variety of patterns where unconditional cooperators coexist with defectors. The figure shows configurations of 100×100 square lattices for seven different parameter regions. There are periodic boundary conditions, which means the edges are wrapped around to generate a toroidal universe. The color code is as follows: blue is a cooperator that was a cooperator in the previous round; red is a defector that was a defector in the previous round; green is a cooperator that was a defector in the previous round; yellow is a defector that was a cooperator in the previous round. The more yellow and green in a picture, the more changes are occurring. An entirely blue and red pattern is completely static. The payoff matrix is given by equation (9.1). The parameter b denotes the advantage for

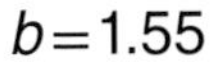
$b=1.55$

$b=1.65$

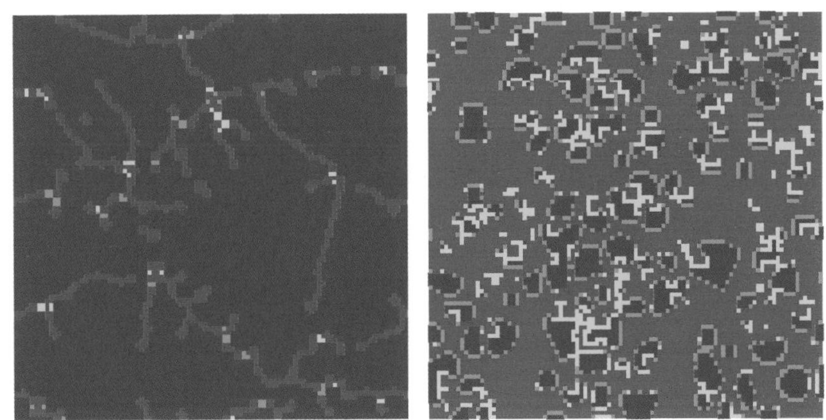

$b=1.70$

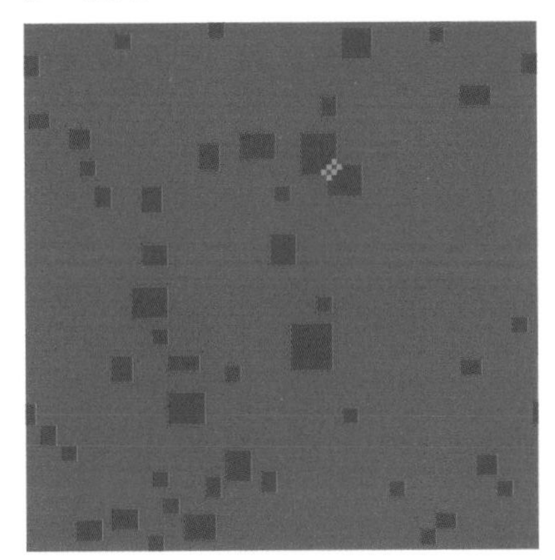

defectors. Cooperators dominate the scene for $b = 1.10$, 1.15, 1.24, 1.35, and 1.55. For these parameter values, there are various (static or pulsating) network structures of defectors in a mostly cooperative world. For $b = 1.65$ there is a dynamic coexistence between cooperators and defectors. Cooperators form clusters that grow, collide, disappear, and fragment to form new clusters. The pattern is always changing, but the average frequency of cooperators is always very close to 0.30. For $b = 1.70$ there are static clusters of cooperators in a frozen world. The initial condition is a random configuration with 10% cooperators; except for $b = 1.70$ when the simulation started with 50% cooperators.

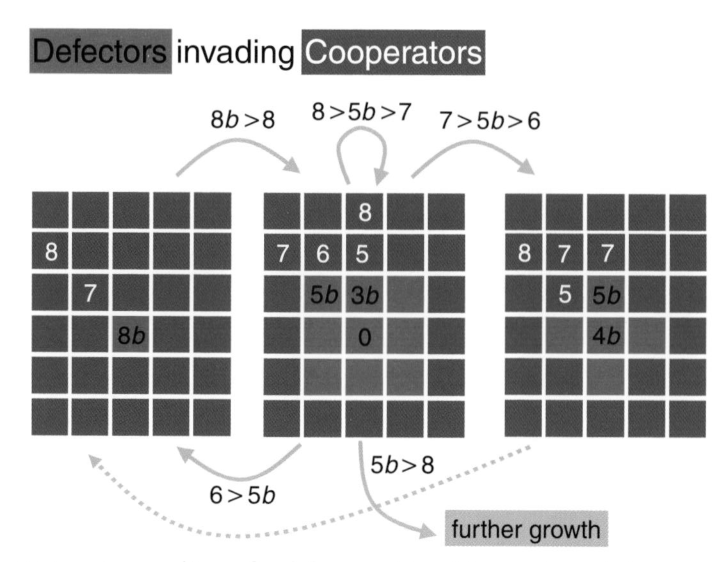

Figure 9.3 Invasion conditions for defectors. If $b > 1$ then a single defector gives rise to a square of 9 defectors, 9D. If $b < 6/5$, this square returns to a single defector. We have a period two oscillator. If $6/5 < b < 7/5$, then the 9D square turns into a 5D cross, which turns into a single defector. We have a period three oscillator. If $7/5 < b < 8/5$, the 9D square is stable. If $8/5 < b$, the 9D square can expand.

9.3 INVASION

The standard procedure for analyzing evolutionary games is to explore the conditions for invasion. When does natural selection favor the spread of a new mutant? Let us start with defectors invading cooperators.

9.3.1 Defectors Invading Cooperators

Figure 9.3 illustrates the conditions for a single defector to invade a population of cooperators. The defector has payoff $8b$. All of its immediate neighbors are cooperators and have payoff 7. All of their neighbors have payoff 8. Therefore, if $8b > 8$, which means $b > 1$, then the defector will take over all its neighbors.

In the square cluster of 9 defectors, 9D, the central defector has payoff 0, the four defectors at the corners have payoff $5b$, the four remaining defectors have

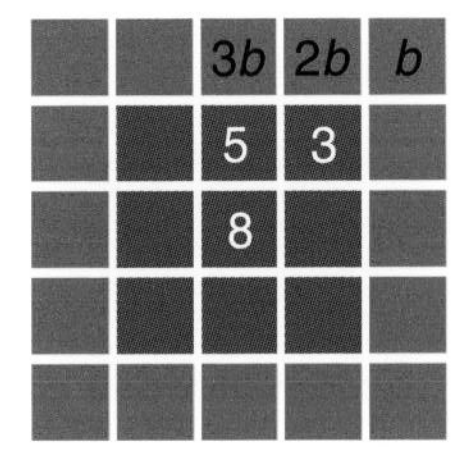
Cooperators invading Defectors
3b
2b
b
5
3
8
If b<3/2, this cluster will expand uniformly
If 3/2<b<5/3, this cluster will expand along lines but not along diagonals
If 5/3<b<8/3, this cluster is stable
If 8/3<b, this cluster will disappear

Figure 9.4 Cooperators can invade defectors when starting from a small cluster. Here we analyze a cluster of 9 cooperators, 9C. If $b < 3/2$, this cluster will expand uniformly. If $3/2 < b < 5/3$, the cluster can grow along the lines but not along the diagonals. In the next generation, there will be 12 new cooperators; all cells in a 5×5 square except the 4 corner cells will be cooperators.

payoff $3b$. The cluster is surrounded by cooperators with payoffs 5, 6 and 7. The second row of cooperators all have payoff 8. There are four possibilities:

(i) If $b < 6/5$, then the 9D square will return to a single defector.

(ii) If $6/5 < b < 7/5$, then the 9D square will turn into a cross consisting of 5 defectors which will subsequently turn into a single defector. There is a period three oscillator: 1D to 9D to 5D and back to 1D.

(iii) If $7/5 < b < 8/5$, then the 9D square will not change.

(iv) If $8/5 < b$, then the 9D square will grow into a square consisting of 25 defectors, which will continue to expand.

9.3.2 Cooperators Invading Defectors

Let us now analyze the conditions for cooperators to invade defectors (Figure 9.4). First, we note that a single cooperator can never survive or expand, but is always doomed to become eliminated in one step. In this deterministic game, cooperators only have a chance if they arise in clusters.

If $b < 3/2$, then a square of 4 cooperators will expand to a square of 16 cooperators, then to 36 cooperators, and so on. There will be ever-increasing squares of cooperators. If $b > 3/2$, then a square of 4 cooperators will be eliminated.

A square of 9 cooperators will also grow into bigger and bigger squares if $b < 3/2$. If $3/2 < b < 5/3$, then the 9C square can expand along the side lines, but not diagonally. It will give rise to a cross-like structure of 21 cooperators. This structure will continue to grow. If $5/3 < b < 8/3$, then the 9C square is stable; it will neither expand nor decline. If $b > 8/3$, the 9C square will be eliminated in two steps.

9.3.3 Three Classes of Parameter Regions

In summary, the above analysis suggests the existence of three classes of parameter regions.

(i) If $b < 8/5$, then only C clusters can keep growing.

(ii) If $b > 5/3$, then only D clusters can keep growing.

(iii) If $8/5 < b < 5/3$, then both C and D clusters can keep growing.

The various dynamical behaviors observed in Figure 9.2 fall into these three broad classes. As long as $b < 8/5$, the world is dominated by cooperators. If $b > 5/3$, defectors take over. If $8/5 < b < 5/3$, there is a dynamic balance between cooperators and defectors.

In parameter regions (i) and (ii), the final abundance of cooperators strongly depends on the starting condition. In parameter region (iii), however, most initial conditions converge to the same mixture of the two strategies with roughly 30% cooperators. While the actual pattern of cooperators and defectors is changing all the time, the frequency of cooperators, in a sufficiently large array, is almost constant. We call this behavior a "dynamic equilibrium."

9.4 DYNAMIC FRACTALS AND EVOLUTIONARY KALEIDOSCOPES

An interesting sequence of patterns emerges if a single defector invades a world of cooperators in the parameter region $8/5 < b < 5/3$. The defector grows to form a 3×3 and then a 5×5 square of defectors. The payoff for defectors at the corners of this square is $5b$, which is larger than 9. The payoffs for defectors

The corner-and-line condition

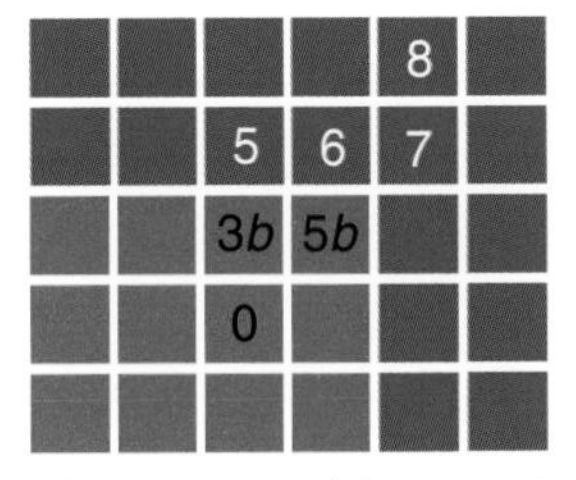

If $5b > 8$, then defectors win at corners

If $3b < 5$, then cooperators win along lines

$5/3 > b > 8/5$ is a clash of titans

Figure 9.5 The corner-and-line condition is responsible for spatial chaos, dynamic fractals, and kaleidoscopes. A large square-shaped cluster of defectors can expand on the corners if $8/5 < b$, but shrink along the lines if $b < 5/3$. Hence in the parameter region $8/5 < b < 5/3$, cooperators win along straight lines, but lose along irregular boundaries.

along the edges of the square is $3b$, which is smaller than 6. Therefore the defectors gain at the corners, but lose along the lines (Figure 9.5). The result is a dynamic fractal that combines symmetry and chaos. Figure 9.6 shows the growing fractal after 64, 124, and 128 time steps before it has encountered any boundaries. There are fractal-like structures that repeat themselves. The growing fractal is square-like at generations that are the powers of 2. The fractal contains many clusters of cooperators, which move around, expand, collide, fragment, and give birth to new clusters of cooperators. The frequency of cooperators within the growing fractal converges to $x \approx 0.30$, which is the same numerical value as in the simulations with random initial conditions.

Figure 9.7 shows a sequence of the "evolutionary kaleidoscope" that is generated by a single defector invading a population of cooperators in a fixed array with periodic boundaries. Each generation shows a new picture. There is an amazing variety. The initial symmetry is never broken, because the rules are symmetrical. The frequency of cooperators oscillates chaotically. These oscillations, however, cannot continue forever, because the total number of possible states is finite. The kaleidoscope must eventually converge to some oscillator with a finite period or a static configuration. Note that this convergence to a periodic orbit also holds, of course, for asymmetric initial conditions.

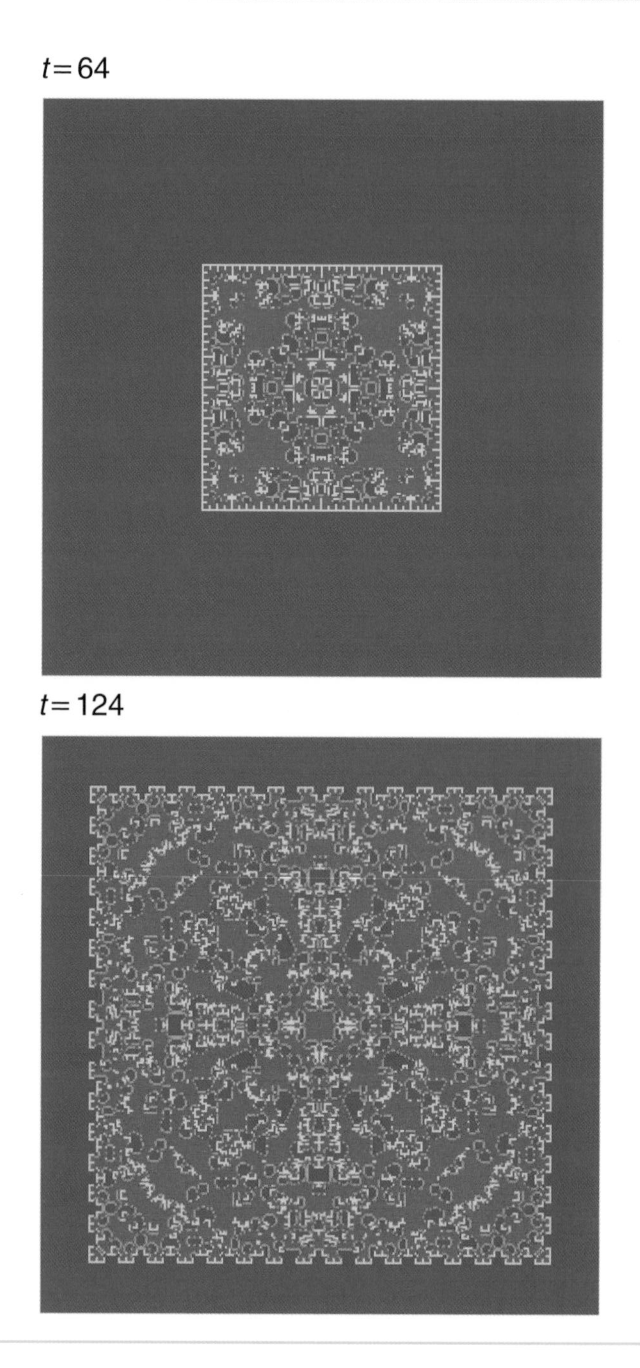
t=64
t=124

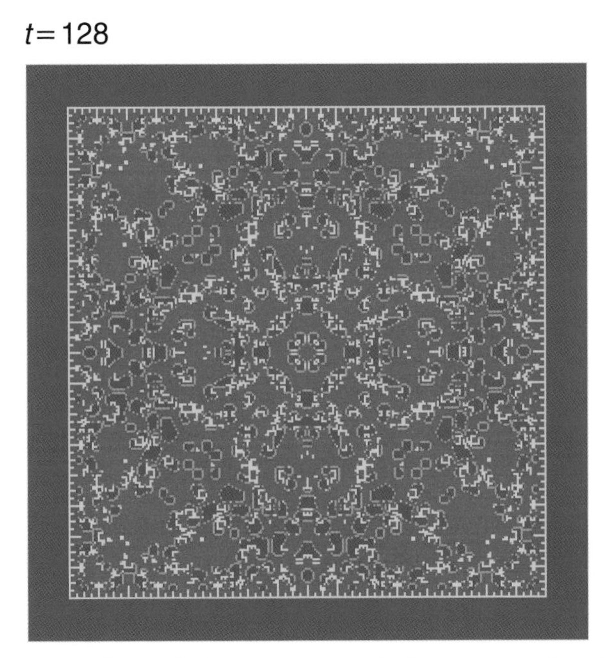

Figure 9.6 Starting with a single defector in a world of cooperators, there is an amazing sequence of ever-growing "Persian carpets." The structure is square-like with straight boundaries at every generation that is a power of 2. Here we show generations 64, 124, and 128. Parameter region $8/5 < b < 5/3$.

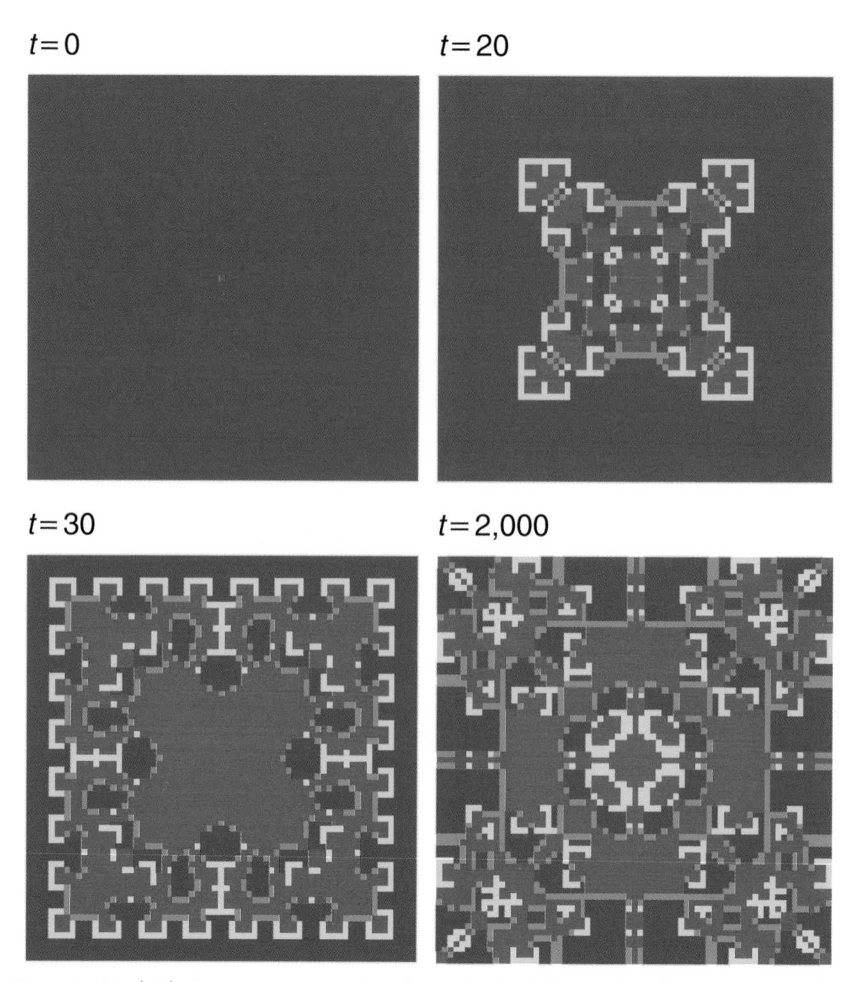

Figure 9.7 Kaleidoscopes are generated by a single defector invading a square of cooperators of fixed size. There can be an amazingly long sequence of always changing symmetric patterns. In the end (but after a very long time) the kaleidoscope must reach a fixed pattern or a cycle, because the number of all possible configurations is finite. Parameter region $8/5 < b < 5/3$. The size of the square is 69×69 with periodic boundary conditions.

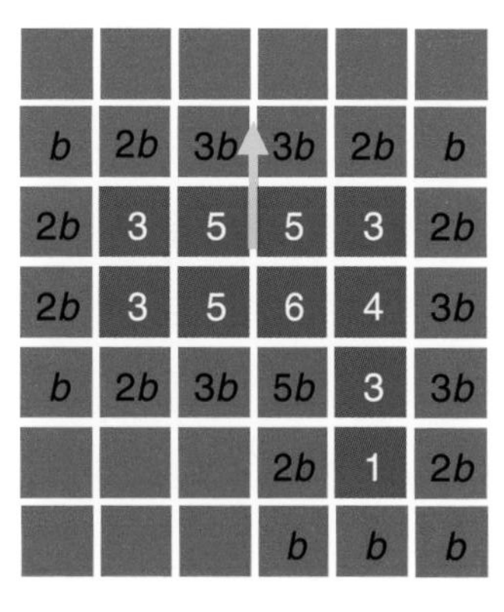

Figure 9.8 A "walker" is a cluster of 10 cooperators. It moves into the direction indicated by the yellow arrow. The leg moves from the right to the left every other generation. When observed on a screen, it appears as if the walker is walking on two legs. Parameter region $3/2 < b < 5/3$.

The interesting mathematical features of the growing fractal and the kaleidoscopes arise from a combination of simplicity (the rules), deterministic unpredictability (the eventual fate), transient chaos (the frequency of cooperators), and symmetry (beauty).

9.5 THE BIG BANG OF COOPERATION

Although the patterns of the previous section are beautiful, there is the disconcerting aspect that they describe the invasion and partial replacement of a world of cooperators by defectors. Fortunately, the reverse invasion is also possible and even more beautiful.

A "walker" is a structure of 10 cooperators (Figure 9.8). For $3/2 < b < 5/3$, this "fellowship of cooperators" moves bravely through an adverse world of defectors. One such walker cannot change the world, but if two walkers collide they can generate a "big bang" of cooperation, exploding into a world of defectors (Figure 9.9).

Less dramatically, but no less beautiful, a big bang can also be initiated by a single square of 9 cooperators or a rectangle of 6 cooperators. Figures 9.10 and 9.11 show big bangs of cooperators for two different parameter values.

A collision of two "walkers" can lead to a "big bang" of cooperation

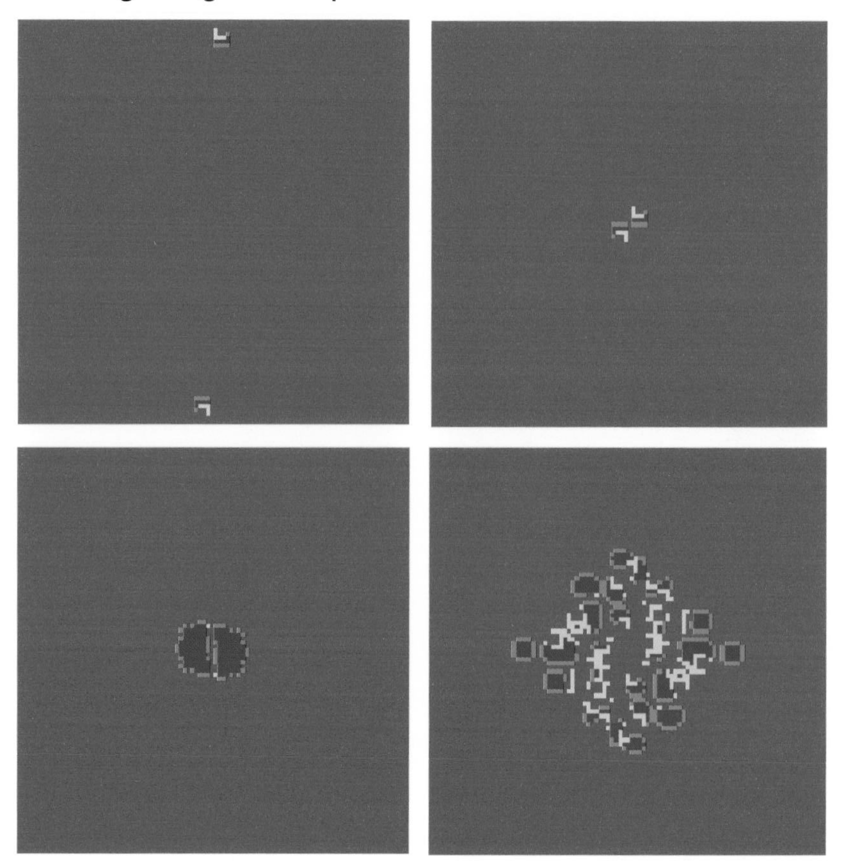

Figure 9.9 Cooperation "walks in." A collision of two walkers in a world of defectors can generate a big bang of cooperation. Four consecutive time points are shown. Parameter region $8/5 < b < 5/3$.

Figure 9.10 A cluster of 3×3 cooperators in the parameter region $8/5 < b < 5/3$ can invade a world of defectors and also generate a fractal-like growth pattern.

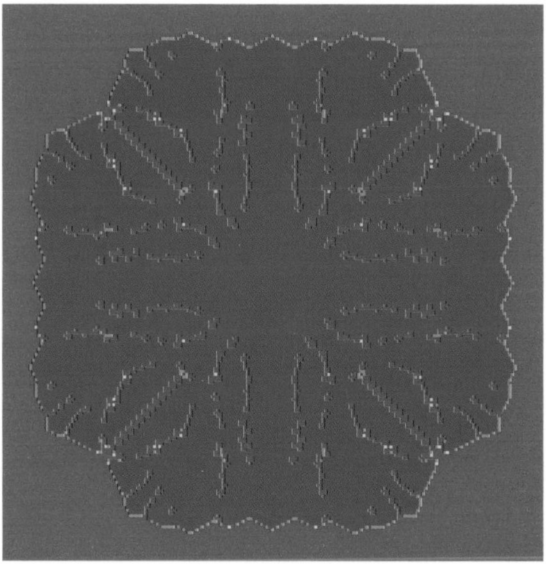

Figure 9.11 Invasion of cooperators starting from a 3×3 cluster in the parameter region $3/2 < b < 8/5$.

9.6 OTHER GEOMETRIES

Spatial games can be studied with many variations on the basic theme. Instead of the Moore neighborhood, we can investigate the "von Neumann" neighborhood, which consists of the 4 adjacent neighbors excluding the diagonals. Again there is a range of different patterns exhibiting coexistence between cooperators and defectors in the nonrepeated Prisoner's Dilemma. For $4/3 < b < 3/2$, we encounter the dynamic equilibrium with kaleidoscopes and fractals of even greater allure (Figure 9.12). For $3/2 < b < 2$, clusters of cooperators can still expand horizontally and vertically generating a rectangular "railway" network.

In a hexagonal lattice, each cell is surrounded by six others. There again different parameter regions allow coexistence between cooperators and defectors, but there is no dynamic equilibrium. The patterns are more static. For other evolutionary games, however, it is possible to obtain a dynamic equilibrium on a hexagonal lattice.

We can also distribute individual cells randomly over a two-dimensional plane. Two individuals are neighbors if their distance is less than a certain "radius of interaction," r. Cells can differ in the number of their neighbors. The resulting random grid is, of course, closer to real-world situations than the symmetrical lattices are. The payoff of an individual is the sum over the interactions with all of its neighbors. As before, a cell is retained by its original owner or given to the most successful neighbor, whoever has the highest payoff. All cells are updated simultaneously. The evolutionary dynamics are deterministic. Cooperators survive up to certain values of r. The equilibrium frequency of cooperators depends on the initial conditions. Population dynamics on random grids are more static than on square lattices. We have not yet found games that generate spatial chaos on irregular grids. Irregularity tends to simplify dynamics.

9.7 OTHER UPDATE RULES

So far we have studied spatial games with entirely deterministic dynamics. Each cell is given to whoever has the highest payoff in the neighborhood, and all cells are updated in synchrony. These assumptions allowed us to study

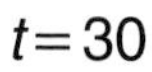

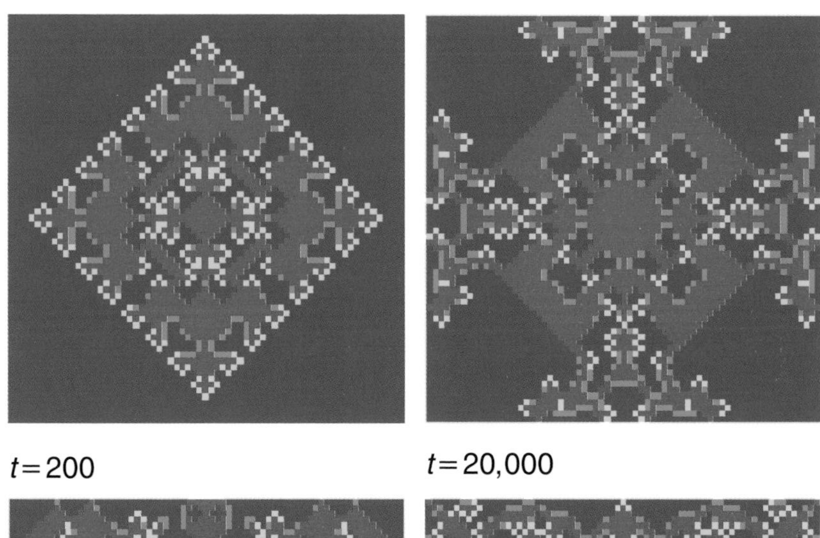

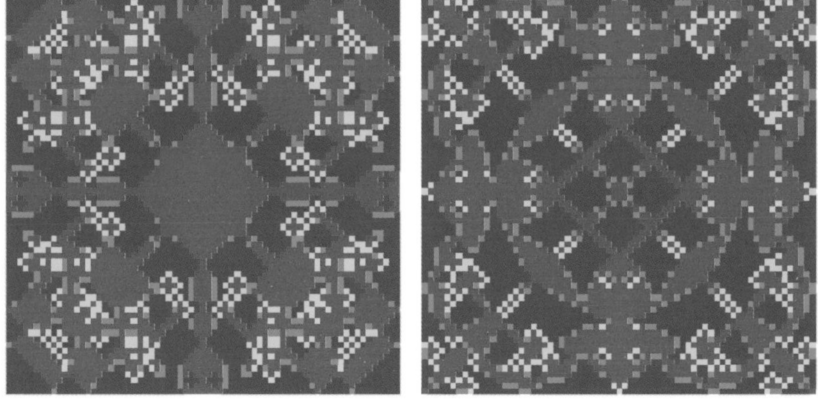

Figure 9.12 A kaleidoscope in the von Neumann neighborhood. On a square lattice, each player interacts with the four nearest neighbors. Parameter region $4/3 < b < 3/2$.

the rich mathematical properties of deterministic spatial dynamics in discrete time. We found fascinating and complicated behavior in terms of spatial chaos and dynamic fractals. Although the fate of each single cell is totally deterministic, the overall population dynamics are highly complicated.

We can also study games with stochastic transition rules. For example, a cell could become a cooperator with a probability that is given by the relative payoff of cooperators in the neighborhood.

Instead of synchronous updating, we can also investigate asynchronous updating: one player is chosen at random; its own payoff and the payoffs of all neighbors are determined. Then the player is updated. Synchronous updating means nonoverlapping generations; asynchronous updating means overlapping generations (with continuous reproduction). Asynchronous updating introduces random choice and therefore stochasticity. Figure 9.13 shows a cluster of cooperators invading a world of defectors for asynchronous updating with Moore neighborhood and $b = 1.59$.

In general, stochastic update rules display less variety in dynamical behaviors. Dynamic fractals and kaleidoscopes are not possible, because the stochastic update rules do not maintain symmetry. Stochasticity disturbs straight lines between cooperators and defectors, and irregular boundaries favor defectors.

If the spatial competition between cooperators and defectors is described by a stochastic process, then, in general, there will be only two absorbing states: all cooperators or all defectors. The system will eventually reach one of these two states, but it can take an extremely long time. In most cases, spatial games with stochastic update rules still allow the coexistence of cooperators and defectors for the lifetime of our universe.

If there are empty sites or more than two competing strategies, then spatial games can lead to spiral waves.

9.8 VIRTUALLABS

Christoph Hauert has written a beautiful programming environment that allows you to study every aspect of evolutionary games, spatial games, and games on graphs. These "VirtualLabs" can be accessed on http://lorax.fas.harvard.edu/virtuallabs/

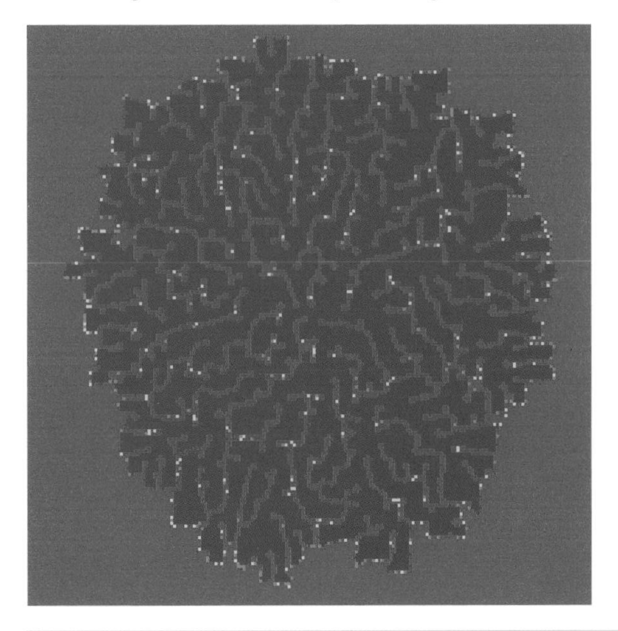

Figure 9.13 Invasion of cooperators with asynchronous updating. At each time point a random cell is chosen to be updated. Its payoff is compared with the payoff of all its neighbors. Then the cell is given to whoever has the highest payoff. This takeover rule is still deterministic, but the growth pattern is stochastic, because the cells updated in each time step are chosen randomly. The starting condition was a 3×3 cluster of cooperators in a world of defectors.

This Web page enables you to retrace the steps we have described here, but you can also make many new discoveries. VirtualLabs represent a "language" for evolutionary dynamics. Many "questions" of this language have not been asked. Many "sentences" have not been spoken. You can use the VirtualLabs to make new discoveries in many settings of evolutionary dynamics. The figures of this chapter were generated with VirtualLabs. VirtualLabs are virtually error free as they are "Swiss made."

SUMMARY

- Evolutionary game dynamics (= frequency-dependent selection) can be studied in a spatial setting.
- In spatial games, players interact with their nearest neighbors.

- A player keeps his current strategy or adopts one of his neighbors' strategies according to who has the highest payoff.
- It is possible to formulate entirely deterministic spatial game dynamics.
- In spatial games, the theory of cellular automata meets game theory.
- In the spatial Prisoner's Dilemma, there is coexistence between cooperators and defectors.
- Cooperators survive in clusters. This principle is called "spatial reciprocity."
- In some parameter regions, we discover spatial chaos, dynamic fractals, and evolutionary kaleidoscopes.
- Cooperators can invade defectors when starting from a small cluster.
- Irregular grids tend to simplify dynamical complexity.
- Asynchronous updating or "proportional winning" introduces stochasticity. Cooperators and defectors can nevertheless coexist for near eternity.

HIV INFECTION 10

THE EMERGENCE of the human immunodeficiency virus (HIV) in the early 1980s demonstrated that infectious diseases represent a major problem for human health all over the world and that newly arising infectious agents can be especially devastating. As the human species becomes more abundant on the globe, having crossed the six billion threshold in 1999, the opportunity increases dramatically for infectious agents of other species to invade the human host. Despite tremendous progress in molecular biology and medicine, our methods to combat infectious diseases are limited. There are successful vaccines against a number of agents, but all attempts have so far failed to construct an HIV vaccine. The reasons for this failure are not clear, but include the virus's ability to infect cells of the immune system and to mutate away from any opposing selective forces.

HIV belongs to the class of retroviruses, which reversely transcribe their RNA genome into DNA (Figure 10.1). Howard Temin and David Baltimore won a Nobel Prize in Medicine for the discovery of the reverse transcriptase enzyme. The viral DNA can be integrated into the genome of the host cell and can remain there for an effectively unlimited time. Anywhere between 2% and 8% of the human genome consists of "burnt-out" retroviruses that

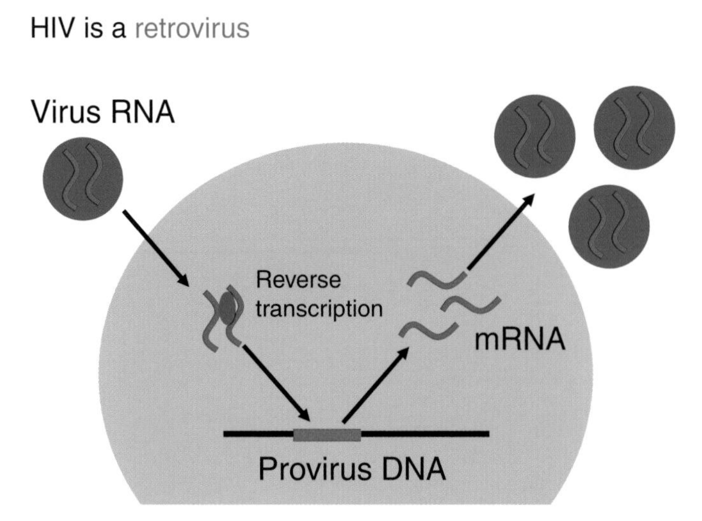

Figure 10.1 The life cycle of the human immunodeficiency virus (HIV). The virion contains two copies of the viral genome in form of single-stranded RNA. After entry into the host cell, the reverse transcriptase that comes with the virion uses both copies of the RNA genome to produce an RNA-DNA hetero-duplex and subsequently double-stranded viral DNA. This viral DNA (the provirus) is integrated into the genome of the host cell. The provirus can remain silent for a long time or can immediately induce the host cell to produce messenger RNA (mRNA). The viral mRNA has a dual purpose: (i) it is used for biosynthesis of viral proteins and (ii) it serves as viral genome for the virions that will eventually leave the host cell.

have integrated at some point in our genomic history and have subsequently received inactivating mutations.

The closest relatives of HIV are the simian immunodeficiency viruses (SIV) that infect many primate species. There are two types of human viruses: HIV-1 is very closely related to SIV from chimpanzees; HIV-2 is very closely related to SIV from sooty mangabeys. Remarkably, the SIV viruses appear not to cause disease in their natural hosts. When transmitted to another host species, however, they can induce a disease very similar to the acquired immunodeficiency syndrome (AIDS) in humans. For example, SIV from African green monkeys can induce AIDS in Asian macaques.

In humans, HIV leads to a primary infection with flu-like symptoms. During the primary infection, there is high virus load, but immune responses

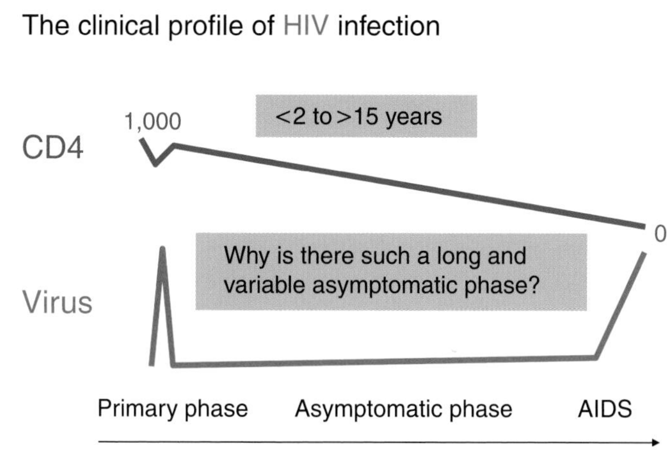

Figure 10.2 The pattern of disease progression of HIV infection requires a mechanistic explanation. There is a short primary phase, with high virus load, followed by a long and variable asymptomatic phase, usually with lower virus load. Eventually patients develop the fatal immunodeficiency disease AIDS. The asymptomatic phase can last less than two years or longer than fifteen years. The average length (without therapy) is about ten years. During disease progression the CD4 cell count drops (as a nearly linear function of time) from 1,000 to essentially 0. AIDS is defined as a CD4 cell count of less than 200. The question is: given that the characteristic time scale of the infection (the generation time of the virus) is on the order of days, why does it take years to develop disease?

against HIV may not yet be detectable. The patient can be negative when tested for HIV antibodies, but is usually positive when tested for the presence of HIV RNA. Subsequently, patients enter an asymptomatic phase that can last for many years (Figure 10.2). Over time there is a linear decline in the number of CD4 cells, the primary target cell of HIV in the human host. When the CD4 cell count drops from about 1,000 initially to below 200, patients enter the final stage of disease called AIDS (acquired immune deficiency syndrome). The failing immune system can no longer control the HIV virus, which replicates to high levels. In addition, the patient is overwhelmed and killed by other opportunistic infections.

CD4 cells represent an important component of the human immune system. CD4 cells are stimulated by the presence of foreign antigens. Once stimulated, they divide and send activation signals to CD8 cells and B cells. CD8 cells recognize and kill virus-infected cells. B cells release antibodies that attack viruses and other infectious agents. By infecting and depleting the CD4 cell population, HIV attacks the immune system that is meant to control it.

There are about twenty anti-HIV drugs. Some of the drugs were originally developed as anticancer drugs and were by coincidence inhibitors of the virus-encoded replication enzyme, reverse transcriptase. Other drugs were specifically designed to inhibit a virus-encoded protease enzyme. All these drugs interfere with virus reproduction. HIV can rapidly evolve resistance to any one of them, but is usually controlled when combinations of three or more drugs are used simultaneously. Successful drug treatment leads to a dramatic reduction in the abundance of virus, the so-called virus load. The drugs, however, do not cure patients, because they cannot eradicate the virus. Drug treatment can nevertheless greatly enhance the life expectancy of patients and can delay the onset of symptoms. Although the anti-HIV drugs constitute an amazing success of biomedicine, it is unclear whether they will have any impact on the global pandemic. The majority of all infected people worldwide are in the poorest countries and have little or no access to expensive drugs. The most effective epidemiological use of anti-HIV drugs may be to prevent mother-to-infant transmission of the virus. In any case, a highly effective vaccine is urgently needed.

Here we will address the following questions: What is the mechanism of HIV disease progression? Given that the virus infects and kills CD4 cells on a time scale of days, why is there such a long and variable asymptomatic period? Without treatment, it takes on average ten years to progress from infection to AIDS. Some people have died within one or two years of infection, while others are still asymptomatic after fifteen or more years. Furthermore, why does HIV cause a fatal disease in humans, while the very closely related SIV virus apparently does not cause disease in its natural hosts?

I will present a model of HIV (and SIV) disease progression. The main idea is that the key mechanism of disease progression is virus evolution in individual patients. During primary infection, there is selection for the fastest-growing virus mutant. Once immune responses emerge, there is selection for virus mutants that escape from these responses. This process is called antigenic variation. The number of different antigenic variants of the virus, the "antigenic diversity," increases over time. The virus evolves more and more successfully to evade any opposing immunological pressure and is finally driven to a point at which the immune system can no longer control it. In this theory, virus evolution in individual patients is responsible for disease progression. The theory can also explain the difference between pathogenic and apathogenic SIV infections.

A central assumption of this theory, which was first proposed in 1990, is that the virus is rapidly replicating in the presence of immune responses. This rapid turnover was confirmed in 1995. Furthermore, the theory assumes that the virus can readily produce escape mutants that evade current immunological attack. Evidence for this fact has been mounting since 1991, but a clearer quantitative picture of viral escape from immune responses has been emerging only in recent years. Finally, the theory requires the virus to impair immune responses. There has been some discussion about the detailed mechanism of how HIV eliminates CD4 cells (some investigators claim that HIV is not directly cytopathic), but there can be no doubt that the CD4 cell population is being destroyed during HIV infection.

We will begin by studying the simplest possible model for the evolutionary dynamics of antigenic variation and subsequently add the HIV-specific property of impairing immune responses.

10.1 ANTIGENIC VARIATION

The simplest model of antigenic variation describes a replicating viral (or other) pathogen that is opposed by strain-specific immune responses. Let v_i denote the population size of virus strain (or mutant) i and x_i the magnitude of the specific immune response against strain i. Consider the following system of ordinary differential equations:

$$\begin{aligned} \dot{v}_i &= r v_i - p x_i v_i \\ \dot{x}_i &= c v_i - b x_i \end{aligned} \qquad i = 1, \ldots, n \tag{10.1}$$

In the absence of immune responses, viral growth is exponential at rate r. Immune responses are stimulated at the rate cv_i, which is proportional to the abundance of virus. Immune responses eliminate virus at the rate px_iv_i. Finally, immune responses decay in the absence of further stimulation at rate bx_i. In this model, there are n virus strains, which are opposed by n specific immune responses. Each immune response, x_i, can only recognize virus strain v_i.

Figure 10.3 shows a computer simulation of the above model. We start with a single virus strain, v_1. Initially virus growth is exponential at rate r, but the virus stimulates the specific immune response, x_1, which reduces virus growth and eventually brings an end to viral expansion. Virus load reaches a maximum value and subsequently starts to decline. Similarly the immune response, x_1, reaches a maximum value and then declines. The system settles in damped oscillations to the equilibrium

$$v_1^* = \frac{br}{cp} \qquad x_1^* = \frac{r}{p}. \tag{10.2}$$

We assume that mutation continuously generates new viral strains, which can escape from the specific immune responses. In the computer simulation of Figure 10.3, the generation of new mutants is a stochastic process. The probability that a new mutant emerges in the time interval $[t, t + dt]$ is given

Figure 10.3 Dynamics of the basic model of antigenic variation as defined by equation (10.1). Each viral strain is only opposed by a strain-specific immune response. There is no cross-reactive immunity. Therefore the dynamics of each strain are independent of all other strains. Each strain rises to high abundance and is subsequently down-regulated in damped oscillations by a specific immune response. Virus load increases in an oscillatory fashion as new strains are being generated. The equilibrium virus load is an increasing function of antigenic diversity. The figure shows the total virus load, v, the abundance of individual strains, v_i, and the strength of specific immune responses, x_i. Parameter values are $r = 2.5$, $p = 2$, $c = 0.1$, and $b = 0.1$. The infection starts with a single strain. The probability that a new mutant arises in the time interval $[t, t + dt]$ is given by Pdt with $P = 0.1$. Reprinted from Nowak and May (2000) by permission of Oxford University Press.

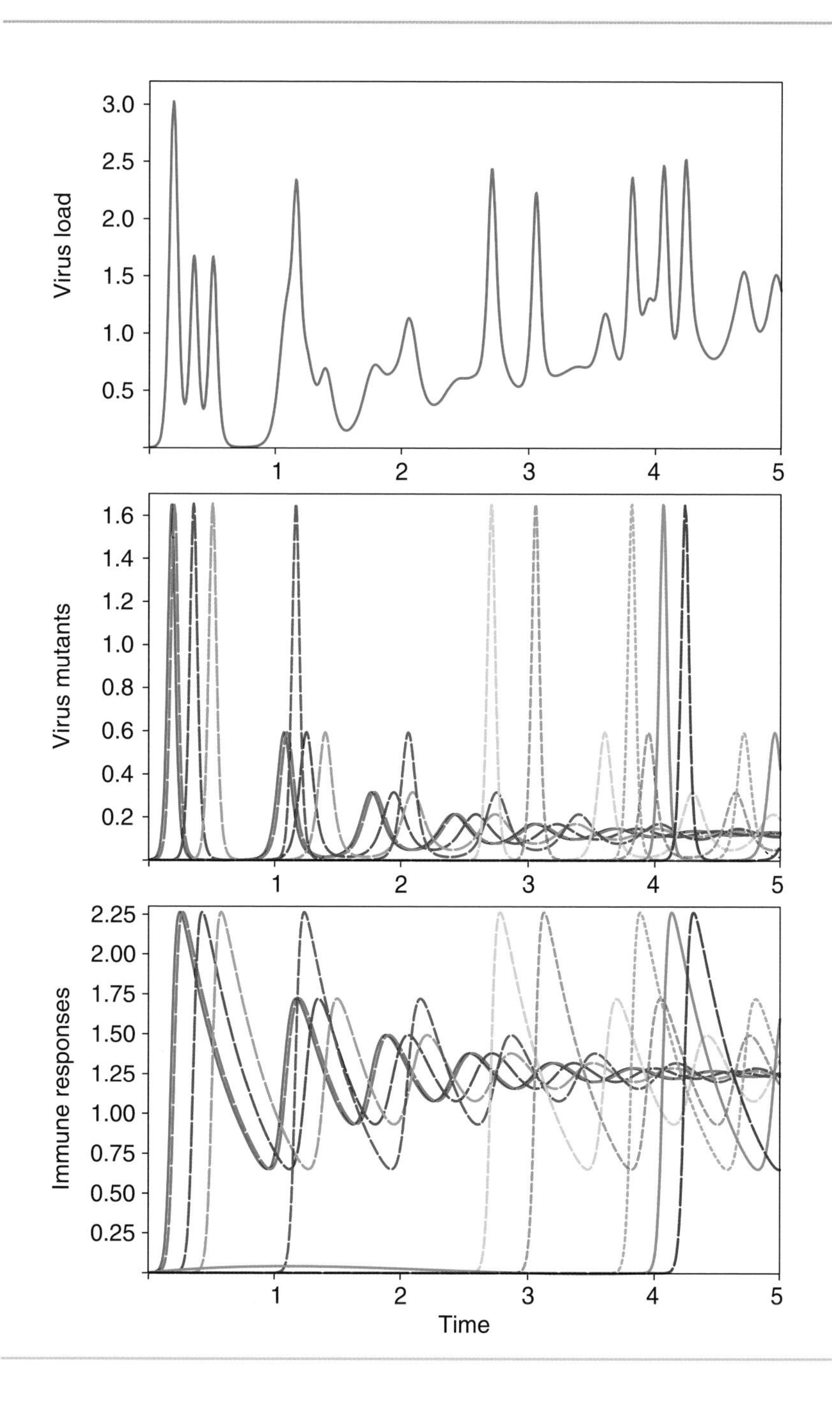

Virus load
3.0
2.5
2.0
1.5
1.0
0.5
1
2
3
4
5
Virus mutants
1.6
1.4
1.2
1.0
0.8
0.6
0.4
0.2
Immune responses
2.25
2.00
1.75
1.50
1.25
1.00
0.75
0.50
0.25
Time

by Pdt, where P is the mutation rate. The simplest assumption is that P is a constant. Alternatively, we can assume that P is proportional to virus load, $v = \sum_{i=1}^{n} v_i$, because the number of mutation events is proportional to the number of replication events.

In Figure 10.3 the new variant, v_2, escapes from the immune response x_1 and grows unchecked initially, but it induces an immune response, x_2, which brings it down after some time. Meanwhile another escape mutant, v_3, has emerged, and so on. The result is a sequence of antigenically different variants that all grow for some time before being controlled by immune responses. If there are n viral variants present in the system and if all are at the equilibrium value given by (10.2), then the total virus abundance is given by

$$v = \frac{brn}{cp}. \tag{10.3}$$

We observe that virus load, v, is an increasing function of antigenic diversity, n.

Equation (10.1) defines the simplest model of antigenic variation. Each virus strain, v_i, is only controlled by one specific immune response, x_i. This means that the dynamics of any one strain are independent of all other strains; the two differential equations describing the dynamics of one strain and its specific immune response are decoupled from the equations for other viral strains.

10.1.1 Strain-Specific and Cross-Reactive Immunity

The most obvious extension of the basic model is to include a cross-reactive immune response that can recognize several (or all) virus mutants. Thus new antigenic variants escape from all existing strain-specific responses but are still recognized by the cross-reactive response.

Denote by z the strength of a cross-reactive immune response that is active against all virus mutants. This leads to the system of equations

$$\begin{aligned} \dot{v}_i &= v_i(r - px_i - qz) \qquad i = 1, \ldots, n \\ \dot{x}_i &= cv_i - bx_i \qquad i = 1, \ldots, n \\ \dot{z} &= kv - bz \end{aligned} \tag{10.4}$$

The cross-reactive immune response, z, is stimulated by all virus mutants at rate kv_j and decays at the rate bz. The main consequence of this model extension is that new antigenic variants do not completely escape from all existing immune responses, and therefore do not grow to the same abundance as the original virus mutant. The dynamics of individual strains are no longer independent of one another. A computer simulation is shown in Figure 10.4.

For n virus mutants, the equilibrium virus load is now given by

$$v = \frac{brn}{cp + kqn}. \tag{10.5}$$

Once again viral load is an increasing function of antigenic diversity, n, but saturates for high values of n. The maximum possible equilibrium virus load is $v_{max} = (br)/(kq)$, which represents the equilibrium virus load in the presence of the cross-reactive immune response alone. Thus increasing antigenic diversity eliminates the effect of strain-specific immunity.

10.2 DIVERSITY THRESHOLD

In the previous section we analyzed general models of antigenic variation. They can in principle describe any virus or other infectious agent that establishes a persistent infection in its host and can mutate to escape from immune responses. Let us now add another feature, which makes the model more specific for HIV. We will assume that the virus can impair immune responses:

$$\begin{aligned} \dot{v}_i &= v_i(r - px_i - qz) \qquad i = 1, \ldots, n \\ \dot{x}_i &= cv_i - bx_i - uvx_i \qquad i = 1, \ldots, n \\ \dot{z} &= kv - bz - uvz \end{aligned} \tag{10.6}$$

As before, v_i denotes the population size of virus mutant i, while x_i denotes the immune response specifically directed against virus strain i. The cross-reactive immune response directed against all different virus strains is given by z. Mutational events occur throughout the infection and increase the number of virus strains, n, as time goes by. The total virus load is given by $v = \sum_i v_i$.

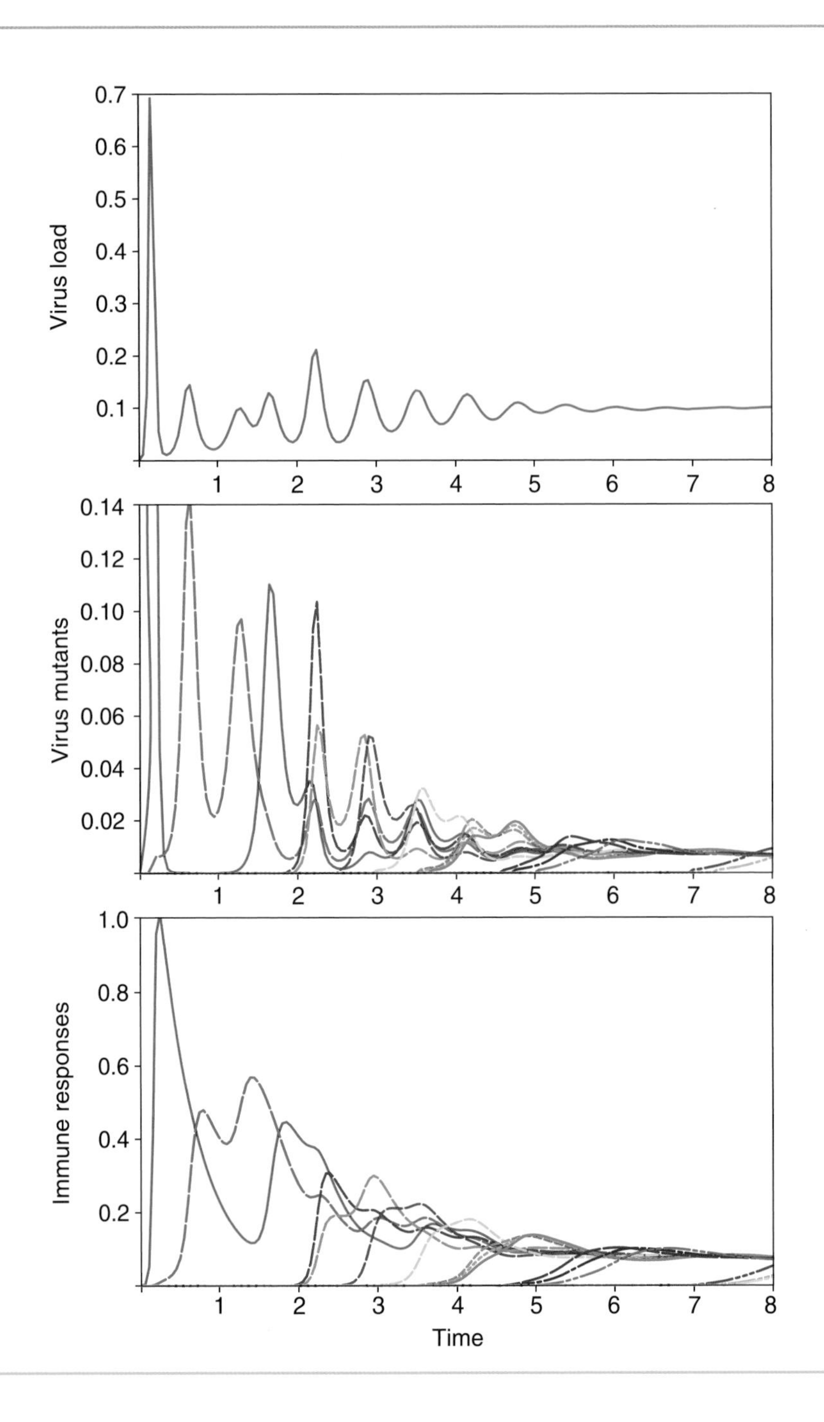
Virus load
0.7
0.6
0.5
0.4
0.3
0.2
0.1
1
2
3
4
5
6
7
8
Virus mutants
0.14
0.12
0.10
0.08
0.06
0.04
0.02
Immune responses
1.0
0.8
0.6
0.4
0.2
Time

The parameter r denotes the average rate of replication of all different virus strains; p specifies the efficacy of the strain-specific immune responses and c specifies the rate at which they are evoked; similarly q denotes the efficacy of cross-reactive immune responses and k the rate at which they are induced. In the absence of further activation, immune responses decay at rate b. HIV and other lentiviruses can impair immune responses by killing CD4-positive cells, which help B cells and cytotoxic T cells to mount immune responses against the virus. We summarize this effect of HIV in the loss terms $-uvx_i$ and $-uvz$. Thus the parameter u characterizes the ability of the virus to impair immune responses; by depleting CD4 cells the virus indirectly impairs immune responses mediated by B cells and cytotoxic T cells. The immune responses converge to

$$x_i^* = \frac{cv_i}{b + uv} \qquad i = 1, \ldots, n \tag{10.7}$$

and

$$z^* = \frac{kv}{b + uv}. \tag{10.8}$$

Once the immune responses have reached these levels, the total virus population changes as

$$\dot{v} = \frac{v}{b + uv}[rb - v(cpD + kq - ru)]. \tag{10.9}$$

Figure 10.4 Antigenic variation in the presence of strain-specific and cross-reactive immune responses. The simulation begins with a single viral strain, which induces both strain-specific and cross-reactive immune responses. Subsequent strains escape from the strain-specific response, but not from the cross-reactive response. The equilibrium virus load is an increasing function of antigenic diversity, but saturates for high levels of antigenic diversity. The simulation is based on equation (10.4), with the parameter values $r = 2.5$, $p = 2$, $q = 2.4$, $c = k = 1$, and $b = 0.1$. The probability that a new mutant arises in the time interval $[t, t + dt]$ is given by Pdt with $P = 0.1$. Reprinted from Nowak and May (2000) by permission of Oxford University Press.

The variable D denotes the Simpson index,

$$D = \sum_{i=1}^{n} (v_i / v)^2. \tag{10.10}$$

This quantity is a number between 0 and 1 and represents an inverse measure of diversity. The Simpson index denotes the probability that two virus particles chosen at random belong to the same strain. If only one virus strain is present, then $D = 1$. If n virus strains present, all with the same frequency, then $D = 1/n$.

The product kq specifies the efficacy of the cross-reactive immune responses. The product cpD denotes strain-specific immune responses. The efficacy of these strain-specific responses depends on the antigenic diversity of the virus population. Equation (10.9) shows that increasing diversity (decreasing D) increases the total population size of the virus.

The model has three distinct parameter regions, which correspond to three qualitatively different courses of infection (Figure 10.5).

i. Immediate Disease

If $ru > kq + cp$, then a single virus strain can outrun the combined effect of strain-specific and cross-reactive immunity. In this case, there is no asymptomatic phase; the virus population immediately replicates to high levels and causes disease and death. Viral replication, r, and/or cytopathic effects, u, are large compared with the combined effects of cross-reactive and strain-specific immune responses, $kq + cp$. The immune response is unable to control the virus, which replicates to high levels within a short time. No antigenic variation is necessary.

An example for this type of behavior is the rapid progression to disease and death of pig-tailed macaques infected with the acutely lethal variant SIVsmm-pbj14, which was isolated from sooty mangabeys. The virus kills within two weeks of infection. The primary manifestation of disease, and the cause of death, is diarrhea and its sequelae (rather than immunodeficiency). Our model only predicts that the immune responses are unable to control the virus, which in turn replicates to very high levels and thereby causes disease. To validate our model in this particular example, one would have to check if

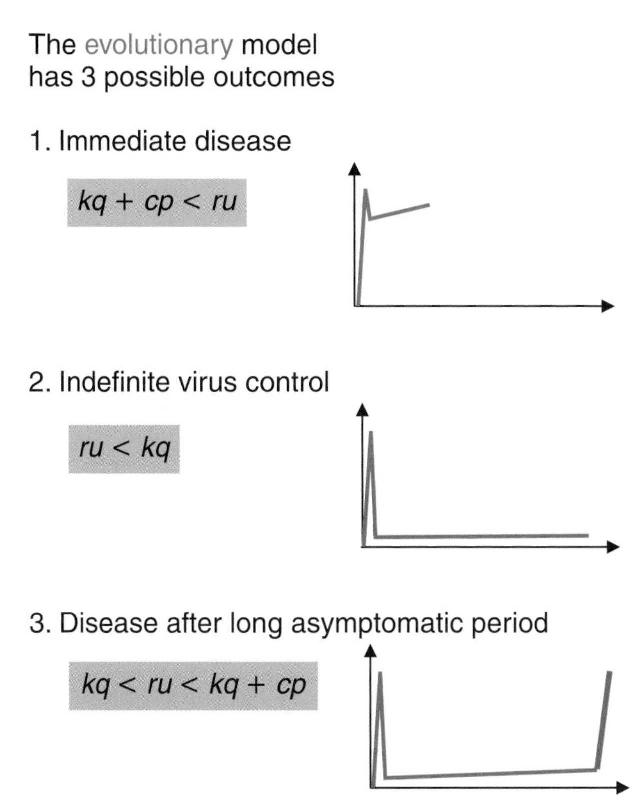

The evolutionary model
has 3 possible outcomes
1. Immediate disease
kq + cp < ru
2. Indefinite virus control
ru < kq
3. Disease after long asymptomatic period
kq < ru < kq + cp

Figure 10.5 The mathematical model of HIV infection (given by equation 10.6) has three possible outcomes, which correspond to observed patterns of lentivirus infection. (i) If the virus cannot be contained despite the combined effort of strain-specific and cross-reactive immune responses, then there is immediate development of disease without need for viral diversification and evolution. This pattern is observed in some cases of very fast disease progression in HIV infection and experimental SIV infection. (ii) If the virus can be contained by cross-reactive immune responses alone, then there is asymptomatic infection without development of disease. This pattern is observed in most natural SIV infections. (iii) If viral replication and cytopathicity can be contained by the combined effort of strain-specific and cross-reactive immune responses, but not by cross-reactive responses alone, then there is evolution toward disease over a long and variable period of infection. This pattern corresponds to human HIV infection or many experimental SIV infections.

virus concentrations are very high in the sick animals. Another test would be to construct an SIVsmm-pbj variant with a reduced replication rate. This may cause not immediate disease and death but a chronic infection (maybe with slow development of immunodeficiency disease).

ii. Chronic Infection without Disease

If $kq > ru$, then the effect of cross-reactive immunity alone is sufficient to control the virus. Antigenic variation will occur and will increase virus load over time, but the immune system is able to control the virus indefinitely (Figure 10.6).

This parameter regime corresponds to infections that SIV viruses cause in their natural hosts. For example, a large proportion of African green monkeys (AGMs) are infected with SIVagm, but do not succumb to immunodeficiency disease. The functional immune response of AGMs to SIVagm seems to be similar to the response of humans to HIV. There is also a productive infection of CD4 cells, and SIVagm has a viral load equivalent to that in asymptomatic HIV-1–infected humans. There is a similar degree of genetic variation. All these observations are consistent with our model. The parameter regions (ii) and (iii) can give rise to similar viral loads and similar antigenic diversities.

The critical difference is that in parameter region (ii) the virus population is effectively controlled by the cross-reactive immune responses alone, and so there is no diversity threshold. The difference between SIV in AGMs and HIV in humans could be caused by a slightly smaller replication rate of SIVagm in AGMs or, more likely, by a more effective cross-reactive immune response.

Figure 10.6 A strong cross-reactive immune response (directed at the conserved epitopes of the virus) can lead to a chronic infection without development of disease. This situation occurs if the cross-reactive response alone is sufficiently strong to control the virus population, that is, if $kq > ru$. The computer simulation is based on equation (10.6), with the parameter values $r = 2.3$, $p = 2$, $q = 2.4$, $c = k = u = 1$, and $b = 0.01$. The infection starts with a single strain. The probability that a new mutant arises in the time interval $[t, t + dt]$ is given by Pdt with $P = 0.1$. Reprinted from Nowak and May (2000) by permission of Oxford University Press.

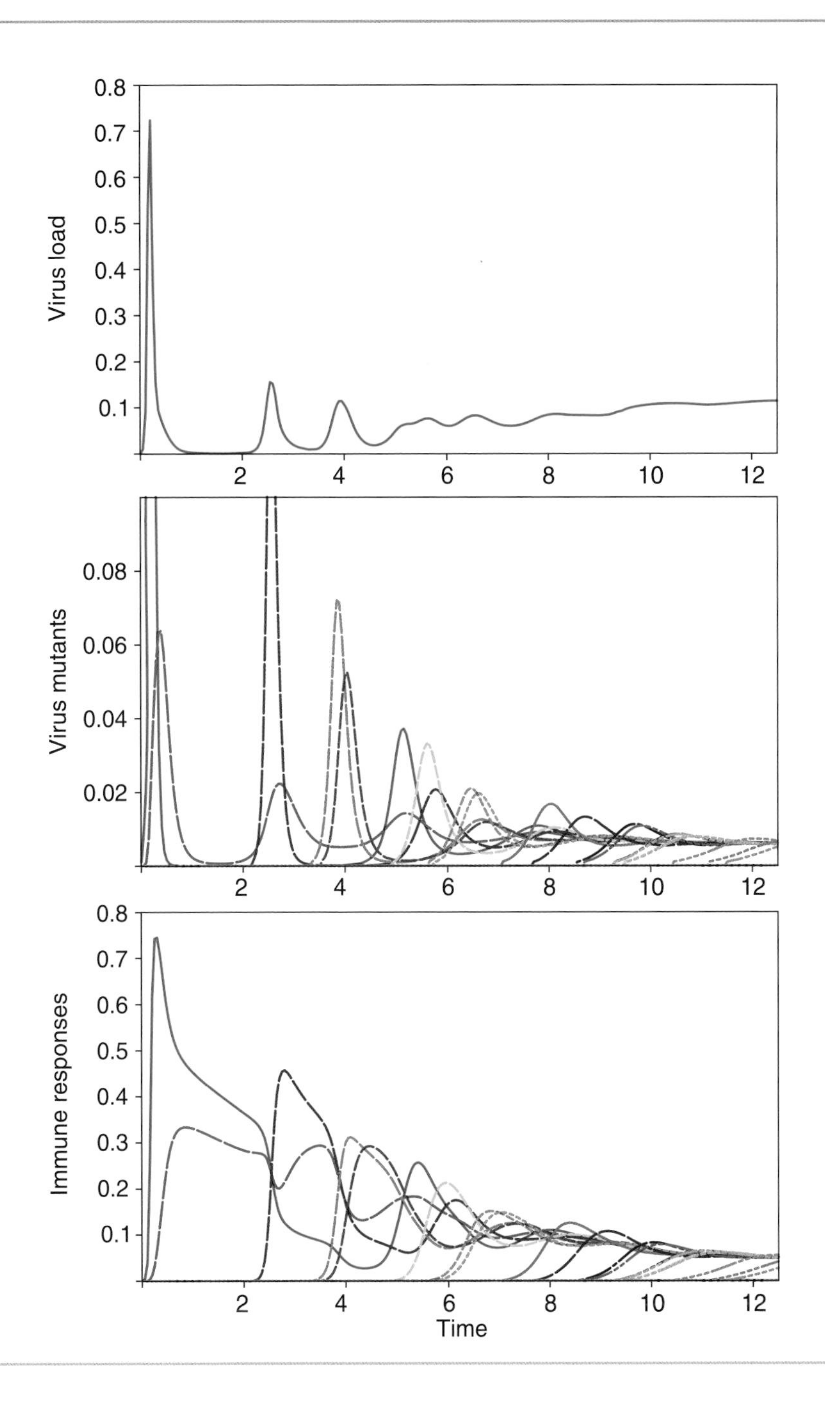

0.8
0.7
0.6
0.5
0.4
0.3
0.2
0.1
Virus load
2
4
6
8
10
12
0.08
0.06
0.04
0.02
Virus mutants
Immune responses
Time

This is plausible: the long-established interaction between SIVagm and its natural host should have selected for efficient cross-reactive immune response, which is directed at parts of the virus that cannot mutate (or only mutate with substantial fitness reduction).

iii. Chronic Infection and Disease after a Long Incubation Period

If $kq + cp > ru > kq$, then the combined effect of cross-reactive and strain-specific immune responses can control any one strain, but cross-reactive immunity alone is not sufficient to control the virus (Figure 10.7). Over time, antigenic variation reduces the effect of strain-specific immunity. In the beginning antigenic diversity is low, and the total virus population size is regulated to some equilibrium value. Antigenic diversity increases over time. Eventually antigenic diversity is so high that equation (10.9) no longer has a steady state for virus load. Now virus load increases without control. Remember that D is an inverse measure for antigenic diversity. In the beginning D is large. D decreases during infection as antigenic diversity increases. The immune system loses control when D drops below a critical value given by

$$D < \frac{ru - kq}{cp}. \tag{10.11}$$

This inequality defines the "antigenic diversity threshold." Once the threshold is exceeded, the virus population escapes from control by the immune response and tends to arbitrarily high values. This process may be interpreted

Figure 10.7 A diversity threshold occurs if the cross-reactive immune response by itself is unable to control the virus population, but a combination between cross-reactive and strain-specific responses can control any one strain. In mathematical terms, this means $kq + cp > ru > kq$. Increasing antigenic diversity enables the virus population to escape from the immune response after a long incubation period. The computer simulation is based on equation (10.6) with the parameter values $r = 2.5$, $p = 2$, $q = 2.4$, $c = k = u = 1$, and $b = 0.01$. The infection starts with a single strain. The probability that a new mutant arises in the time interval $[t, t + dt]$ is given by Pdt with $P = 0.1$. Reprinted from Nowak and May (2000) by permission of Oxford University Press.

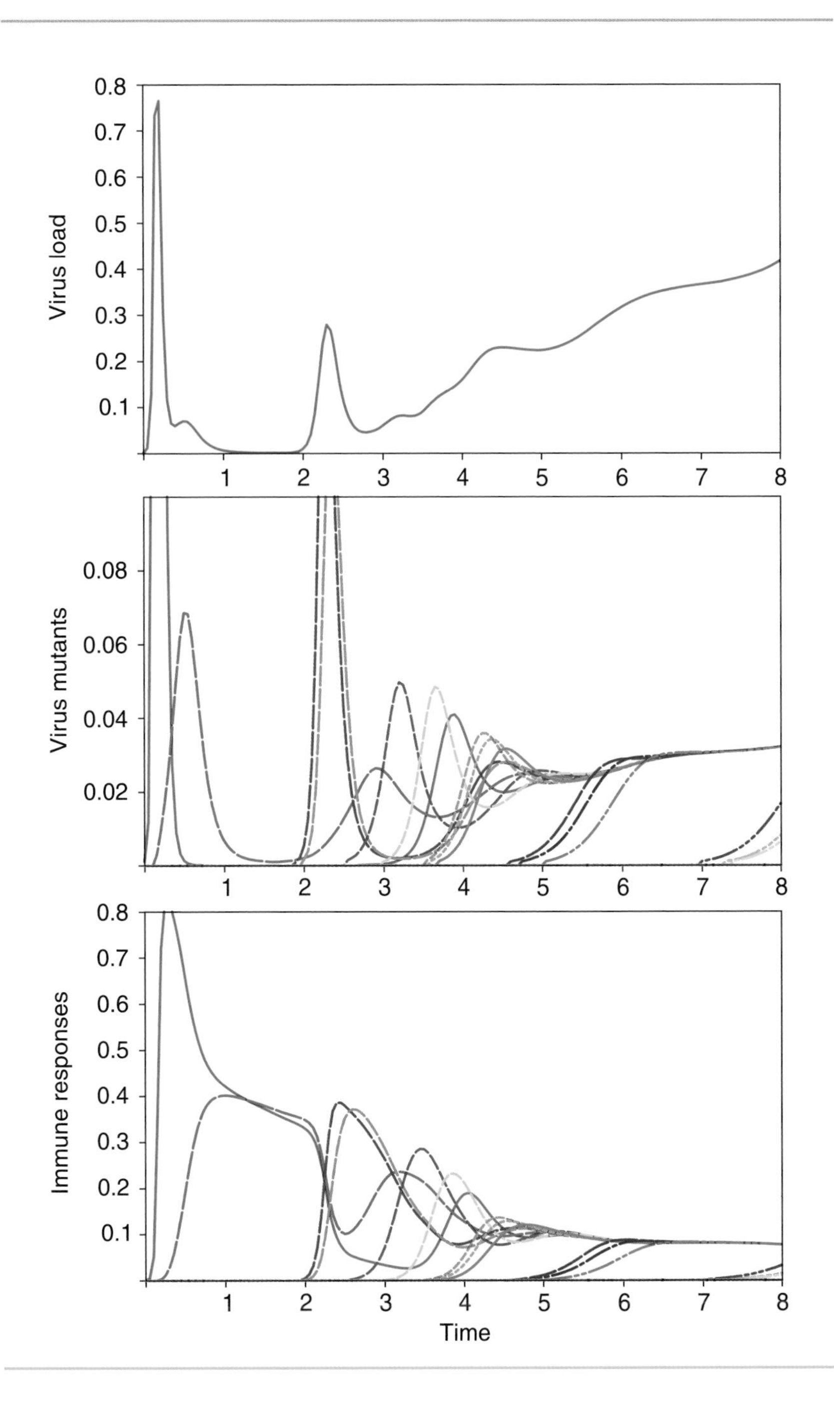

Virus load
0.8
0.7
0.6
0.5
0.4
0.3
0.2
0.1
1
2
3
4
5
6
7
8
Virus mutants
0.08
0.06
0.04
0.02
Immune responses
Time

as the development of immunodeficiency disease, which is characterized by high virus load and almost total depletion of CD4-positive cells.

Parameter region (iii) corresponds to the typical HIV-1 or HIV-2 infection in humans and also to experimental SIV infection, where an animal is infected with a virus from another species.

The "antigenic diversity threshold" is an intuitive concept. In a natural setting, the original infection occurs with a heterogeneous virus population. But the immune system in the newly infected patient has not yet been activated. There is exponential expansion of the invading virus, selecting for the fastest-growing strain without consideration for immunological escape. This initial phase will lead to a virus population with very low genetic and antigenic diversity. Subsequently the immune system becomes activated and selects for antigenic variation in those epitopes that are recognized by relevant immune responses. The increasing antigenic diversity makes it more and more difficult for the immune system to down-regulate all the various mutants simultaneously. The reason for this loss of control is the asymmetric interaction between immunological and viral diversity. Each virus strain can impair all immune responses by cutting off their CD4 help, but individual strain-specific immune responses can only attack specific virus strains. In more heterogeneous virus populations, the ratio between immune response–induced killing of virus and virus-induced killing of immune cells is shifted in favor of the virus. This shift eventually leads to a complete breakdown of the immune system and uncontrolled virus replication.

I call this phenomenon the "diversity threshold," because in the simplest mathematical model there is a critical number of antigenically distinct variants that can be controlled simultaneously by the immune system. In more realistic and more complicated versions of the model, this "diversity threshold" condition takes a more general form, and indicates the point at which the immune system fails to control the virus population. These complications arise, for example, when one acknowledges that different virus strains have different replication rates or immunological properties, or that the basic parameters of the model are not constant but change during the course of infection (such as increasing virus replication rates, resulting from increasing CD4 cell activation). Many different versions of the model have been studied, including responses to multiple epitopes, deletion of epitopes, cost of escape, and target cell limitation.

It is important to note that the model does not predict that patients with higher genetic or antigenic diversity must necessarily progress faster than patients with lower diversity. First of all, observe that the fastest progression occurs in parameter region (i) without any antigenic variation. Second, patients will differ in the strength of their immune response to HIV. Patients with a weak strain-specific response will tend to allow higher virus load without selecting for large antigenic diversity. In contrast, patients with strong strain-specific responses will reduce virus load to low levels, but also select for high antigenic diversity. Hence there need not be a simple relation between viral diversity and rate of disease progression in a comparison between different patients.

Finally, we note that the model does explain the difference between pathogenic and apathogenic SIV and HIV infections. Apathogenic infections correspond to parameter region (ii), where the cross-reactive immune responses suffice to control the virus. Pathogenic infections correspond to parameter region (iii), where cross-reactive and strain-specific responses are needed to control the virus; in this case, virus evolution will allow escape from strain-specific responses and lead to disease progression over time. Therefore understanding the reason why natural SIV infections seem to be apathogenic, while HIV causes a fatal disease in humans, requires a quantitative measurement of the virological and immunological parameters of the infection that make the difference between parameter regions (ii) and (iii).

We have seen that HIV evolution in individual infections provides a plausible mechanism for disease progression. The virus is initially controlled by immune responses, but continuously evolves to escape from these responses. Virus evolution can lead to increasing antigenic diversity, more efficient avoidance of immune responses, faster replication rates, and broader cell tropism (meaning that the virus can infect a larger number of different cell types). After some time, virus evolution reaches a threshold above which the immune system can no longer control the virus (Figure 10.8).

The evolutionary mechanism of disease progression has three parameter regions that correspond to three different outcomes of lentivirus infections: (i) there is immediate disease and death if the combination of cross-reactive

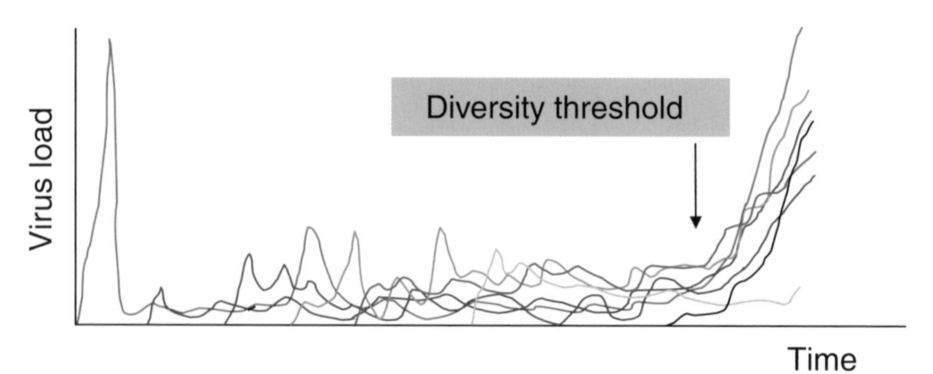

Figure 10.8 HIV disease progression is a consequence of viral evolution that occurs in individual patients. The virus continuously evolves to escape from antiviral immune responses (and drug treatment). In addition, the virus can generate mutants that replicate faster in a particular patient and can infect a wider variety of cells (increased cell tropism). Eventually the virus evolution reaches a point, called the "diversity threshold," at which the immune system loses control.

and strain-specific immune responses fails to control the invading virus strain; (ii) there is asymptomatic infection without disease if the cross-reactive immune responses alone can control the virus; (iii) there is progression to disease after a long and variable asymptomatic period if the combined effect of cross-reactive and strain-specific immunity can control any one strain, but cross-reactive immunity alone is not enough; in this case, virus evolution will eventually lead to AIDS.

Although the theory was received with skepticism originally and misunderstood at times, in the meantime every confirmed fact of HIV biology is in agreement with an evolutionary model of disease progression. It is therefore likely that this model provides the correct mechanism of HIV disease progression.

SUMMARY

- HIV infects CD4 cells, which represent a crucial component of the human immune system. CD4 cells help other cells, such as CD8 cells and B cells, to mount immune responses.
- CD4 cell numbers decline during HIV infection.
- The generation time of HIV in an infected patient is about one to two days. Yet it takes on average ten years for HIV to destroy the entire CD4 cell population. The crucial question arises: what is the mechanism for the slow disease progression in HIV infection?
- HIV evolves during individual infections.
- Antigenic variation allows HIV to escape from immune responses that are meant to control it.
- Antigenic variation leads to increasing viral diversity, increasing viral load (= abundance) and declining immunological control.
- Eventually this evolutionary process reaches a point (a "diversity threshold") above which the immune system can no longer control the virus.
- The diversity threshold is a consequence of the asymmetric interaction between HIV and the immune system: different virus mutants can kill CD4 cells irrespective of the mutants' specificity, but specific immune responses are only active against specific virus mutants.
- The model has three different parameter regions, which correspond to the observed patterns of HIV and SIV infection: (i) rapid progression to disease and death, (ii) asymptomatic infection without disease, and (iii) development of disease after a long and variable asymptomatic period.
- According to the proposed model, HIV disease progression is caused by the evolutionary dynamics in individual infections.

EVOLUTION OF VIRULENCE 11

EVOLUTIONARY CONSIDERATIONS of host–parasite interactions provide a fascinating topic for experimental and theoretical biologists. I use the term "parasite" to denote anything that lives and multiplies inside another organism and usually causes some harm. Phages are parasites of bacteria. Many viruses and bacteria are parasites of humans. There are many single and multicellular eukaryotic parasites that cause infectious diseases in humans and other animals. Our genome contains "parasitic" DNA that simply wants to increase its own abundance without much concern for other genes.

Parasites are as old as life itself. As soon as there were self-replicating machines, there were parasites to exploit them. Much of the design of individual cells and higher organisms can be explained as an adaptation to defend against parasites and limit the damage that is associated with infection. Bacteria have enzymes to cut viral genomes into pieces. Plants produce a vast library of chemicals in self-defense. The vertebrate immune system is a highly complicated, costly organ with the task of protecting against infectious agents. Even sexual reproduction has been explained as an adaptation to maintain genetic diversity and to help evolve away from parasites. In return, sexually transmitted parasites use this mode of reproduction of their hosts to their own advantage.

The conventional wisdom of many medical textbooks has been that well-adapted parasites are harmless to their hosts. This notion is based on the argument that killing its host does not help a parasite that relies on its host for reproduction. Some well-known observations seem to support this view. A much-cited example is the evolution toward reduced virulence of the myxoma virus in Australian rabbit populations. A more recent example, which we encountered in the previous chapter, is the observation that long-standing primate lentivirus associations seem to be apathogenic. Simian immunodeficiency viruses (SIV) apparently do not cause disease in their natural hosts. These viruses and their hosts have been coevolving for millions of years. In contrast, the human immunodeficiency virus (HIV) entered the human population only a few decades ago and causes a fatal disease.

There are also many counterexamples, however, where long-standing host–parasite systems have not evolved to become harmless. A major example is human malaria, which is estimated to have caused more human death than any other infectious disease. Another well-known example is provided by nematodes in fig wasps. These nematodes have a strong detrimental effect on their host, despite the observation that fig wasps preserved in twenty-million-year-old amber have already been infected by nematodes.

Mathematical epidemiology is one of the oldest disciplines of theoretical biology. In 1760 Daniel Bernoulli, hoping to influence public health policy, developed a mathematical model to evaluate the effectiveness of variolation against smallpox. In 1840 William Farr performed a statistical analysis of deaths from smallpox in England and Wales. In 1908 Ronald Ross, who had discovered that malaria was transmitted by mosquitoes, formulated a simple mathematical model to explore the relationship between the prevalence of mosquitoes and the incidence of malaria. William Ogilvy Kermack and Anderson Gray McKendrick, in 1927, established the important "threshold theory": introducing a few infected individuals into a population will cause an epidemic only if the density of susceptibles is above a certain threshold. In 1979 Roy Anderson and Robert May formulated many new approaches for theoretical epidemiology and laid the foundation for much subsequent work.

They developed simple mathematical models in order to explain laboratory experiments or epidemiological data. They also studied ecological questions by analyzing how infectious agents regulate the population size of their hosts. They emphasized the importance of the "basic reproductive ratio" and its consequences for vaccination programs.

May and Anderson also point out that parasite evolution does not necessarily lead to avirulence, but instead selection works to increase the parasite's basic reproductive ratio, R_0. If the rate of transmission is linked to virulence, then selection can favor increasing virulence. They reanalyzed the classical myxoma virus infection of Australian rabbits and argued that evolution had led to intermediate levels of virulence. The data actually suggest an equilibrium distribution of viruses with different levels of virulence; after many years, both the most virulent and the least virulent virus strains are still present in the virus population. Most infections are caused by virus strains with intermediate levels of virulence.

In this chapter we will study the evolutionary dynamics of parasites, but will assume that the host does not evolve on the time scale that is under consideration. This is a good assumption because, in general, parasites evolve much faster than their hosts. We begin by investigating the basic model of epidemiology, where parasite evolution maximizes the basic reproductive ratio. This result is based on the assumption that an already infected host cannot be superinfected by another parasite strain. We will subsequently remove this constraint and explore the evolutionary dynamics of superinfection. Superinfection means that an already infected host can be infected and taken over by another parasite strain.

In the classification of Anderson and May, this whole chapter deals with "microparasites," which typically include viruses, bacteria, and protozoans. They have small sizes, short generation times (compared with those of their hosts), and high rates of direct reproduction within their hosts. In contrast "macroparasites," which comprise parasitic helminths and arthropods, have longer generation times than microparasites and reproduce only very slowly within a host individual. Mathematical models for microparasites are typically formulated in terms of infected and uninfected (and immune/recovered) hosts. Models for macroparasites, in contrast, must keep track of the number of parasites in individual hosts.

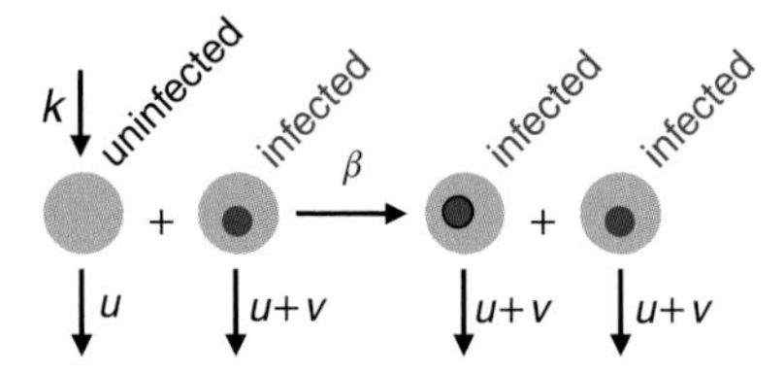

Figure 11.1 The basic model of infection dynamics describes the spread of an infectious agent (a parasite) in a population of hosts. An infected host meets an uninfected host and passes on the infection. It is often useful to think of biological dynamics as chemical kinetics: here an infected host "reacts" with an uninfected host to produce two new infected hosts. The rate constant of this reaction, β, denotes the infectivity of the parasite. The normal mortality of hosts is described by the death rate u. The disease-induced mortality (virulence) is given by v. Uninfected hosts enter the population at a constant rate, k.

11.1 THE BASIC MODEL OF INFECTION BIOLOGY

The basic epidemiological dynamics of a host–parasite interaction (Figure 11.1) can be described by the following system of ordinary differential equations

$$\begin{aligned}\dot{x} &= k - ux - \beta xy \\ \dot{y} &= y(\beta x - u - v)\end{aligned} \qquad (11.1)$$

Uninfected and infected hosts are denoted by x and y, respectively. In the absence of the parasite, the host population is regulated by a simple immigration-death process, with k specifying the constant immigration rate of uninfected hosts and u their natural death rate. This represents a simple, if somewhat artificial, way of attaining a stable host population in the absence of infection. Infected hosts transmit the parasite to uninfected hosts at the rate βxy, where β is the rate constant characterizing the parasite's infectivity. Infected hosts die at the increased rate $u + v$. The parameter v defines the virulence of the infection; it is the excess mortality associated with infection. More generally, virulence can be defined as the parasite's effect on reducing the fitness of infected hosts.

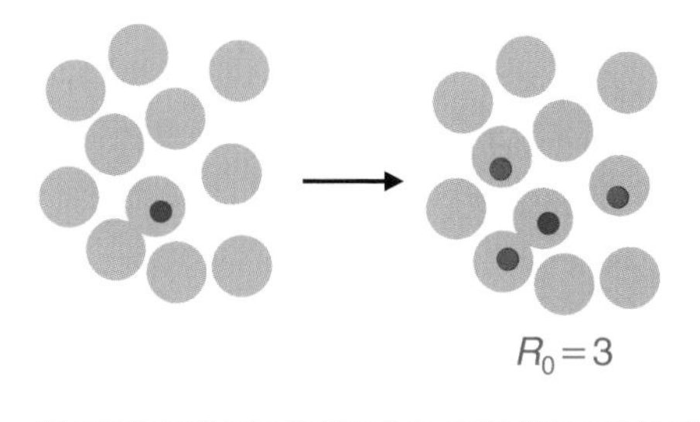

Figure 11.2 The basic reproductive ratio, R_0, of an infectious agent is given by the number of secondary infections that are caused by one infection that is introduced into an uninfected population of hosts. R_0 is a crucial quantity that determines whether or not a parasite can spread in a host population. If $R_0 < 1$, then the parasite will die out after a few rounds of infection. If $R_0 > 1$, then an explosive increase in the number of infections (an epidemic) will occur.

The basic reproductive rate of the parasite is defined as the number of new infections caused by a single infected host if introduced in a population of uninfected hosts (Figure 11.2). For system (11.1), the basic reproductive ratio is given by

$$R_0 = \frac{\beta}{u+v}\frac{k}{u}. \tag{11.2}$$

This can be understood as follows. The average lifetime of an infected host is $1/(u+v)$. The rate at which one infected host produces new infections is βx. The product of these two quantities is the average number of new infections caused by a single infected host in its lifetime if there are x uninfected hosts. The equilibrium abundance of uninfected hosts prior to the arrival of the infection is given by $x = k/u$. Hence equation (11.2) represents the basic reproductive ratio, R_0, which is a crucial concept of epidemiology.

If R_0 is less than one, then the parasite cannot spread. The "chain reaction" is sub-critical: a single case might cause a few additional cases, but then the transmission chain will die out again. An epidemic cannot take place.

If R_0 is greater than one, then the chain reaction is super-critical. There will be an exponential increase in the number of infected hosts. An epidemic will occur. After some time, the number of infected individuals will peak and then start to decline. Damped oscillations lead to a stable equilibrium given by

$$x^* = \frac{u+v}{\beta} \qquad y^* = \frac{\beta k - u(u+v)}{\beta(u+v)} \tag{11.3}$$

A successful vaccination program must reduce the population size of susceptible hosts such that the basic reproductive ratio is below one. If $R_0 = 5$, then more than 80% of the population must be vaccinated to prevent an epidemic. If $R_0 = 50$, then more than 98% of the population must be vaccinated. In general, successful vaccines are those that are directed against infectious agents with low reproductive ratios.

I call system (11.1) the "basic model of infection biology," because it describes not only the dynamics of an infectious agent in a population of hosts but also the dynamics of a virus within a single infected host. In the latter case, x and y denote, respectively, uninfected and infected cells. The application of this model to HIV infection is described in my book *Virus Dynamics*, coauthored with Robert May.

11.2 SELECTION MAXIMIZES THE BASIC REPRODUCTIVE RATIO

To understand parasite evolution, we have to study the epidemiological dynamics of at least two parasite strains competing for the same host. Extending equation (11.1), we obtain

$$\begin{aligned} \dot{x} &= k - ux - x(\beta_1 y_1 + \beta_2 y_2) \\ \dot{y}_1 &= y_1(\beta_1 x - u - v_1) \\ \dot{y}_2 &= y_2(\beta_2 x - u - v_2) \end{aligned} \tag{11.4}$$

The two parasite strains differ in their infectivity, β_1 and β_2, and in their degree of virulence, v_1 and v_2. The basic reproductive ratios of strains 1 and 2 are, respectively, given by

$$R_1 = \frac{\beta_1}{u + v_1} \frac{k}{u} \tag{11.5}$$

and

$$R_2 = \frac{\beta_2}{u + v_2} \frac{k}{u}. \tag{11.6}$$

Coexistence between the two parasite strains is only possible if $R_1 = R_2$, which is ungeneric. At equilibrium, the time derivatives, $\dot{x}$, $\dot{y}_1$, and $\dot{y}_2$, must be zero.

Furthermore, stable coexistence between strain 1 and strain 2 requires that both y_1 and y_2 are positive at equilibrium. From $\dot{y}_1 = 0$ and $y_1 > 0$, we obtain $x = (u + v_1)/\beta_1$. But from $\dot{y}_2 = 0$ and $y_2 > 0$, we obtain $x = (u + v_2)/\beta_2$. Both conditions can only hold simultaneously if $R_1 = R_2$. Generically, however, we expect that $R_1 \neq R_2$, in which case coexistence is not possible.

If both basic reproductive ratios are less than one, $R_1 < 1$ and $R_2 < 1$, then the only stable equilibrium is the uninfected population,

$$E_0: \qquad x = \frac{k}{u} \qquad y_1 = 0 \qquad y_2 = 0 \tag{11.7}$$

If $R_1 > 1 > R_2$, then strain 2 becomes extinct and the only stable equilibrium is

$$E_1: \qquad x^* = \frac{u + v_1}{\beta_1} \qquad y_1^* = \frac{\beta_1 - u(u + v_1)}{\beta_1(u + v_1)} \qquad y_2^* = 0 \tag{11.8}$$

If $R_1 < 1 < R_2$, then strain 1 becomes extinct and the only stable equilibrium is

$$E_2: \qquad x^* = \frac{u + v_2}{\beta_2} \qquad y_1^* = 0 \qquad y_2^* = \frac{\beta_2 - u(u + v_2)}{\beta_2(u + v_2)} \tag{11.9}$$

If both basic reproductive ratios exceed one, $R_1 > 1$ and $R_2 > 1$, then the strain with the higher basic reproductive ratio will outcompete the strain with the lower basic reproductive ratio. If $R_2 > R_1$, then all infected individuals will eventually carry strain 2, while strain 1 becomes extinct. The system will converge to equilibrium E_2.

Note that $R_2 > R_1$ is precisely the condition that strain 2 can invade equilibrium E_1. This means that the derivative $\partial \dot{y}_2 / \partial y_2$, evaluated at equilibrium E_1, is positive. $R_2 > R_1$ is also the condition that strain 1 cannot invade equilibrium E_2. This means that the derivative $\partial \dot{y}_1 / \partial y_1$, evaluated at equilibrium E_2, is negative. These derivatives characterize the growth rate of an infinitesimal amount of the invading strain at a particular equilibrium point. We conclude that E_1 is unstable, while E_2 is stable. Coexistence between the two strains is not possible. Therefore strain 2 outcompetes strain 1.

Therefore evolution will tend to maximize the basic reproductive ratio (Figure 11.3). If there is no constraint between infectivity and virulence, then the

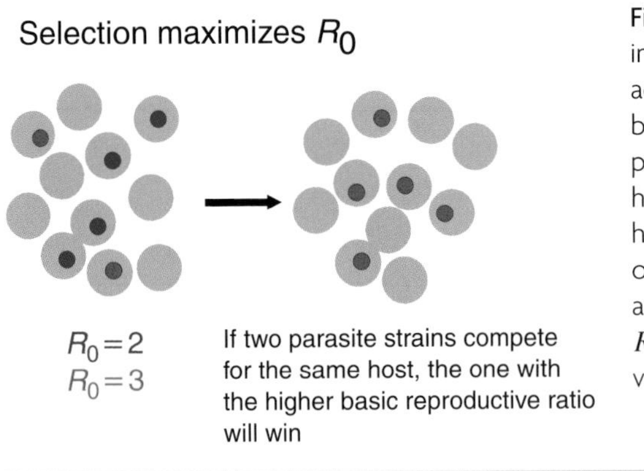

Selection maximizes R_0
$R_0=2$
$R_0=3$
If two parasite strains compete for the same host, the one with the higher basic reproductive ratio will win

Figure 11.3 In simple models of infection dynamics, selection acts to maximize the parasite's basic reproductive ratio. If two parasites compete for the same host, then the parasite with higher R_0 will outcompete the other parasite. Therefore well-adapted parasites have a high R_0, but not necessarily low virulence.

evolutionary dynamics will increase β and reduce v. This represents the conventional wisdom that infectious diseases will evolve to become less virulent.

In general, however, we expect an association between virulence v and infectivity β; usually the harm done to hosts (v) is associated with the production of transmission stages (β). For certain functional relations between v and β there is an evolutionarily stable degree of virulence, corresponding to the maximum value of R_0. Other situations allow evolution toward the extreme values of very high or low virulences. The detailed dynamics depend on the shape of β as a function of v. It is interesting to note that along some trajectories where virulence increases, parasite evolution can lead to lower and lower parasite population sizes (in terms of total number of infected hosts).

If the infectivity is proportional to virulence, $\beta = av$, where a is some constant, then the basic reproductive ratio, R_0, is an increasing function of virulence, v. In this case selection will always favor more virulent (and therefore more infectious) strains.

If the infectivity is a saturating function of virulence, $\beta = av/(c + v)$, then the basic reproductive ratio, R_0, is a one-humped function of virulence. The maximum R_0 is achieved at an intermediate optimum level of virulence given by $v_{\text{opt}} = \sqrt{cu}$. If the virulence of a parasite population is greater than v_{opt}, then selection will reduce virulence. If it is less than v_{opt}, then selection will increase virulence.

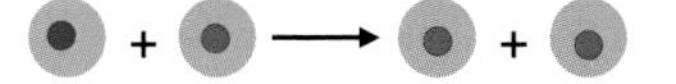
Superinfection means that one strain can take over a host already infected by another strain
+
+
If there is superinfection, then selection does not maximize the basic reproductive ratio

Figure 11.4 Superinfection means that an already-infected host can be infected by another parasite strain. There is competition between the two parasite strains in the superinfected individual; one parasite strain may win this competition and outcompete the other. A consequence of superinfection is that natural selection no longer maximizes the basic reproductive ratio. Instead there can be coexistence of different parasite strains with different levels of virulence. In general, superinfection leads to increased virulence beyond what would be optimum for the parasite. Superinfection introduces competition on two levels: within an infected host and in the population of hosts.

11.3 SUPERINFECTION

The analysis of the previous section did not include the possibility of superinfection. An infected host is not susceptible to another infection. We will now remove this limitation and allow for an infected host to be superinfected by another parasite strain (Figure 11.4).

We will consider a heterogeneous parasite population with a range of different virulences, and assume that more virulent strains outcompete less virulent strains within an infected individual. Thus increased virulence provides a competitive advantage over other parasites in the same host.

For simplicity, we assume that the infection of a single host is always dominated by one parasite strain. Therefore superinfection means that a more virulent strain takes over a host infected by a less virulent strain. This can be described by the following system of ordinary differential equations:

$$\dot{x} = k - ux - x\sum_{i=1}^{n}\beta_i y_i$$
$$\dot{y}_i = y_i\Big(\beta_i x - u - v_i + s\beta_i \sum_{j=1}^{i-1} y_j - s\sum_{j=i+1}^{n}\beta_j y_j\Big) \qquad i = 1,\ldots,n \tag{11.10}$$

Here v_i denotes the virulence of strain i. We order the strains such that $v_1 < v_2 < \ldots < v_n$. A more virulent strain can superinfect a host already infected with a less virulent strain. The parameter s describes the rate at which superinfection occurs relative to infection of uninfected hosts. If either the host or the parasite has evolved mechanisms to make superinfection more difficult, then s is smaller than one. If already-infected hosts are more susceptible to acquiring a second infection, then s is greater than one, which means superinfection occurs at increased rates.

For the numerical simulations shown in Figure 11.5, we assume a functional relation between virulence and infectivity given by

$$\beta_i = \frac{av_i}{c + v_i}. \tag{11.11}$$

For low virulence, infectivity increases linearly with virulence. For high virulence, there is a saturation of infectivity at a maximum level. The basic reproductive ratio is given by

$$R_{0,i} = \frac{akv_i}{u(c + v_i)(u + v_i)}. \tag{11.12}$$

The optimal virulence, which maximizes R_0, is given by

$$v_{opt} = \sqrt{cu}. \tag{11.13}$$

Figure 11.5 shows the equilibrium population structure of the parasite for various values of s between 0 and 2. We have assumed $k = 1$, $u = 1$, and $\beta_i =$

Figure 11.5 The equilibrium distribution of parasite strains with different levels of virulence. The simulation is performed according to equation (11.10) with $k = 1$, $u = 1$, $n = 50$, $\beta_i = 8v_i/(1 + v_i)$, and $s = 0, 0.2, 1, 2$ as indicated. The individual v_i are randomly distributed between 0 and 5. In the absence of superinfection, $s = 0$, the strain with the maximum basic reproductive rate, R_0, is selected. With superinfection, $s > 0$, we find the coexistence of many different strains with different virulences, v_i, within a range v_{min} and v_{max}, but the strain with the largest R_0 is not selected. Superinfection does not optimize parasite reproduction. For increasing s, the values of v_{min} and v_{max} increase, as well. The x-axis denotes virulence, the y-axes indicate equilibrium frequencies (always scaled to the same largest value).

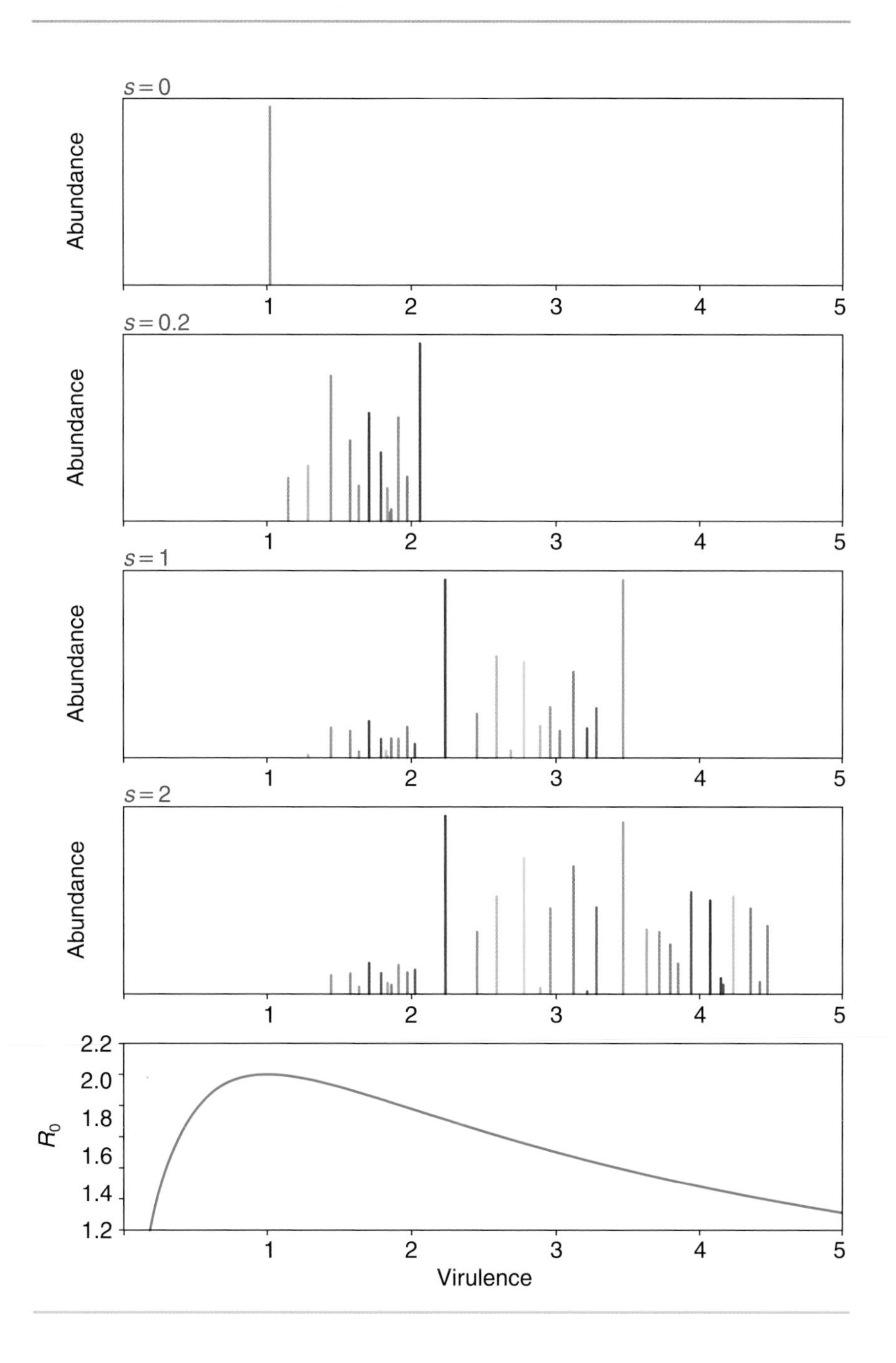
s = 0
Abundance
1
2
3
4
5
s = 0.2
Abundance
1
2
3
4
5
s = 1
Abundance
1
2
3
4
5
s = 2
Abundance
1
2
3
4
5
2.2
2.0
1.8
1.6
1.4
1.2
R_0
1
2
3
4
5
Virulence

$8v_i/(1+v_i)$. We simulated $n = 100$ strains of parasites with virulences randomly distributed between 0 and 5. For this choice of parameters, the strain with a virulence closest to 1 has the largest R_0. Indeed we find that this strain is selected in the absence of superinfection, $s = 0$. If superinfection is possible ($s > 0$), then there is selection of an ensemble of strains with a range of virulences between two boundaries, v_{min} and v_{max}, with $v_{min} > v_{opt}$. Thus superinfection has two important effects: (i) it shifts parasite virulence to higher levels, beyond the level that would maximize the parasite's reproductive rate; and (ii) it leads to a coexistence between a number of different parasite strains with a range of virulences. There are amusing ups and downs in the equilibrium densities of strains. A strain has a high equilibrium frequency if the strain with a slightly larger virulence has low frequency, and vice versa. Only a subset of strains survive at equilibrium. What determines this complicated and unexpected equilibrium structure?

11.4 AN ANALYTICAL MODEL OF SUPERINFECTION

Let us now derive an analytical understanding of the complexities introduced by superinfection. Instead of using a constant immigration rate k for uninfected hosts, we choose a variable immigration rate that balances exactly the death of uninfected and infected hosts. This can be done by setting

$$k = ux + uy + \sum v_i y_i \tag{11.14}$$

in equation (11.10). The total number of infected hosts is given by $y = \sum_{i=1}^{n} y_i$. The sum $x + y$ remains constant and without loss of generality we choose $x + y = 1$. We obtain the following system of n equations

$$\dot{y}_i = y_i \left[\beta_i(1-y) - u - v_i + s\left(\beta_i \sum_{j=1}^{i-1} y_j - \sum_{j=i+1}^{n} \beta_j y_j \right) \right] \tag{11.15}$$

$$i = 1, \ldots, n$$

Note that y remains in the closed interval $[0, 1]$.

System (11.15) is a Lotka-Volterra equation. It can be written in the form

$$\dot{y}_i = y_i(R_i + \sum_{j=1}^{n} A_{ij}y_j) \qquad i = 1, \ldots, n \tag{11.16}$$

Here $R_i = \beta_i - v_i - u$. The matrix is given by

$$A = -\begin{pmatrix} \beta_1 & \beta_1 + s\beta_2 & \beta_1 + s\beta_3 & \ldots & \beta_1 + s\beta_n \\ \beta_2(1-s) & \beta_2 & \beta_2 + s\beta_3 & \ldots & \beta_2 + s\beta_n \\ \beta_3(1-s) & \beta_3(1-s) & \beta_3 & \ldots & \beta_3 + s\beta_n \\ \vdots & \vdots & \vdots & \ddots & \vdots \\ \beta_n(1-s) & \beta_n(1-s) & \beta_n(1-s) & \ldots & \beta_n \end{pmatrix} \tag{11.17}$$

For an analytic understanding, we take the limit $c \to 0$ in our expression for $\beta_i = av_i/(c + v_i)$. Now all parasite strains have the same infectivity, β, and differ only in their degree of virulence, v_i. We obtain

$$\dot{y}_i = y_i\beta[1 - y - \frac{v_i + u}{\beta} + s(\sum_{j=1}^{i-1} y_j - \sum_{j=i+1}^{n} y_j)] \qquad i = 1, \ldots, n \tag{11.18}$$

This is a Lotka-Volterra equation with $R_i = \beta - v_i - u$ and

$$A = -\beta\begin{pmatrix} 1 & 1+s & 1+s & \ldots & 1+s \\ 1-s & 1 & 1+s & \ldots & 1+s \\ 1-s & 1-s & 1 & \ldots & 1+s \\ \vdots & \vdots & \vdots & \ddots & \vdots \\ 1-s & 1-s & 1-s & \ldots & 1 \end{pmatrix} \tag{11.19}$$

This system belongs to a class of Lotka-Volterra equations for which Josef Hofbauer and Karl Sigmund have shown the existence of a unique globally stable equilibrium. This equilibrium attracts all orbits from the interior of the positive orthant. If this equilibrium lies on a face of the positive orthant, then it also attracts all orbits from the interior of this face.

Equation (11.18) can be rewritten as

$$\dot{y}_i = y_i\beta[f_i - sy_i]. \tag{11.20}$$

Here

$$f_i = 1 - \frac{v_i + u}{\beta} - (1 - s)y + 2s \sum_{j=i+1}^{n} y_j. \tag{11.21}$$

All equilibrium points of equation (11.20) are given by the following relations:

$$\begin{aligned} y_1 &= 0 \quad \text{or} \quad y_1 = f_1/s \\ y_2 &= 0 \quad \text{or} \quad y_2 = f_2/s \\ &\vdots \\ y_n &= 0 \quad \text{or} \quad y_n = f_n/s \end{aligned} \tag{11.22}$$

Note that each f_i only depends on the total sum y and all y_j with $j > i$. Suppose we know y; then we can construct a specific equilibrium point in a recursive "top-down" way:

$$\begin{aligned} y_n &= \max\{0, f_n/s\} \\ y_{n-1} &= \max\{0, f_{n-1}/s\} \\ y_{n-2} &= \max\{0, f_{n-2}/s)\} \\ &\vdots \\ y_1 &= \max\{0, f_1/s\} \end{aligned} \tag{11.23}$$

The notation max{., .} simply denotes the larger of the two numbers. This equilibrium point has to be stable, because either $f_i < 0$ and hence $y_i \to 0$, or $f_i > 0$ and $y_i \to f_i/s$.

11.4.1 The Case s = 1

The case $s = 1$ offers a quick solution, because y drops out of equation (11.23). Hence the unique stable equilibrium distribution is given recursively in the following way:

$$
\begin{aligned}
y_n &= \max\{0, 1 - \frac{v_n + u}{\beta}\} \\
y_{n-1} &= \max\{0, 1 - \frac{v_{n-1} + u}{\beta} - 2y_n\} \\
y_{n-2} &= \max\{0, 1 - \frac{v_{n-2} + u}{\beta} - 2(y_n + y_{n-1})\} \\
&\vdots \\
y_1 &= \max\{0, 1 - \frac{v_1 + u}{\beta} - 2(y_n + y_{n-1} + \cdots + y_2)\}
\end{aligned}
\tag{11.24}
$$

This is the only stable equilibrium. For each parasite strain i with equilibrium frequency $y_i = 0$, we have $\partial \dot{y}_i / \partial y_i < 0$ for a generic choice of parameters. Moreover, equation (11.24) corresponds to a simple and elegant geometric method for constructing the equilibrium configuration of the population (Figure 11.6).

11.4.2 The General Case $s > 0$

Let us consider an equilibrium distribution with $y_i > 0$ for $i = 1, \ldots, n$, which means we count only those strains that are present at equilibrium. From equation (11.15) we can write $\sum_{j=1}^{i-1} y_j = y - y_i - \sum_{j=i+1}^{n} y_j$, to get

$$
y_i = B_i - 2 \sum_{j=i+1}^{n} y_j \tag{11.25}
$$

with $B_i = [1 - \frac{v_i+u}{\beta} - (1-s)y]/s$. We obtain

$$
\begin{aligned}
y_n &= B_n \\
y_{n-1} &= -2B_n + B_{n-1} \\
y_{n-2} &= 2B_n - 2B_{n-1} + B_{n-2}
\end{aligned}
\tag{11.26}
$$

For even n we obtain $y = B_1 - B_2 + B_3 - \cdots - B_n = (v_n - v_{n-1} + \cdots - v_1)/\beta s$. For odd n we obtain $y = B_1 - B_2 + B_3 - \cdots + B_n$ and hence $y = (\beta - u - v_n + v_{n-1} - \cdots - v_1)/\beta$. At first sight the expressions for odd and

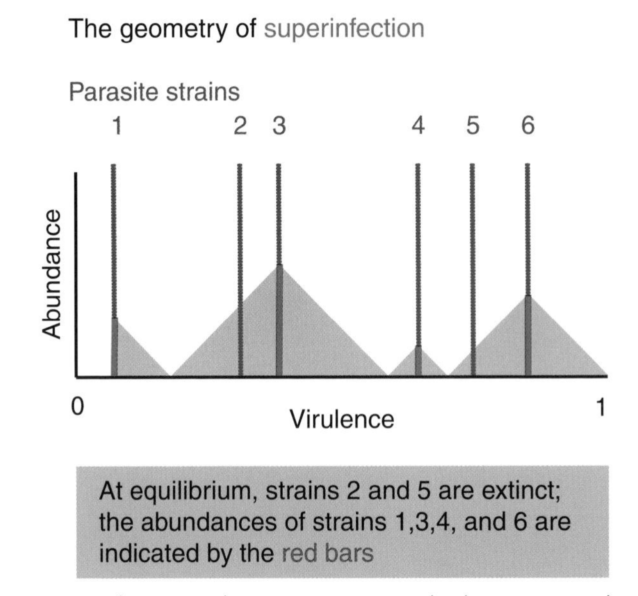
The geometry of superinfection
Parasite strains
1
2 3
4 5 6
Abundance
0
Virulence
1
At equilibrium, strains 2 and 5 are extinct; the abundances of strains 1,3,4, and 6 are indicated by the red bars

Figure 11.6 For $s = 1$, there is an elegant geometric method to construct the equilibrium distribution of the parasite population. Suppose there are n strains with virulences v_1 to v_n, all between 0 and 1. Start by drawing vertical lines at v_1 to v_n (shown in blue). Draw a 45-degree line running up to the left from $v = 1$; the intersection with the vertical line at v_n determines the abundance y_n. This corresponds to $y_n = 1 - v_n$. Now mirror the construction triangle (shaded in blue) at the axis given by the vertical line at v_n. The intersection with the downward-pointing 45-degree line with the baseline determines the point $v = 1 - 2y_n$. Now there are two possibilities: (i) either $v_{n-1} < v$, in which case draw a new 45-degree line up to the left from v; the intersection with the vertical line at v_{n-1} gives y_{n-1}; this corresponds to $y_{n-1} = v_n - v_{n-1} - (1 - v_n)$; or (ii) $v_{n-1} > v$, in which case the strain $n - 1$ will not be present at equilibrium and the construction method proceeds directly with strain v_{n-2}, and so on. The figure is self-explanatory. We choose $n = 6$ strains. Four of those strains are present at equilibrium; their abundances are indicated by red bars. Two strains are extinct.

even n look quite different. We want to calculate $v_{\max}$, the maximum level of virulence present in an equilibrium distribution for a given s. Assuming equal spacing (on average), that is, $v_k = kv_1$, leads to $y = v_n/2\beta s$ for n even and to $y = 1 - u/\beta - v_n/2\beta$ for n odd. (For n odd we have used the approximation $n - 1 \approx n$.) From $y_n \geq 0$ we derive in both cases

$$v_{max} = \frac{2s(\beta - u)}{1+s}. \quad (11.27)$$

This is the maximum level of virulence that can be maintained in an equilibrium distribution. For $s = 0$, this is simply $v_{max} = 0$, that is, the strain with the lowest virulence, which for our choice of parameters is also the strain with the highest basic reproductive ratio. For $s > 1$, strains can be maintained with virulences above $\beta - u$. These are strains that are by themselves unable to invade an uninfected host population, because their basic reproductive ratio is smaller than one.

Finally resolving the even- and oddities, we insert v_{max} for v_n into the two different expressions for y and find in both cases

$$y = \frac{\beta - u}{\beta(1+s)}. \quad (11.28)$$

This is the equilibrium frequency of infected hosts. The more superinfection, the fewer infected hosts.

11.5 DYNAMICAL COMPLEXITIES

Let us now return to the model with different strains having different infectivities, β_i, as given by equation (11.15). Here the solutions need not converge to a stable equilibrium. Equation (11.15) can lead to very complex dynamics.

For two strains of parasite ($n = 2$) we may find coexistence (that is, a stable equilibrium between the two strains) or a bistable situation, where either one or the other strain wins, depending on the initial conditions. An interesting situation can occur if $s > 1$ and strain 1 has a virulence too high to sustain itself in a population of uninfected hosts ($R_0 < 1$), whereas strain 2 has a lower virulence but an $R_0 > 1$. Since $s > 1$, infected hosts are more susceptible to superinfection, and thus the presence of strain 2 can effectively shift the reproductive rate of strain 1 above one. Superinfection can stabilize parasite strains with extremely high levels of virulence.

For three or more strains of parasite, we may observe oscillations with increasing amplitude and period, tending toward a heteroclinic cycle. Imagine three parasite strains, each of which by itself is capable of establishing an equilibrium between uninfected and infected hosts (that is, all have $R_0 > 1$). The system in which these three strains occur simultaneously has three boundary equilibria, where two strains always have frequency 0 and the population consists of uninfected hosts and hosts infected by the third strain only. There is also one unstable interior equilibrium with all three strains present. The system converges toward the boundary equilibria and cycles from the first one to the second to the third and back to the first. The period of such cycles gets larger and larger. There will be long times where the infection is just dominated by one parasite strain (and hence only one level of virulence), and then suddenly another strain takes over. Such a dynamic can, for example, explain sudden upheavals of pathogens with dramatically altered levels of virulence. If we wait long enough, one of the parasite strains may become extinct by some fluctuation when its frequency is low. Then one of the two remaining strains will outcompete the other.

For small values of s all elements of matrix (11.17) will be negative. Such a Lotka-Volterra system is called "competitive," and all trajectories will converge to an $n-1$-dimensional subspace, which reduces the dynamical complexities. This implies that for $n = 2$ there are no damped oscillations, and for $n = 3$ one can exclude chaos.

SUMMARY

- The basic reproductive ratio of an infectious agent (parasite) is the number of secondary infections caused by one infected individual that has been introduced into a population of uninfected individuals.
- Parasite evolution tends to maximize the basic reproductive ratio.
- If there is a functional relationship between infectivity and virulence, then well-adapted parasites need not be harmless. Parasite evolution can lead to intermediate levels of virulence.
- Superinfection means that an already infected host can be infected by another parasite strain.

- Superinfection triggers intrahost competition for increased levels of virulence and reduced transmission rates.
- Superinfection increases the average level of virulence above what would be optimum for the parasite population.
- Superinfection does not maximize the basic reproductive ratio. Even the strain with the highest R_0 can become extinct.
- Superinfection leads to a coexistence of parasite strains with many different levels of virulence within a well-defined range.
- Superinfection can maintain strains with very high levels of virulence, including strains that are so virulent that they themselves could not persist alone in an otherwise uninfected host population.
- Superinfection can lead to very complicated dynamics, such as heteroclinic cycles, with sudden and dramatic changes in the average level of virulence.
- The higher the rate of superinfection, the smaller the number of infected hosts. Hence superinfection is not advantageous for the parasite population as a whole.

12 EVOLUTIONARY DYNAMICS OF CANCER

CANCER IS THE CONSEQUENCE of an evolutionary process. Usually we think of evolution as leading to improvement and innovation. But cancer is an evolution that leads backward to selfish cellular proliferation, often destroying the organism in which it occurs. Cancer is the unfortunate byproduct of our design. We are built of individual cells that have their own reproductive machinery and that can sometimes revert to a primitive program of uncontrolled self-replication. Cancer is the "evolution of defection" among our cells.

Computers have viruses, but they do not get cancer. Computers can copy information, and therefore viruses can manipulate this process and reproduce. But computers are not built of smaller self-reproducing units. Therefore silicon-based cancer does not yet exist.

Cancer is a disease of multicellular organisms. For the development of multicellularity, the main obstacle that needed to be overcome was how to establish and maintain cooperation among many individual cells. Cancer is a breakdown of cellular cooperation. Cells must divide whenever needed for the developmental program, but not otherwise. A complicated genetic control network had to evolve to achieve this task. Many of our genes work to ensure that cancer does not happen too early. These genes are involved (i) in maintaining the integrity of the genome; (ii) in performing error-free cell division; (iii) in determining the developmental program that tells cells when to divide; and (iv) in monitoring the status of the cell and if necessary inducing programmed cell death (apoptosis).

Most cells in the body are continuously listening to reassuring signals from other cells telling them that they are doing all right. If these signals fail to arrive, then the default program for a cell is to commit suicide. Apoptosis is a defense mechanism against cancer. If things go wrong, then cells kill themselves. Cancer cells have escaped from apoptotic control.

The evolutionary process that leads to cancer is different from most other evolutionary processes, because there are so many genes that can be inactivated or modified without any fitness loss for the cell and in many cases even with a fitness gain. Therefore cancer progression can be seen as a "destructive evolution," getting rid of mechanisms that are implemented to protect against cancer.

A relatively large fraction of all possible mutations that can occur in a precancerous cell will increase its somatic fitness (reproductive rate). In contrast, a vanishingly small fraction of all possible mutations that can occur in a well-adapted organism (such as a mouse or a rabbit) will increase its fitness. Hence there should be strong selection pressure on cancer cells to increase their mutation rates. The optimum mutation rate for a cell on the way to cancer is much higher than the normal somatic mutation rate. Throughout this chapter, the term "mutation" is used to include any genetic modification such as point mutations, insertions, deletions, chromosome rearrangements, mitotic recombination, or loss or gain of whole chromosomes or arms of chromosomes.

The idea that cancer is a genetic disease caused by somatic evolution has emerged over the last one hundred years. In 1890 the German physician David von Hansemann noted that cancer cells had abnormal cell division events. In 1914 Theodor Boveri observed that something was wrong with the chromosomes of cancer cells. Today we know that most cancer cells are aneuploid, that is, they do not have the normal number of chromosomes (Figure 12.1). Ernest Tyzzer in 1916 was the first to apply the term "somatic mutation" to cancer. Herman Muller discovered in 1927 that ionizing radiation, which was known to be carcinogenic, was also mutagenic. This observation presented further evidence for the association between somatic mutation and cancer. In 1951 Muller proposed that for cancer to occur, a single cell had to receive multiple mutations. A few years later, the mathematical modeling of cancer began with a statistical analysis of age-incidence patterns. Work by C. O. Nordling in 1953 as well as Peter Armitage and Richard Doll in 1954 led to the important insight that the emergence of cancer requires multiple probabilistic events.

In 1971 Alfred Knudson discovered that the age incidence of retinoblastoma, a childhood cancer, grows as a linear function of time in the group of children who have multiple cancers in both eyes, but as a slower quadratic function of time in the group of children who have only one cancer. Knudson proposed the concept of a tumor suppressor gene (TSG). Cancer emerges if both alleles are inactivated. In the first group of children, one allele is already inactivated in their germ line, while the second allele is inactivated by a somatic mutation. In the second group of children, both alleles are inactivated by somatic mutations. This is known as Knudson's two-hit hypothesis: it takes two hits to inactivate a TSG (Figure 12.2). On the basis of this discovery, Suresh Moolgavkar and Alfred Knudson developed probabilistic models to describe cancer initiation and progression.

In 1986 the retinoblastoma tumor suppressor gene was identified. In the meantime, about thirty tumor suppressor genes have been found to be associated with human cancers. They have the property that the somatic mutations are recessive: inactivating the first allele is neutral (or nearly neutral), while inactivating the second allele changes the phenotype of the cell and usually increases its net reproductive rate, which constitutes a step toward cancer.

An important TSG is p53, which is mutated in more than half of all human cancers. This gene is located at the center of a control network that monitors genetic damage (including double-strand breaks of the DNA). If there is a certain amount of damage, then cell division will be paused and the cell will be given time to repair itself. If there is too much damage, then the cell will undergo apoptosis. In many cancer cells, the function of p53 is inactivated, which allows these cell to divide in the presence of substantial genetic damage.

Oncogenes represent another class of genes that are involved in cancer. Oncogenes increase cellular proliferation if one allele is mutated or inappropriately expressed (Figure 12.3). The concept of oncogenes was introduced by Michael Bishop and Harold Varmus in 1976; in 1989 they shared the Nobel

Prize in Medicine for this work. Some viruses carry oncogenes that induce the infected cell to proliferate like a cancer cell. In the last three decades, many oncogenes have been discovered that are involved in various stages of human cancers including tumor initiation, progression, angiogenesis (the process of attracting blood vessels to growing tumors), and metastasis formation.

While mutations of tumor suppressor genes and oncogenes tend to increase the net reproductive rate (somatic fitness) of a cell, mutations in genetic instability genes increase the mutation rate. For example, mutations in mismatch repair genes lead to a 50–1,000-fold increase of the point mutation rate, which manifests itself primarily in the accumulation of mutations in the microsatellite regions of the genome. Therefore this type of genetic instability is called microsatellite instability (MIN). About 15% of colon cancers have MIN, but 85% of colon cancers and most other cancers have chromosomal instability (CIN).

CIN is defined as an increased rate of gaining or losing whole chromosomes or large fractions of chromosomes during cell division (Figure 12.4). Often the first allele of a TSG is inactivated by a point mutation, while the second allele is inactivated by loss of heterozygosity (LOH). LOH can be caused by somatic recombination or by loss of (the part of) the chromosome that contains the unmutated allele. In both cases, we obtain a cell that has lost both alleles of the TSG. Christoph Lengauer and Bert Vogelstein have determined that the rate of

Figure 12.1 Normal cells are diploid, with two copies of each autosomal chromosome. The autosomal human chromosomes are labeled 1 to 22 according to decreasing size. In addition, there are two sex chromosomes: in females XX, in males XY. Therefore in a normal human cell there are 46 chromosomes, as shown in the top spectral karyotype. (A karyotype depicts the complete set of chromosomes within a cell.) In most cancer cells, especially in solid tumors, there is aneuploidy, which means the total number of chromosomes is not 46. Some chromosomes exist in more than two copies, others only in one. Moreover, some chromosomes in cancers can be fusions of two or more chromosomes. The middle karyotype is from the breast cancer line HCC1937, which has a BRCA1 mutation. The bottom karyotype shows the chromosomal abnormality in a Capan1 cell (pancreatic cancer) with a BRCA2 mutation. Images courtesy of Joanne M. Staines and Paul Edwards, Cancer Genomics Project, University of Cambridge. See also Davidson et al. (2000).

1
2
3
4
5
6
7
8
9
10
11
12
13
14
15
16
17
18
19
20
21
22
X
Y
1
2
3
4
5
6
7
8
9
10
11
12
13
14
15
16
17
18
19
20
21
22
X
Y
1
2
3
4
5
6
7
8
9
10
11
12
13
14
15
16
17
18
19
20
21
22
X
Y

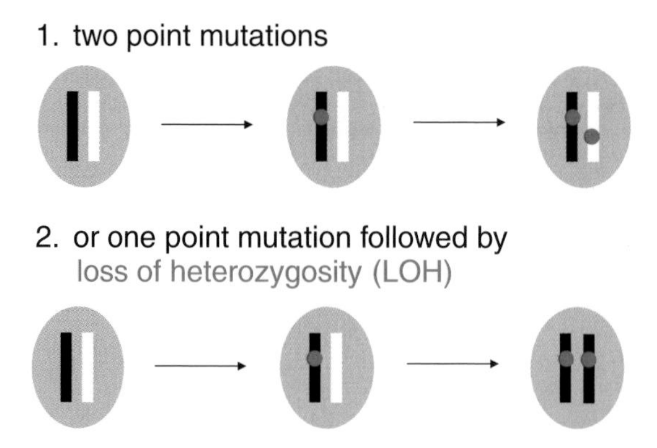
Tumor suppressor genes are inactivated by ...
1. two point mutations
2. or one point mutation followed by
loss of heterozygosity (LOH)

Figure 12.2 A tumor suppressor gene (TSG) is inactivated by two mutational events. The first hit is often caused by a point mutation. The second hit can be another point mutation or a loss of heterozygosity (LOH). There are various mechanisms of LOH, including somatic recombination or loss of a whole chromosome or chromosome arm. If one copy of the chromosome is lost, sometimes the other copy is duplicated. TSGs play central roles in regulatory networks that determine the rate of cell cycling. Inactivation of the function of a TSG modifies the regulatory network and can lead to increased cell proliferation. The basic idea of a TSG is that inactivation of one allele has no (or only a minor) effect, while inactivation of both alleles represents a step toward cancer.

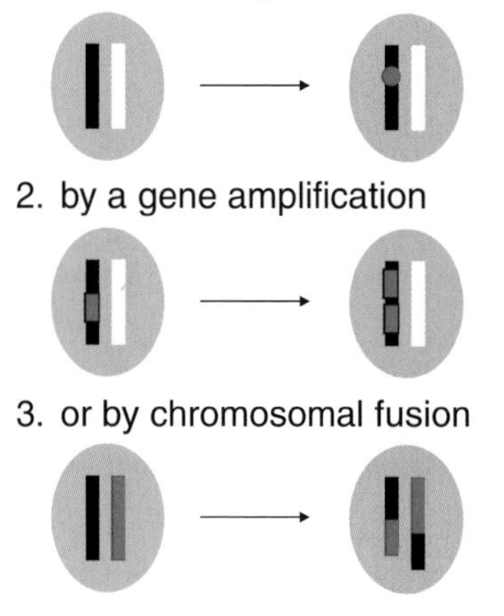
Oncogenes are activated by ...
1. one specific point mutation
2. by a gene amplification
3. or by chromosomal fusion

Figure 12.3 Oncogenes must be activated to induce a step toward cancer. It is usually sufficient to mutate one of the two alleles. The activation can occur by a specific point mutation, a gene amplification, or a chromosomal rearrangement. The latter can lead to a fusion gene, where the first half comes from one gene, the second half from another. Activated oncogenes increase cell proliferation.

Chromosomal instability (CIN):
mutations in CIN genes increase the rate of gaining or losing whole chromosomes

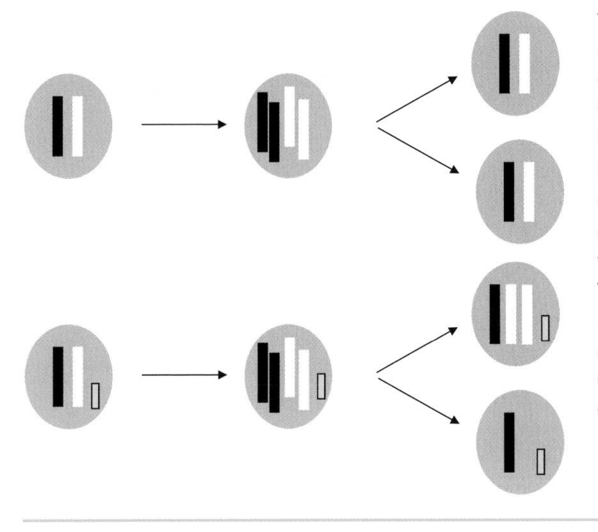

Figure 12.4 Many hundreds of genes work together during cell division to ensure that all chromosomes are properly duplicated and distributed to each of the two daughter cells. Mutations in such genes can lead to chromosomal instability (CIN). The CIN phenotype of a cell is defined by an increased rate of gaining or losing whole chromosomes or arms of chromosomes. The figure shows a normal cell division (top) and one that leads to aneuploidy (bottom). The black and white bars indicate maternal and paternal copies of a chromosome. The yellow bar illustrates a CIN mutation (somewhere in the genome).

losing a chromosome in CIN cancer cells is about 10^{-2} per chromosome per cell division. In contrast, the rate of LOH in non-CIN cells is thought to be on the order of 10^{-7} to 10^{-6}. Therefore CIN leads to a dramatic acceleration in the inactivation of TSGs.

The molecular basis for CIN is just beginning to be understood. A large number of genes that trigger CIN when mutated have been discovered in yeast. These so-called CIN genes are involved in chromosome condensation, sister-chromatid cohesion, kinetochore structure and function, microtubule formation, and cell cycle checkpoints. By comparison with yeast, we expect several hundred human CIN genes, but only a few have been identified so far. These genes include MAD2, hBUB1, BRCA2, and hCDC4.

The classification of CIN genes is based on the mutational events that are required to trigger CIN. Class I CIN genes, such as MAD2, trigger CIN if one allele of the gene is mutated or lost. Class II CIN genes, such as hBUB1, trigger CIN if one allele is mutated in a dominant-negative fashion; this means the

3 classes of CIN genes

Onco-CIN genes

Class I CIN genes trigger CIN if one allele is mutated or lost.
Example: MAD2

Class II CIN genes trigger CIN if one allele is mutated in a dominant negative fashion.
Example: hBUB1

Class III CIN genes trigger CIN if both alleles are mutated.
Example: BRCA2

CIN suppressor genes

Figure 12.5 CIN is caused by mutations in genes that participate in maintaining genomic integrity during cell division. We can classify CIN genes according to the number and types of mutational events that are needed to activate the CIN phenotype. Class I genes lead to CIN if one allele is mutated or lost. Class II genes lead to CIN if one allele is mutated. Class III genes lead to CIN if both alleles are mutated. There exist examples among human CIN genes for all three classes. Class I and II genes can be called onco-CIN genes, while class III genes are CIN suppressor genes.

mutated allele interferes with the normal function of the unmutated allele. Class III CIN genes, such as BRCA2, trigger CIN if both alleles are mutated. Class I and II CIN genes could be called "onco-CIN genes," while class III genes are "CIN suppressor genes" (Figure 12.5).

Because of the brilliant work of Bert Vogelstein and his friends, colon cancer has one of the best understood evolutionary trajectories. The epithelial layer of the colon has an enormous cellular turnover: a large number of cells are produced and discarded each day. This massive amount of cell division events implies a higher risk of mutations that might lead to cancer. The geometric organization of the colon, however, reduces this risk. The colon is organized into about 10^7 crypts. Each crypt contains thousands of cells. At the bottom of the crypt there is a small number of stem cells that slowly divide to produce differentiated cells. Differentiated cells divide a few times while migrating to the top of the crypt, where they undergo apoptosis. The architecture of the crypt implies that only a small subset of the cells are at risk of acquiring mutations that will become fixed in permanent cellular lineages. Many mutations that arise in the differentiated cells will be washed out. Therefore the number of cells that are at risk of receiving mutations toward cancer is greatly reduced by this architecture (Figure 12.6).

Colon cancer arises in a crypt

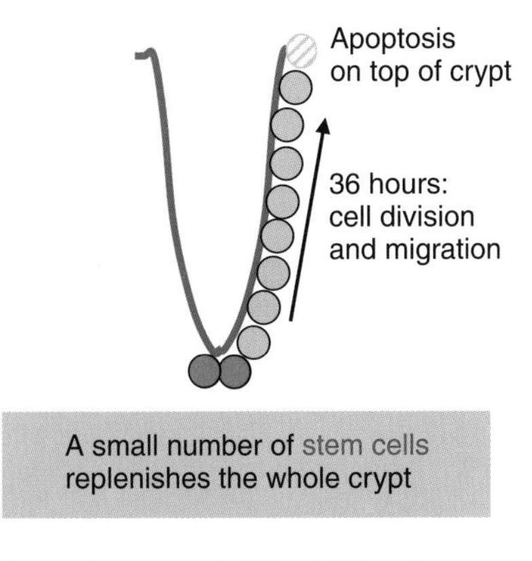

A crypt contains 1,000–4,000 cells.
The colon contains 10^7 crypts.

Figure 12.6 The epithelial layers of many tissues seem to require large numbers of rapidly dividing cells. This cell turnover represents a main risk for the emergence of cancer. The colon is organized into roughly 10^7 crypts. Each crypt contains about 1,000 to 4,000 cells. At the bottom of the crypt there are a small number of stem cells (maybe 1 to 4) which divide slowly and differentiate into cells that divide rapidly and migrate to the top of the crypt, where they undergo apoptosis (programmed cell death). This design helps to reduce the risk of cancer, because most cell division events (and therefore most somatic mutations) occur in cells that are short lived. When designing models for the dynamics of colon cancer initiation, we have to take into account that the epithelial layer is subdivided into crypts, and that each crypt has a small effective population size.

Colorectal cancer is thought to be initiated by a mutation that inactivates the adenomatous polyposis coli (APC) tumor suppressor gene pathway. In about 95% of cases, the APC tumor suppressor gene is mutated. In the remaining cases, there are other mutations that affect the same pathway. The crypt in which the APC mutant cell arises becomes dysplastic. The slow accumulation of abnormal cells produces a polyp. The emergence of a large polyp seems to require further mutations, for example, activation of the oncogenes RAS or BRAF. Subsequently 10–20% of these large polyps progress to cancer

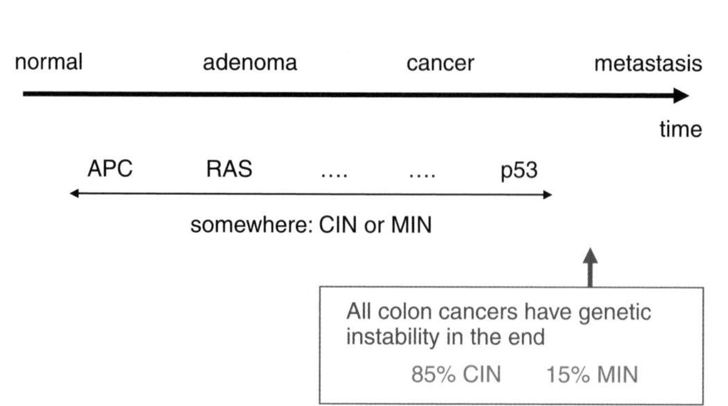

Figure 12.7 The "Vogelgram" illustrates the sequence of mutations that lead to colon cancer. Usually the first step is inactivation of the APC tumor suppressor gene pathway, followed by activation of the RAS or BRAF oncogene. There are one or two additional mutations that are not yet clearly characterized. Eventually there is inactivation of the p53 tumor suppressor gene. While normal colon tissue is genetically stable, all colon cancers have genetic instability in the end. About 85% of sporadic colon cancers have chromosomal instability (CIN). The remaining 15% have microsatellite instability (MIN). At some stage genetic instability must arise. A major question is whether genetic instability is an early event and a driving force of tumorigenesis or a late stage by-product.

by acquiring mutations in genes that are part of the TGF-β pathway, the p53 pathway, or other pathways (Figure 12.7).

In the end, all colon cancers are genetically unstable. About 15% of colon cancers are diploid and have MIN. The remaining 85% are aneuploid and have CIN. A major question is whether genetic instability arises early or late during the progression to colon cancer.

A quantitative understanding of cancer biology requires a mathematical framework to describe the fundamental principles of population genetics and evolution that govern tumor initiation and progression. Mutation, selection, and tissue organization determine the dynamics of tumorigenesis and should be studied quantitatively in terms of both experiment and theory.

Here I will address the following questions: What are the fundamental principles that determine the dynamics of activating oncogenes and inactivating tumor suppressor genes? How do mutation, selection, and tissue architecture influence the rate of tumor initiation and progression? And how do quantitative approaches help to investigate the role of genetic instability in tumorigenesis?

12.1 ONCOGENES

Oncogenes contribute to cancer progression if one allele is mutated or inappropriately expressed. Let us explore the basic aspects of the evolutionary dynamics of oncogene activation.

Most tissues of multicellular organisms are subdivided into compartments, which contain populations of cells that proliferate to fulfill organ-specific tasks. Compartments are subject to homeostatic mechanisms that ensure that the cell number remains approximately constant over time. Whenever a cell divides, another cell has to die to keep the total population size constant. Cancer results if the equilibrium between cell birth and death is shifted toward uncontrolled proliferation. Not all cells of a compartment, however, may be at risk of becoming cancerous. Differentiated cells, for example, may not have the capacity to divide often enough to accumulate the necessary number of mutations in cancer susceptibility genes (these are genes that may lead to cancer when mutated, such as tumor suppressor genes, oncogenes, or genetic instability genes). The effective population size of a compartment describes those cells that are at risk of becoming cancer cells. In the following, compartment size will be used synonymously with effective population size within a compartment.

Consider a compartment of replicating cells. During each cell division, there is a small probability that a mistake will be made during DNA replication; in this case, a mutated daughter cell is produced. The mutation might confer a fitness advantage to the cell by ameliorating an existing function or inducing a new function. Then the mutation is "advantageous" in terms of somatic selection. Alternatively, the mutation might impair an important cellular function and confer a fitness disadvantage to the cell. Then the cell proliferates more slowly or dies more quickly than its neighbors. The net reproductive rate is decreased, and the mutation is "deleterious" in terms of somatic selection. (Somatic selection describes the process of natural selection among cells of the soma of a multicellular organism. Somatic selection leads to cancer.) Finally, the mutation might not change the reproductive rate of the cell. Then the cell proliferates at the same rate as its neighbors, and the mutation is "neutral" in terms of somatic selection. All of these mutations can represent steps toward cancer and are therefore disadvantageous for the organism.

Let us discuss the dynamics of a particular mutation within a compartment. Initially, all cells are unmutated. What is the probability that a single mutated cell has arisen by time t? We measure time, t, in cell cycles. If the relevant cells divide once per day, then the unit of time is one day. Denote by N the number of cells in a compartment, and denote by u the mutation rate per gene per cell division. The probability that at least one mutated cell has arisen by time t is given by

$$P(t) = 1 - e^{-Nut}. \tag{12.1}$$

What is the fate of a single mutated cell? In the simplest scenario, there is a constant probability, q, that this cell will not die, but will initiate a neoplasia. Hence the probability that a compartment has initiated a neoplasia by time t is given by

$$P(t) = 1 - e^{-Nuqt}. \tag{12.2}$$

Alternatively, consider a scenario in which the mutated cell has a relative fitness r compared with a wild-type cell with fitness 1 ("wild type" means unmutated). If $r > 1$, the mutation is advantageous; if $r < 1$, the mutation is disadvantageous; if $r = 1$, the mutation is neutral. Normally, we expect mutations in oncogenes to cause increased net growth rates, $r > 1$; however, a mutation in an oncogene could be kept in check by apoptotic defense mechanisms, and therefore r could be less than one.

What is the probability that such a mutation takes over the compartment? In order to calculate this probability, we consider the Moran process (Chapter 6). The fixation probability of a single mutant with relative fitness r is given by

$$\rho = \frac{1 - 1/r}{1 - 1/r^N}. \tag{12.3}$$

For a neutral mutant, $r = 1$, we have $\rho = 1/N$. An advantageous mutation has a higher fixation probability than a neutral mutation, which has a higher fixation probability than a deleterious mutation. The events in a small compartment, however, are dominated by random drift: if N is small, then even a deleterious mutation can have a fairly high probability of reaching fixation due to chance events.

The probability that a mutation has been fixed by time t is given by

$$P(t) = 1 - e^{-Nu\rho t}. \tag{12.4}$$

Note that any mutation has a higher fixation probability, ρ, in a small compartment than in a large compartment, but $P(t)$ is an increasing function of N for $r > 1$ and a decreasing function of N for $r < 1$. Thus large compartments accelerate the accumulation of advantageous mutations, but slow down the accumulation of deleterious mutations. Conversely, small compartments slow down advantageous mutations, but accelerate deleterious ones. Therefore the compartment size is important in determining the types of mutations that are likely to occur.

We can argue that the most dangerous steps toward cancer are those that lead to cells with increased net reproductive rate, such as mutations in oncogenes or tumor suppressor genes. The best tissue architecture for containing those mutations uses a large number of small compartments. It seems that this is the dominant tissue architecture adopted by human organs that require fast cell division. The mutant cell that carries a fitness advantage is likely to reach fixation in the compartment, but its further spread is at least initially limited by the compartment boundaries. It turns out, however, that this architecture is especially vulnerable to cancer initiation via mutations that lead to genetic instability.

The difference between equations (12.2) and (12.4) is the following: in (12.2) there is a fixed probability q that a mutated cell initiates a neoplasia; in (12.4) the corresponding probability, ρ, depends on the selective advantage of the cell, r, and the effective population size, N, of the compartment. Equation (12.4) describes a situation where the mutated cell has a fitness advantage and must reach fixation in the compartment to initiate cancer progression. Equation (12.2) refers to a situation where a mutated cell might induce a clonal expansion, which is not subject to the constraints of the compartment.

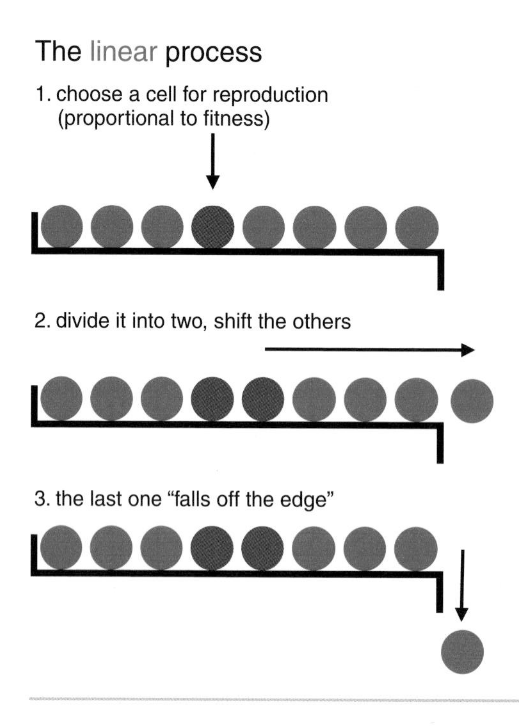

The linear process
1. choose a cell for reproduction (proportional to fitness)
2. divide it into two, shift the others
3. the last one "falls off the edge"

Figure 12.8 The linear process represents the simplest stochastic model that describes the subdivision of somatic tissue into stem cells and differentiated cells. At each time step, a cell is chosen for reproduction proportional to fitness. The cell is replaced by the two daughter cells. All cells to the right are shifted one place. The right-most cell falls off the edge (= undergoes apoptosis). What is the probability of fixation of a mutant cell with relative fitness r that arises in a random position?

12.2 THE LINEAR PROCESS

So far we have considered the evolutionary dynamics of a mutation that arises in a well-mixed compartment. This approach describes a tissue compartment in which all relevant cells are in equivalent positions and in direct reproductive competition with one another—there are no spatial effects. However, we can also envisage theories in which cellular differentiation and spatial structure are explicitly modeled. One simple approach considers N cells in a linear array. At each time step, a cell is chosen at random, but proportional to fitness. The cell is replaced by two daughter cells, and all cells to its right are shifted by one place to the right. The cell at the far right undergoes apoptosis. The cell at the far left acts as a stem cell (Figure 12.8).

Let us now assume that a mutated cell arises that has relative fitness r. The fixation probability of this mutant cell is given by $\rho = 1/N$, irrespective of r, because only a mutation in the left-most cell can reach fixation in the compartment. A mutation arising in any other cell will eventually be "washed out" of the compartment by the continuous production of cells and their migration from the stem cell to differentiation and apoptosis. The probability that all cells of the compartment are mutated at time t is given by

$$P(t) = 1 - e^{-ut}. \tag{12.5}$$

Here time is measured in units of stem cell divisions. If the stem cell divides more slowly than the other cells, then the accumulation of mutated cells is decelerated.

This "linear process" of cancer initiation has the important feature of balancing out fitness differences between mutations. Advantageous, deleterious, and neutral mutations all have the same fixation probability, $\rho = 1/N$. This is in contrast to a well-mixed compartment, in which the fittest mutation has the highest probability of fixation. In comparison with a well-mixed compartment, a linear compartment delays the development of cancers that are initiated by advantageous mutations, such as mutations in oncogenes and tumor suppressor genes (Figure 12.9).

Incidentally, the linear process prompted the idea of evolutionary graph theory, but its shifting cell population is not covered by the mathematical formalism developed in Chapter 8.

12.3 NUMERICAL EXAMPLES

Three simple numerical examples illustrate how tissue architecture can affect the rate of cancer progression.

(i) Suppose an organ consists of $M = 10^7$ compartments. Each compartment has $N = 10^3$ cells that divide once per day. Let us assume that the rate of activating the oncogene per cell division is $u = 10^{-9}$ and that this activation confers a 10% growth advantage to the cell, which means $r = 1.1$. Then the probability of fixation is $\rho = (1 - 1/r)/(1 - 1/r^N) \approx 0.09$. The probability that a compartment has been taken over by mutated cells after 70 years ($t = 70 \times 365.25$ days) is $P(t) = 1 - \exp(-Nu\rho t) \approx 0.0023$. The expected number of mutated compartments at this age is $M \cdot P(t) \approx 23{,}000$.

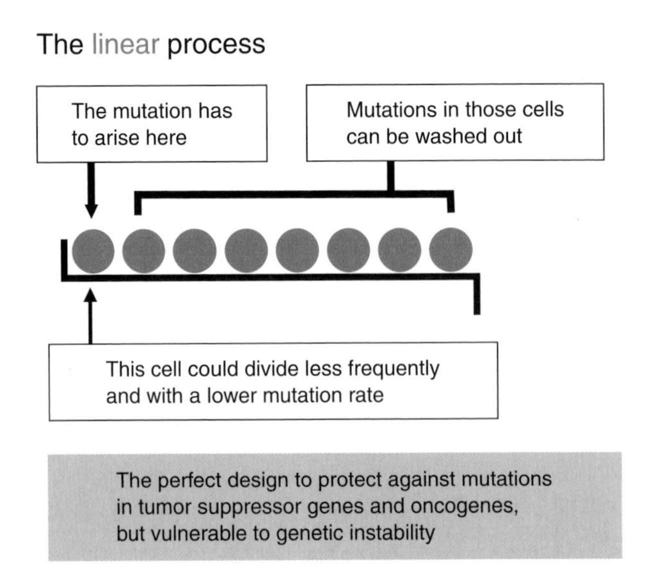

Figure 12.9 The linear process acts as a powerful suppressor of selection. A mutant can only take over the whole compartment if it arises in the left-most cell, which acts like a stem cell. Only the lineage arising from this cell is here to stay. All lineages descending from other cells are transient; mutations that occur in those cells will be washed out. The probability of fixation of a randomly placed mutant is $1/N$, irrespective of its relative fitness. Hence the selective advantage of an activated oncogene or inactivated tumor suppressor gene is negated by the population structure of the linear process. Furthermore, the stem cell can divide more slowly and with a smaller mutation rate than other cells. This effect can further reduce the rate of somatic evolution that might lead to cancer.

(ii) Let us now assume a linear tissue architecture for each compartment. As before, there are $M = 10^7$ compartments consisting of 10^3 cells each. But each compartment is now fed by one stem cell that divides every 10 days. The probability that a compartment has been taken over by mutated cells at time $t = 70$ years is reduced to $P(t) \approx 2.6 \cdot 10^{-6}$. The expected number of mutated compartments at this age is only 26. With these numerical values, the linear architecture reduces the rate of cancer progression about 1,000-fold.

(iii) Finally, consider a population of $N = 10^7$ cells that divide every day. This population size describes, for example, a lesion that has already accumulated mutations in one or a few cancer susceptibility genes. The probability is $P \approx 0.28$ that a mutated cell with relative fitness $r = 1.1$ arises and takes over the whole population within one year. The time until this probability is 1/2 is given by $T_{1/2} = 2.1$ years.

12.4 TUMOR SUPPRESSOR GENES

A normal cell has two alleles of a tumor suppressor gene. The standard idea is that inactivation of the first allele does not lead to a phenotypic change, but inactivation of both alleles increases the net reproductive rate of the cell and therefore represents a step toward cancer.

While an oncogene is activated by one or a few specific mutations, a TSG can typically be inactivated by any mutation that disrupts the function of the gene. Therefore the mutation rate for inactivating an allele of a TSG is much higher than the mutation rate that results in activation of an oncogene. But two events are required to eliminate a TSG, while one event suffices to activate an oncogene.

Consider a TSG, A. Let us introduce the following nomenclature: A "type 0" cell, $A^{+/+}$, is a normal cell with two functioning (wild-type) alleles of the TSG. A "type 1" cell, $A^{+/-}$, has only one functioning allele of the TSG. A "type 2" cell, $A^{-/-}$, has no functioning allele of the TSG.

Let us now ask the most basic question concerning the evolutionary dynamics of TSGs. What is the probability that a single cell with two inactivated TSG alleles has arisen by time t in a population of reproducing cells? The answer, which turns out to be surprisingly difficult, will be presented in subsections 12.4.1–12.4.4. We note that the whole system has only three parameters: the population size, N; the mutation rate for the first hit, u_1; and the mutation rate for the second hit, u_2 (Figure 12.10).

We will assume that u_1 is smaller than u_2, because certain mutational mechanisms, such as mitotic recombination, can only constitute the second hit. In CIN cells, u_1 is much smaller than u_2, because of the dramatically increased rate of losing whole chromosomes.

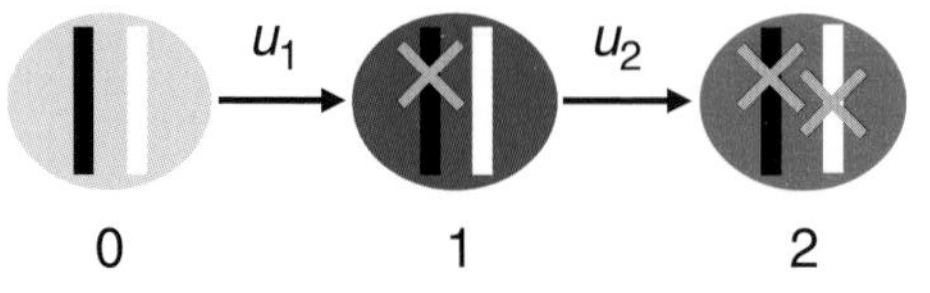
Evolutionary dynamics of tumor suppressor genes:
given a population of N reproducing cells,
what is the probability that at least one cell
has received two mutations by time t?
u1
u2
0
1
2

Figure 12.10 How long does it take for a population of reproducing cells to inactivate a tumor suppressor gene? If we assume that the first hit is neutral, then the answer will depend on three parameters: the population size, N; and the mutation rates for the first and second allele, u_1 and u_2.

12.4.1 The Exact Markov Process

The evolutionary dynamics can be described by a Markov process with $N + 2$ states. The states $i = 0, \ldots, N$ are transient and indicate the presence of i cells of type 1 and $N - i$ cells of type 0. The state $N + 1$ is the only absorbing state and indicates that a cell of type 2 has been produced. The transition probabilities for this Markov process are as follows:

$$\begin{aligned}
P_{i,i-1} &= \frac{i}{N}\frac{N-i}{N}(1-u_1) \qquad i = 1, \ldots, N \\
P_{i,i} &= \frac{i}{N}\Big[\frac{i}{N}(1-u_2) + \frac{N-i}{N}u_1\Big] + (\frac{N-i}{N})^2(1-u_1) \quad i = 0, \ldots, N \\
P_{i,i+1} &= \frac{N-i}{N}\Big[\frac{i}{N}(1-u_2) + \frac{N-i}{N}u_1\Big] \qquad i = 0, \ldots, N-1 \\
P_{i,N+1} &= \frac{i}{N}u_2 \qquad i = 0, \ldots, N \\
P_{N+1,i} &= 0 \qquad i = 0, \ldots, N \\
P_{N+1,N+1} &= 1
\end{aligned} \tag{12.6}$$

All other entries of the transition matrix are 0.

We are interested in the expected time until absorption into state $N+1$ starting from state $i=0$. Denote by t_i the expected absorption time from state i. We have

$$t_0 = 1 + P_{0,0}t_0 + P_{0,1}t_1$$

$$t_i = 1 + P_{i,i-1}t_{i-1} + P_{i,i}t_i + P_{i,i+1}t_{i+1} \qquad i = 1, \ldots, N \qquad (12.7)$$

$$t_{N+1} = 0$$

This linear system can be solved numerically to get the precise value for t_0. The analytic expressions are complicated. In the following, we will derive excellent approximations for small, intermediate, and large population size.

12.4.2 Small Population Size

In a small population, a type 1 cell reaches fixation before a type 2 cell arises. "Small population" means

$$N \ll 1/\sqrt{u_2}. \qquad (12.8)$$

This can be understood as follows. The average fixation time of the first mutation is of order $\tau_1 = N$. This is the expected time it takes to proceed from one mutant cell to N mutant cells in the Moran process, given that fixation does occur. The average waiting time for the second mutation is $\tau_2 = 1/(Nu_2)$. If $\tau_1 \ll \tau_2$, then it is likely that the first mutation reaches fixation before the second mutation arises. From $\tau_1 \ll \tau_2$, we obtain inequality (12.8). Note that each cell divides on average once every time unit. If the population size is N, there are N cell divisions per time unit. Consequently there are N^2 cell divisions in N time units.

In the parameter region given by (12.8), the evolutionary dynamics can be described as transition among three states. State 0 means that all cells are of type 0. State 1 means that all cells are of type 1. State 2 means that at least one type 2 cell has been generated. Denote by $X_0(t)$, $X_1(t)$, and $X_2(t)$, respectively, the probability of being in state 0, 1, or 2 at time t. At time $t=0$, all cells are unmutated. Therefore $X_0(0)=1$, while $X_1(0)=X_2(0)=0$. State 2 is the only absorbing state. For $t \to \infty$, the system converges to $X_0(t)=X_1(t)=0$ and $X_2(t)=1$.

The time derivatives of the three probabilities are given by

$$\dot{X}_0 = -u_1 X_0$$
$$\dot{X}_1 = u_1 X_0 - N u_2 X_1 \tag{12.9}$$
$$\dot{X}_2 = N u_2 X_1$$

In state 0, the rate of producing type 1 cells is Nu_1. The probability that such a cell reaches fixation is $1/N$. Therefore the transition rate from state 0 to state 1 is given by the mutation rate u_1. If the population is in state 1, then the rate of producing type 2 cells is given by Nu_2.

Equation (12.9) is a linear system of ordinary differential equations, which can be solved analytically. The probability that at least one cell of type 2 has been produced by time t is given by

$$P(t) = X_2(t) = 1 - \frac{N u_2 e^{-u_1 t} - u_1 e^{-N u_2 t}}{N u_2 - u_1}. \tag{12.10}$$

For short times, $t \ll 1/(Nu_2)$, we have

$$P(t) \approx N u_1 u_2 t^2/2. \tag{12.11}$$

Therefore the probability accumulates as a second order of time. The 2 in the exponent is the same as in Knudson's two-hit hypothesis: it takes two rate-limiting hits to inactivate a TSG in a small population of cells (Figure 12.11).

In contrast, for long times, $t > 1/(Nu_2)$, we have

$$P(t) \approx 1 - e^{-u_1 t}. \tag{12.12}$$

On this time scale, the second hit is fast and can be neglected. Only the first hit is rate limiting.

The exact definition for the number of rate-limiting hits is given by the slope of $\log P(t)$ versus $\log t$. If there is one rate-limiting hit, then $P(t)$ is a linear function of time. Two rate-limiting hits mean that $P(t)$ is a quadratic function of time. If $P(t)$ is constant (on a certain time scale), then there are zero rate-limiting hits. The number of rate-limiting hits depends on the time

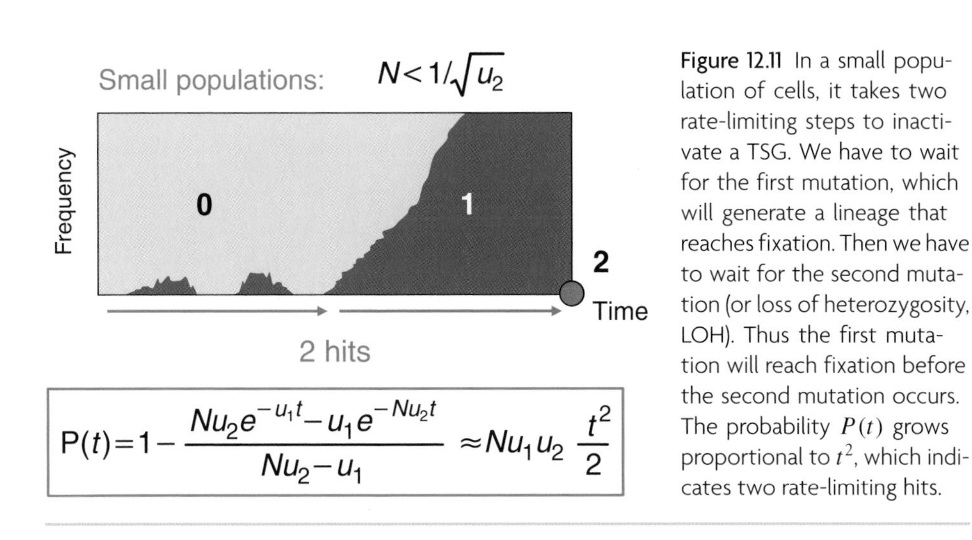

Figure 12.11 In a small population of cells, it takes two rate-limiting steps to inactivate a TSG. We have to wait for the first mutation, which will generate a lineage that reaches fixation. Then we have to wait for the second mutation (or loss of heterozygosity, LOH). Thus the first mutation will reach fixation before the second mutation occurs. The probability $P(t)$ grows proportional to t^2, which indicates two rate-limiting hits.

scale. For example, equation (12.10) has two rate-limiting hits on the short time scale, $t \ll 1/(Nu2)$, but only one rate-limiting hit on the longer time scale, $t > 1/(Nu_2)$. On the very long time scale (too long for human life), $t \gg 1/u_1$, equation (12.10) has zero rate-limiting hits.

12.4.3 Intermediate Population Size

In populations of intermediate size, we still have to wait considerable time until the first type 1 cell has been produced. The lineage that arises from such a cell can die out or produce a type 2 cell. In the latter case, the type 2 cell normally arises before the lineage of type 1 cells has reached fixation. "Intermediate population size" means

$$1/\sqrt{u_2} \ll N \ll 1/u_1. \tag{12.13}$$

The average waiting time for a type 1 cell is $1/(Nu_1)$. If $N < 1/u_1$, then this waiting time is longer than the characteristic time scale of cell division. Hence we have to wait for a "long" time until the first type 1 cell is generated. If $N > 1/\sqrt{u_2}$, then a type 2 cell will be generated before the lineage of type 1

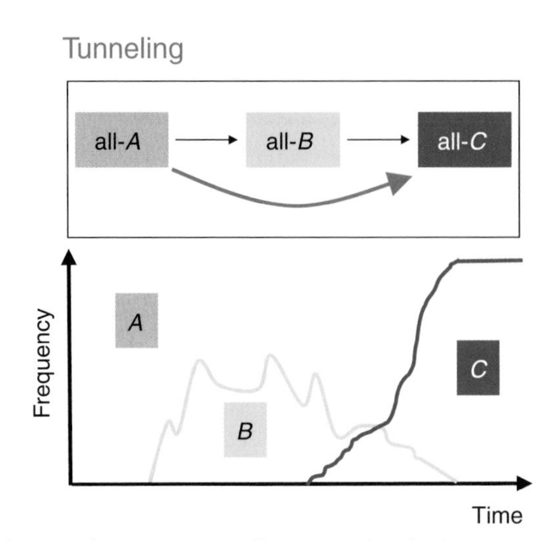

Figure 12.12 If the population size is small compared with the inverse of the mutation rates, then the evolutionary dynamics can be described as a stochastic transition between homogeneous states. Consider two consecutive mutations leading from A cells to B cells and C cells. Initially the population is in state all-A. If a B cell is produced and reaches fixation, the population is in state all-B. Subsequently, if a C cell is produced and reaches fixation, the population is in state all-C. It is possible, however, that a C cell is produced before B cells have reached fixation. In this case, the population moves from all-A to all-C without ever visiting all-B. This phenomenon is called "evolutionary tunneling."

cells has taken over the whole population. We say the population "tunnels" from state 0 to state 2 without ever reaching state 1 (Figure 12.12).

The probability that at least one cell with two hits has arisen before time t is

$$P(t) = 1 - \exp(-Nu_1\sqrt{u_2}t). \tag{12.14}$$

This probability accumulates as a first order of time: it takes only one rate-limiting hit to inactivate a TSG in a population of intermediate size. This one rate-limiting hit is characterized by the waiting time until a type 1 cell emerges, which gives rise to a lineage that will generate a type 2 cell (Figure 12.13). Derivations of equation (12.14) are not trivial and can be found in Komarova et al. (2003) and Iwasa et al. (2005).

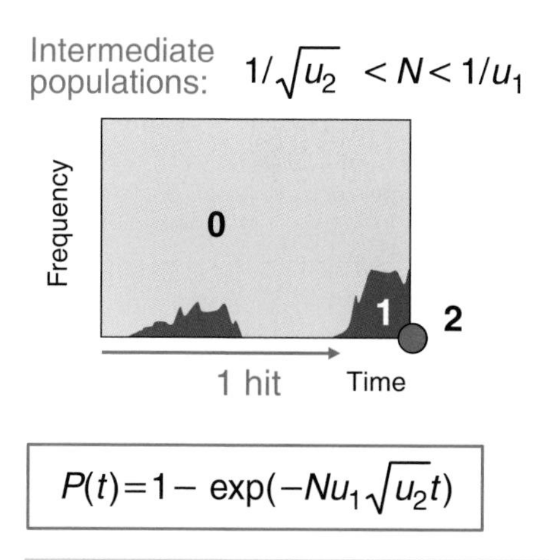

Figure 12.13 For intermediate population size, it takes only one rate-limiting step to inactivate a TSG. We have to wait for the first mutation that gives rise to a lineage that will produce the second mutation. Hence the second mutation occurs before the first mutation has reached fixation. For small t, the probability $P(t)$ grows proportional to t, which indicates one rate-limiting hit.

12.4.4 Large Population Size

In large populations, type 1 cells arise immediately, and their abundance grows as a linear function of time. We simply have to calculate the probability that this growing cell population will generate a type 2 cell. "Large population size" means

$$N \gg 1/u_1. \tag{12.15}$$

In this case, the waiting time for producing type 1 cells, $1/(Nu_1)$, is less than one time unit. Hence type 1 cells arise immediately. The abundance of these cells, x_1, grows according to

$$x_1(t) = Nu_1t. \tag{12.16}$$

The probability of having produced a type 2 cell at time t is

$$P(t) = 1 - e^{-u_2\int_0^t x_1(\tau)d\tau}. \tag{12.17}$$

Solving the integral, we obtain

$$P(t) = 1 - \exp(-Nu_1u_2t^2/2). \tag{12.18}$$

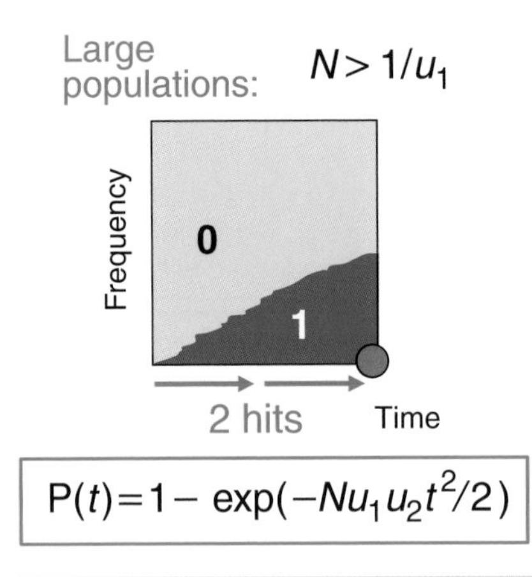

Figure 12.14 In large populations, the first mutation will arise immediately. The abundance of cells with one mutation increases as a linear function of time. We have to wait for these cells to produce the second mutation. For small t, the probability $P(t)$ grows proportional to t^2, which means that it takes two hits to eliminate a TSG in a large population of cells. But neither of these hits will be rate limiting in the overall process of cancer progression, because the events in very large populations of cells occur on a much faster time scale.

This probability again accumulates as a second order of time (Figure 12.14). Eliminating a TSG in a large population of cells is, however, most likely not rate limiting for the overall process of cancer progression. The time it takes to eliminate a TSG in a large population of cells is negligible compared with the time it takes to wait for a mutation in a small population.

12.4.5 Three Dynamic Laws for Tumor Suppressor Gene Inactivation

The three dynamic laws (12.10, 12.14, 12.18) provide a complete description of TSG inactivation. In a normal tissue consisting of small compartments of cells, a TSG is eliminated by two rate-limiting hits. The overall rate of inactivation is proportional to the second order of time. For intermediate population size, only one rate-limiting hit is needed to inactivate a TSG. The rate of inactivation is proportional to the first order of time. In large tumors, it again takes two hits to inactivate a TSG, but neither of them is rate limiting for the overall process of tumorigenesis. Therefore as the population size, N, increases, a TSG is inactivated in two, one, or zero rate-limiting steps (Figures 12.15 and 12.16).

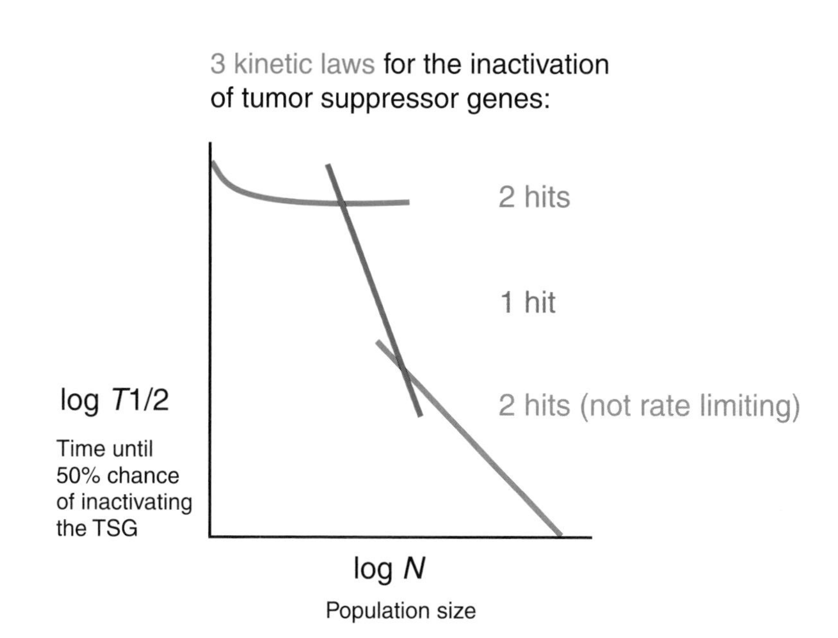

Figure 12.15 There are three kinetic laws for the inactivation of a tumor suppressor gene (TSG). In small, intermediate, and large populations, inactivation takes two, one, and zero rate-limiting hits, respectively. The diagram shows the "half-life of a TSG" versus the population size of cells in a log-log plot. The "half-life of a TSG" is defined by the time until the probability is 1/2 that the TSG has been inactivated.

12.5 GENETIC INSTABILITY

An important question in oncology is to what extent genetic instability is a driving force of cancer progression. Normal cells are genetically stable, but all solid tumors seem to have some form of genetic instability in the end. Therefore genetic instability must arise at some stage of tumorigenesis. One possibility is that genetic instability arises early and accelerates the somatic evolution of cancer via increased mutation rates. Another possibility is that genetic instability is a by-product of the final stages of cancer development.

The idea of a "mutator phenotype" in cancer genetics was first proposed by Larry Loeb in 1974. He argues that somatic selection should favor cells with increased mutation rates, because those cells will more rapidly accumulate other

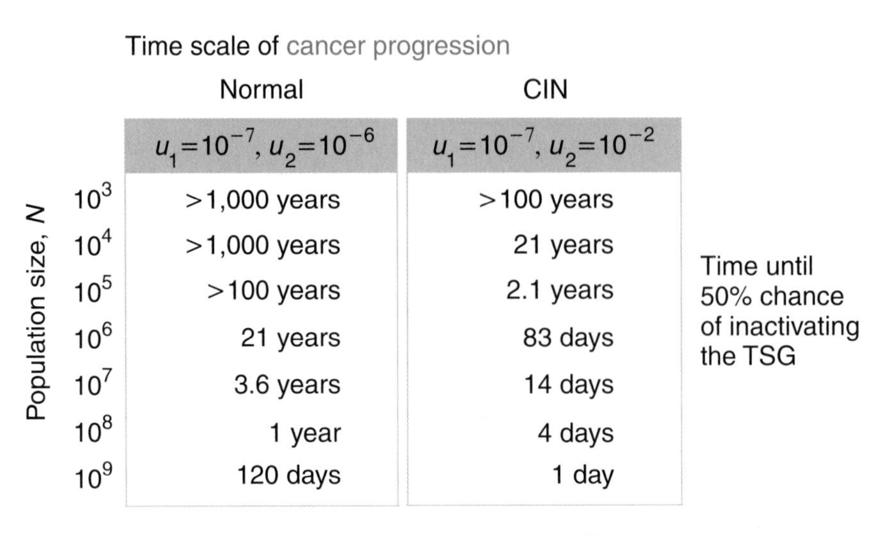

Time scale of cancer progression

Population size, N	Normal $u_1=10^{-7}, u_2=10^{-6}$	CIN $u_1=10^{-7}, u_2=10^{-2}$
10^3	>1,000 years	>100 years
10^4	>1,000 years	21 years
10^5	>100 years	2.1 years
10^6	21 years	83 days
10^7	3.6 years	14 days
10^8	1 year	4 days
10^9	120 days	1 day

Time until 50% chance of inactivating the TSG

Small lesions need genetic instability

Figure 12.16 Chromosomal instability (CIN) can dramatically accelerate cancer progression. The table shows the elapsed time until there is a 50% chance of inactivating the "next" tumor suppressor gene. We assume that for stable cells the mutation rates for inactivating the first and second allele are $u_1 = 10^{-7}$ and $u_2 = 10^{-6}$, respectively. For CIN cells, the rate of inactivating the second allele (due to loss of heterozygosity, LOH) is $u_2 = 10^{-2}$. For example, a lesion of $N = 10^5$ cells takes 2.1 years with CIN and more than 100 years without CIN to inactivate a TSG. We can also see the dramatic reduction in time scale as the population size (= effective number of cancer cells) increases.

mutations that are necessary for cancer progression. There exists a large literature in evolutionary biology regarding what the optimum mutation rate is in a given situation, but cancer progression represents a rather unique evolutionary scenario. A large number of our genes contribute to the goal that cancer does not happen too early. All these genes are targets for possible advantageous mutations in the somatic evolutionary process that leads to cancer. Since cancer cells have many possibilities for advantageous hits, their optimum mutation rate should be much higher than the normal somatic mutation rate.

In the following, we will explore the most radical possibility of all. We will calculate the probability that chromosomal instability comes before the

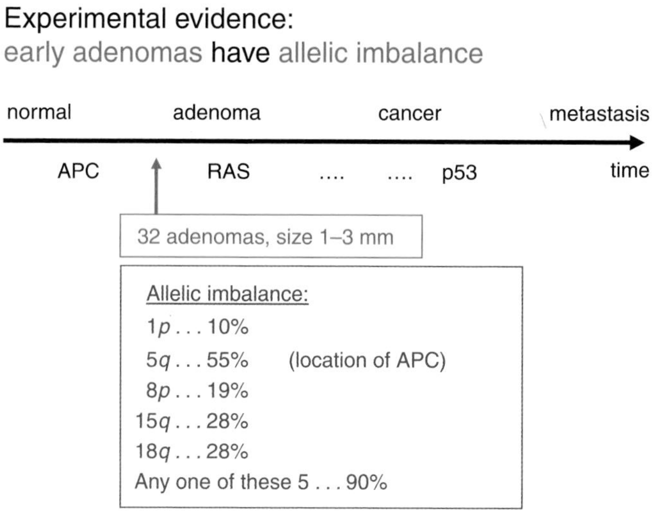

Experimental evidence:
early adenomas **have** allelic imbalance

normal	adenoma	cancer	metastasis
APC	RAS	.... p53	time

32 adenomas, size 1–3 mm

Allelic imbalance:
1*p* . . . 10%
5*q* . . . 55% (location of APC)
8*p* . . . 19%
15*q* . . . 28%
18*q* . . . 28%
Any one of these 5 . . . 90%

Figure 12.17 Early adenomas have allelic imbalance. Shih et al. (2001) analyzed 32 adenomas 1–3 mm in size. They looked for allelic imbalance (which means either less than two copies or more than two copies) at five different places in the genome: 1*p*, 5*q*, 8*p*, 15*q*, and 18*q*. They found that 90% of the adenomas had allelic imbalance in at least one of those five locations.

inactivation of the first tumor suppressor gene and therefore initiates cancer progression. For colon cancer there is indirect evidence that CIN is an early event: most small adenomas, 1 to 3 mm in size, already have allelic imbalance (Figure 12.17).

12.5.1 Neutral CIN before One TSG

Let us study a case where tumorigenesis starts with the inactivation of a TSG, A, in a small compartment of cells. An appropriate example is the inactivation of the APC gene in a colonic crypt. Initially all cells have two active alleles of the TSG, $A^{+/+}$. One of the two alleles can become inactivated at mutation rate u_1 to generate a cell of type $A^{+/-}$. The second allele can become inactivated at mutation rate u_2 to generate a cell of type $A^{-/-}$. $A^{+/+}$ cells can also receive mutations that trigger the CIN phenotype. This happens at rate u_c and the resulting cell is of type $A^{+/+}CIN$. Such a cell can inactivate the first allele of

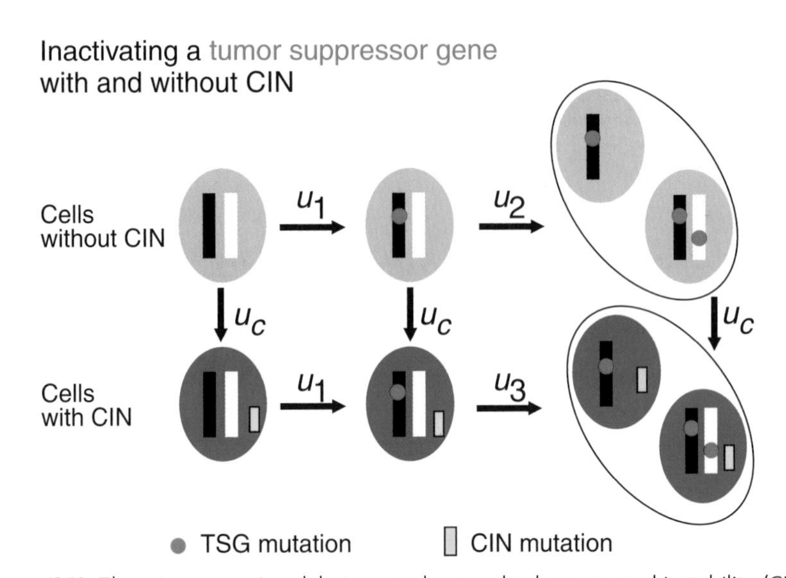

Figure 12.18 There is an ongoing debate as to how early chromosomal instability (CIN) arises during tumor progression. The most radical (and most interesting) suggestion is that CIN precedes the inactivation of the first tumor suppressor gene (TSG) and thus induces the first phenotypic change on the way to cancer. The figure shows the mutational pathway that needs to be analyzed. Inactivation of the first allele is usually caused by a point mutation. Inactivation of the second allele can be caused either by a point mutation or by loss of heterozygosity (LOH). Denote by u_1 and u_2 the inactivation rates of the first and second allele, respectively. Denote by u_c the mutation rate for triggering CIN. If there are many CIN genes, then u_c can be much larger than u_1 or u_2. For a CIN cell, there is rapid loss of heterozygosity (LOH). Hence the rate u_3 is the fastest of all.

the TSG with normal mutation rate u_1 to produce a cell of type $A^{+/-}CIN$. This cell type also arises when an $A^{+/-}$ cell receives a CIN mutation. The $A^{+/-}CIN$ cell rapidly undergoes LOH, at rate u_3, to produce a cell of type $A^{-/-}CIN$ (Figure 12.18).

Therefore, we have the following parameters: u_1 is the mutation rate for inactivating the first allele of the TSG; u_2 is the mutation rate for inactivating the second allele of the TSG in a cell without CIN; u_3 is the mutation rate for inactivating the second allele of the TSG in a cell with CIN; u_c is the mutation rate for triggering CIN; the effective population size of the compartment is given by N.

The first allele is normally inactivated by a point mutation. The mutation rate per gene per cell division is estimated to be around $u_1 \approx 10^{-7}$. In a normal cell, the second allele can be eliminated by another point mutation or by an LOH event.

If the effective compartment size, N, is much less than the inverse of the mutation rates u_1, u_2, and u_c, then the actual evolutionary dynamics can be approximated by stochastic transitions among homogeneous states. A lineage arising from a mutated cell will usually reach fixation or become extinct before another mutated cell arises. In this case, we can consider a stochastic process with six states (Figure 12.19):

(i) In state X_0, all cells are of type $A^{+/+}$.

(ii) In state X_1, all cells are of type $A^{+/-}$.

(iii) In state X_2, all cells are of type $A^{-/-}$.

(iv) In state Y_0, all cells are of type $A^{+/+}CIN$.

(v) In state Y_1, all cells are of type $A^{+/-}CIN$.

(vi) In state Y_2, all cells are of type $A^{-/-}CIN$.

Let us also denote by $X_0(t)$, $X_1(t)$, $X_2(t)$, $Y_0(t)$, $Y_1(t)$, $Y_2(t)$ the probabilities that the stochastic process is in the corresponding state at time t. The evolutionary dynamics of cancer initiation are given by the following system of linear differential equations:

$$\begin{aligned}
\dot{X}_0 &= -(u_1 + u_c)X_0 \\
\dot{X}_1 &= u_1 X_0 - (u_c + N u_2)X_1 \\
\dot{X}_2 &= N u_2 X_1 \\
\dot{Y}_0 &= u_c X_0 - u_1 Y_0 \\
\dot{Y}_1 &= u_c X_1 + u_1 Y_0 - N u_3 Y_1 \\
\dot{Y}_2 &= N u_3 Y_1
\end{aligned} \tag{12.19}$$

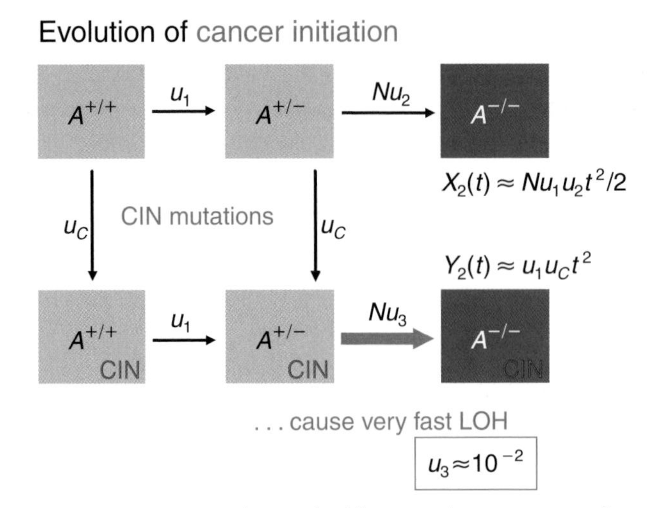

Figure 12.19 Cancer initiation can be studied by a stochastic process that describes transitions between homogeneous compartments. Mutation of the first allele is neutral. Hence the rate of evolution from $A^{+/+}$ to $A^{+/-}$ is given by the mutation rate u_1. Mutation of the second allele leads to a selective advantage. If this advantage is large, then the rate of evolution from $A^{+/-}$ to $A^{-/-}$ is approximately given by Nu_2. If chromosomal instability (CIN) is neutral, its rate of evolution is the mutation rate u_c. CIN has no consequence for inactivating the first allele, but greatly accelerates the loss of heterozygosity (LOH) of the second allele. The red arrow indicates that this transition rate is many orders of magnitude faster than all other transition rates. On the relevant time scale, the probability of inactivating a TSG without CIN is given by $X_2(t) = Nu_1u_2t^2/2$, and with CIN it is given by $Y_2(t) = u_1u_ct^2$. If $Y_2(t) > X_2(t)$, which means $u_c > Nu_2/2$, then it is more likely to initiate cancer via CIN.

We have assumed that the probability that a lineage arising from a single $A^{+/-}$ cell takes over the compartment is $1/N$. Therefore inactivation of the first allele of the TSG is neutral. We have also assumed that the CIN mutation is neutral. Finally, we have assumed that the probability for an $A^{-/-}$ cell to take over the compartment is close to 1, which means this mutation has a strong selective advantage.

Initially, at time $t = 0$, we have $X_0 = 1$ and all other probabilities are 0. The time-explicit solutions of this system can be easily obtained with standard techniques. But on the relevant time scale of human life (about one hundred years), these solutions take an even simpler form. If the cells divide once per day, then our time, t, is measured in units of days. If the cells divide once per week, then our time, t, is measured in units of weeks. In both cases, we find that $u_1 t$, $Nu_2 t$, and $u_c t$ are all much less than 1.

On the relevant time scale, the approximate solutions of system (12.19) take the form

$$
\begin{aligned}
X_0(t) &\approx 1 \\
X_1(t) &\approx u_1 t \\
X_2(t) &\approx N u_1 u_2 t^2/2 \\
Y_0(t) &\approx u_c t \\
Y_1(t) &\approx u_1 u_c t^2 \\
Y_2(t) &\approx u_1 u_c t^2
\end{aligned}
\tag{12.20}
$$

For the lifetime of a person, almost all compartments remain wild type, which means that $X_0(t) \approx 1$. Compartments in state X_1 arise as a linear function of time. Compartments in state X_2 arise as a quadratic function of time. Similarly, Y_0 compartments arise as a linear function of time, and Y_1 compartments as a quadratic function. Surprisingly, however, $Y_2(t)$ is equivalent to $Y_1(t)$, and therefore Y_2 compartments also accumulate as a quadratic function of time. The reason is that $Nu_3 t$ is much larger than one on the relevant time scale, and the corresponding step is not rate limiting. As soon as the system has reached state Y_1, it will proceed to Y_2. The waiting time for the LOH event in CIN cells is negligible compared with the waiting time for all other mutations in the system.

Therefore we make the interesting observation that it takes two rate-limiting hits to eliminate a TSG both with CIN and without CIN. The additional CIN mutation adds another rate-limiting step, but the subsequent LOH event is no longer rate limiting (Figure 12.20).

We can now estimate the ratio of cancers that are initiated with CIN versus without CIN. We have

$$
Y_2(t) : X_2(t) = 2u_c : Nu_2. \tag{12.21}
$$

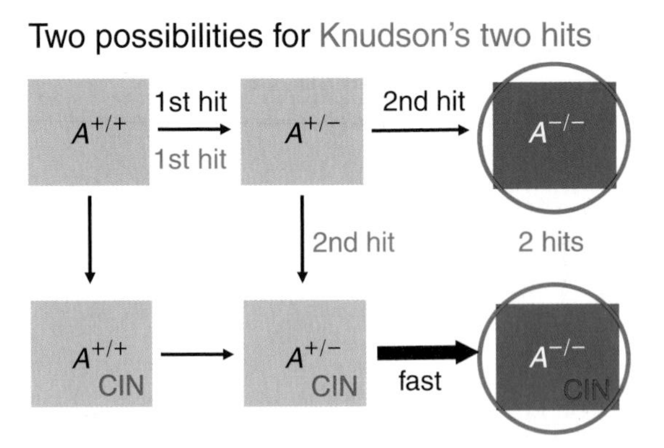

Figure 12.20 Knudson's two-hit hypothesis gave rise to the concept of a tumor suppressor gene (TSG). It takes two rate-limiting hits to inactivate a TSG: one for the first allele and one for the second. These rate-limiting hits were first observed in cancer-incidence data of retinoblastoma. Our analysis shows that Knudson's two hits are entirely compatible with chromosomal instability (CIN). It also takes only two hits to inactivate a TSG with an additional CIN mutation: one hit for the first allele of the TSG and one hit for the CIN mutation. Losing the second allele of the TSG in CIN cells is very fast and therefore not rate limiting.

The ratio is independent of time.

The mutation rate for inactivating the second allele of the TSG in a normal (non-CIN) cell can be written as the sum of the point mutation rate per gene, u, and the rate of LOH, p_0. We have

$$u_2 = u + p_0. \tag{12.22}$$

If there are n_1 class 1 CIN genes and n_2 class 2 CIN genes in the human genome that can be mutated in this particular scenario of cancer initiation, then the mutation rate u_c that triggers the CIN phenotype is given by

$$u_c = 2n_1(u + p_0) + 2n_2u. \tag{12.23}$$

Hence the majority of cancers are initiated by CIN if

$$4n_1(u + p_0) + 4n_2u > N(u + p_0). \tag{12.24}$$

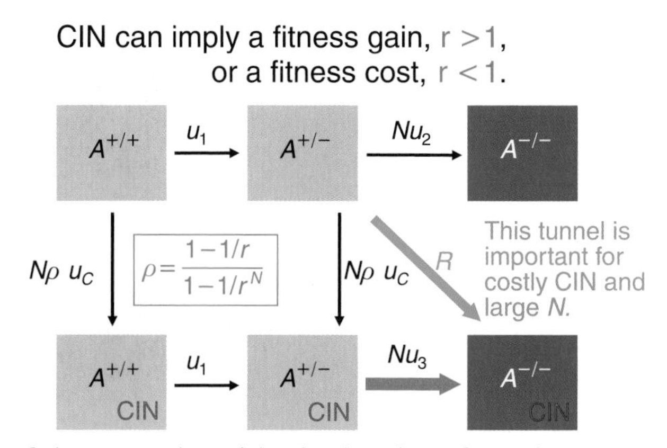

Figure 12.21 If chromosomal instability (CIN) implies a fitness loss or gain, then the transitions from non-CIN to CIN states occur at the rate $Nu\rho$, where ρ is the probability that a newly produced CIN cell takes over the population. But for costly CIN and large N, the tunnel from $A^{+/-}$ to $A^{-/-}$CIN is important. This tunnel has a transition rate $R = Nu_c u_3 r/(1-r)$.

If, for example, $N = 4$ and $u \approx p_0$, then the presence of one class 1 and one class 2 CIN gene suffices to ensure that CIN initiates more than half of all cancers. If there are two class 1 genes and two class 2 genes, then 75% of cancers are initiated by CIN. The calculation so far does not include a possible cost of CIN. Let us do this now.

12.5.2 Costly CIN in Small Compartments

Let us assume that the relative fitness of a CIN cell is $r < 1$; the cost of CIN is $1 - r$. It could be that the CIN phenotype is recognized by apoptotic defense mechanisms, which lead to higher death rates and consequently lower r. Moreover, CIN cells might have a lower fitness, because they accumulate deleterious mutations (Figure 12.21).

In a Moran process with population size N, the probability that the lineage arising from a CIN mutant takes over the population is given by

$$\rho = \frac{1 - 1/r}{1 - 1/r^N}. \tag{12.25}$$

The transition rate from a non-CIN state to a CIN state of our stochastic process is now given by $N\rho u_c$. We obtain the same equations as before, but u_c is replaced by $N\rho u_c$:

$$\begin{aligned}
\dot{X}_0 &= -(u_1 + N\rho u_c)X_0 \\
\dot{X}_1 &= u_1 X_0 - N(\rho u_c + u_2)X_1 \\
\dot{X}_2 &= N u_2 X_1 \\
\dot{Y}_0 &= N\rho u_c X_0 - u_1 Y_0 \\
\dot{Y}_1 &= N\rho u_c X_1 + u_1 Y_0 - N u_3 Y_1 \\
\dot{Y}_2 &= N u_3 Y_1
\end{aligned} \tag{12.26}$$

On the relevant time scale, the solutions are given by

$$\begin{aligned}
X_0(t) &\approx 1 \\
X_1(t) &\approx u_1 t \\
X_2(t) &\approx N u_1 u_2 t^2/2 \\
Y_0(t) &\approx N\rho u_c t \\
Y_1(t) &\approx N\rho u_1 u_c t^2 \\
Y_2(t) &\approx N\rho u_1 u_c t^2
\end{aligned} \tag{12.27}$$

The ratio of cancers that are initiated with CIN versus without CIN is given by

$$Y_2 : X_2 = 2\rho u_c : u_2. \tag{12.28}$$

As before, we assume

$$u_2 = u + p_0 \tag{12.29}$$

and

$$u_c = 2n_1(u + p_0) + 2n_2 u. \tag{12.30}$$

If, for example, $N = 4$, $r = 0.8$, $u \approx p_0$, $n_1 = 2$, and $n_2 = 2$, then about 68% of all cancers are initiated by CIN. If the effective compartment size is small, then even substantial costs of CIN can be tolerated, because the evolutionary dynamics in small compartments are dominated by random drift rather than by selection.

12.5.3 Costly CIN in Large Compartments

For larger compartment sizes, however, the ratio $N\rho$ could become vanishingly small for $r < 1$. If, for example, $N = 100$ and $r = 0.7$, then $N\rho \approx 10^{-16}$; a CIN mutation with this cost could never reach fixation in a population with effective size 100.

"Stochastic tunneling" can still allow a significant fraction of cancers to be initiated by CIN. The stochastic process will never reach states Y_0 or Y_1. From state X_1, however, the process can tunnel to Y_2 without ever reaching Y_1. In state X_1, all cells are of type $A^{+/-}$. Cells of type $A^{+/-}CIN$ are being produced at rate Nu_c. These cells do not reach fixation, but remain near a mutation selection balance with average abundance $Nu_c/(1-r)$. From there they produce $A^{-/-}CIN$ cells at rate ru_3. Therefore the rate of the tunnel from state X_1 to Y_2 is given by

$$R = \frac{Nu_c r u_3}{1-r}. \tag{12.31}$$

The stochastic evolution is described by the system

$$\begin{aligned}
\dot{X}_0 &= -u_1 X_0 \\
\dot{X}_1 &= u_1 X_0 - (R + Nu_2) X_1 \\
\dot{X}_2 &= Nu_2 X_1 \\
\dot{Y}_2 &= R X_1
\end{aligned} \tag{12.32}$$

The approximate solutions on the relevant time scale are

$$\begin{aligned} X_0(t) &\approx 1 \\ X_1(t) &\approx u_1 t \\ X_2(t) &\approx N u_1 u_2 t^2/2 \\ Y_0(t) &\approx 0 \\ Y_1(t) &\approx 0 \\ Y_2(t) &\approx R u_1 t^2/2 \end{aligned} \tag{12.33}$$

Remember that the system will never reach states Y_0 and Y_1. Therefore both $Y_0(t)$ and $Y_1(t)$ are zero.

The ratio of cancers that are initiated with CIN versus without CIN is given by

$$Y_2 : X_2 = R : N u_2 = \frac{u_c r u_3}{1-r} : u_2. \tag{12.34}$$

The population size has canceled out of this comparison. If, for example, $r = 0.8$, $u \approx p_0$, $u_3 = 0.01$, $n_1 = 5$, and $n_2 = 5$, then about 38% of all cancers are initiated by CIN.

12.5.4 CIN before Two TSGs

Consider a path to cancer in which two TSGs, A and B, have to be eliminated (Figure 12.22). Initially the compartment consists of N_0 wild type cells, $A^{+/+}B^{+/+}$. Suppose gene A has to be inactivated first. The evolutionary pathway proceeds from $A^{+/+}B^{+/+}$ via $A^{+/-}B^{+/+}$ to $A^{-/-}B^{+/+}$, and subsequently to $A^{-/-}B^{+/-}$ and $A^{-/-}B^{-/-}$. CIN can emerge at any stage of this pathway; once arisen, CIN accelerates the transitions from $A^{+/-}$ to $A^{-/-}$ and from $B^{+/-}$ to $B^{-/-}$. Inactivation of the first TSG induces neoplastic growth. We assume that the $A^{-/-}$ compartment gives rise to a small lesion of N_1 cells. In this lesion, the second TSG has to be inactivated for further tumor progression. Because of the increased compartment size, the evolutionary trajectory will tunnel from $A^{-/-}B^{+/+}$ directly to $A^{-/-}B^{-/-}$. The condition for this tunnel is $1/u_1 > N_1 > 1/\sqrt{u_2}$.

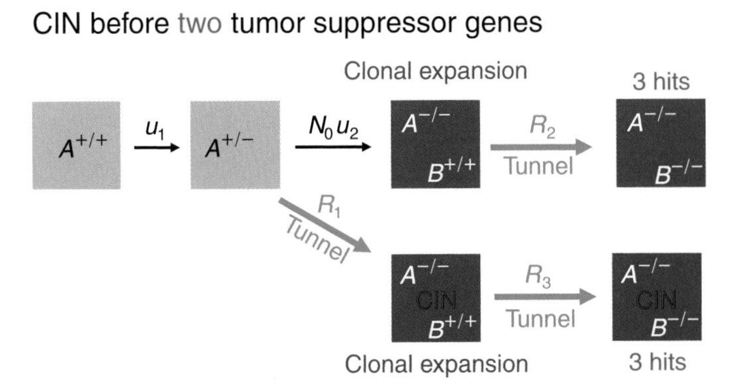

Figure 12.22 Inactivation of the first tumor suppressor gene (TSG) might lead to a moderate clonal expansion. In this example, a second TSG needs to be inactivated for further cancer progression. We can calculate the probability that chromosomal instability (CIN) arises before the inactivation of the first TSG. Even moderate clonal expansion will lead to a population of intermediate size and, therefore, inactivation of the second TSG will occur via a tunnel. Thus it takes three hits to inactivate two TSGs with and without CIN. CIN, however, can accelerate both loss of heterozygosity (LOH) events. We find that one (or a few) costly CIN genes are enough to ensure that CIN comes before the first TSG in a pathway of cancer progression where two TSGs have to be inactivated in rate-limiting situations.

Since we already know that CIN is likely to initiate cancer progression if it has no cost (section 12.5.1) or if inactivation of the first TSG happens in a very small compartment (section 12.5.2), we will only investigate the case where CIN has a substantial cost and the compartment size, N_0, is so large that fixation of a CIN cell is nearly impossible. In this case, evolutionary dynamics can be described as a stochastic process on the following six states:

(i) In state X_0, all cells are of type $A^{+/+}B^{+/+}$.

(ii) In state X_1, all cells are of type $A^{+/-}B^{+/+}$.

(iii) In state X_2, all cells are of type $A^{-/-}B^{+/+}$.

(iv) In state X_3, all cells are of type $A^{-/-}B^{-/-}$.

(v) In state Y_2, all cells are of type $A^{-/-}B^{+/+}CIN$.

(vi) In state Y_3, all cells are of type $A^{-/-}B^{-/-}CIN$.

As before, $X_i(t)$ and $Y_i(t)$ denote the probabilities to be in the corresponding states at time t. Initially, at time $t = 0$, we have $X_0(0) = 1$ and all other states have zero probability. The evolutionary dynamics are given by the following system of linear differential equations:

$$
\begin{aligned}
\dot{X}_0 &= -u_1 X_0 \\
\dot{X}_1 &= u_1 X_0 - (R_1 + N_0 u_2) X_1 \\
\dot{X}_2 &= N_0 u_2 X_1 - R_2 X_2 \\
\dot{X}_3 &= R_2 X_2 \\
\dot{Y}_2 &= R_1 X_1 - R_3 Y_2 \\
\dot{Y}_3 &= R_3 Y_2
\end{aligned}
\tag{12.35}
$$

The rates of the tunnels are given by

$$
\begin{aligned}
R_1 &= \frac{N_0 u_c r u_3}{1 - r} \\
R_2 &= N_1 u_1 \sqrt{u_2} \\
R_3 &= N_1 u_1 \sqrt{u_3}
\end{aligned}
\tag{12.36}
$$

Without CIN, inactivation of two TSGs requires three rate-limiting hits. It takes two hits to inactivate the first TSG. If this leads to a moderate clonal expansion, then the second TSG can be inactivated in one rate-limiting step. With CIN, inactivation of two TSGs also requires three rate-limiting hits. It takes one rate-limiting hit to inactivate one allele of the first TSG, one rate-limiting hit to trigger CIN, and one rate-limiting hit to inactivate both alleles of the second TSG (Figure 12.22).

A numerical analysis of system (12.35) shows that even under the assumption of a substantial cost for CIN and a large compartment size, N_0, only a small number of Class I or II CIN genes are needed to ensure that cancer progression is initiated by a CIN mutation.

The cost of CIN is compensated by an acceleration of every successive TSG inactivation. It is possible that the first TSG, A, is predominantly inactivated in cells without CIN. Thus most lesions that are caused by inactivation of TSG A do not have CIN, but the small fraction of lesions with CIN will eliminate TSG *B* within the time scale of a human life. In such a situation all (or almost all) cancers will derive from lesions in which a CIN mutation preceded inactivation of the first TSG.

If the inactivation of the first TSG leads to a vast clonal expansion, then the inactivation of a further TSG is not rate limiting. In this case, the analysis of the system leads back to the question of whether CIN arises before inactivation of only one TSG.

There is also the possibility that the clonal expansion caused by inactivation of the first TSG is so small that the non-CIN trajectory requires two hits to inactivate the second TSG. In this case, it takes four hits to inactivate *A* and *B* without CIN, but only three hits to inactivate them with CIN. Thus CIN is at an enormous advantage in this situation.

Throughout this section, we have assumed that the clonal expansion caused by inactivation of *A* occurs with certainty once such a cell has been produced. A correction term for a reduced probability can easily be incorporated. We have also made the plausible assumption that the time needed for clonal expansion can be neglected compared with the other transition rates.

We conclude as follows. In a pathway of cancer progression where one TSG is inactivated in a rate-limiting situation, one or a few neutral CIN genes are enough to ensure that CIN occurs before inactivation of the TSG. In a pathway of cancer progression where two TSGs must be inactivated in rate-limiting situations, one or a few costly CIN genes are enough to ensure that CIN occurs before inactivation of the first TSG. By analogy with yeast, there should be hundreds of CIN genes in the human genome (although in any particular tissue only a subset of them might give rise to a CIN phenotype when mutated). Therefore not only does CIN accelerate cancer progression, but there are also many mutations that lead to CIN. The combination of these two effects must imply that CIN is an early event and an important driving force in cancer progression.

SUMMARY

- Cancer is an evolutionary process.
- A quantitative understanding of cancer progression requires a mathematical analysis of the underlying evolutionary dynamics.

- We have calculated the rate of evolution of activating oncogenes and inactivating tumor suppressor genes. The key parameters include the population size of reproducing cells, the mutation rates, and fitness values.
- Tissue architecture can affect the rate of cancer initiation and the types of mutations that are likely to occur. Small compartments confer protection against mutations in oncogenes and tumor suppressor genes, but are vulnerable to genetic instability.
- The "linear process" is an effective tissue design to delay the onset of cancer.
- In small, intermediate, and large populations of reproducing cells, a tumor suppressor gene is eliminated in 2, 1, and 0 rate-limiting steps, respectively.
- In a small compartment (such as a colonic crypt), it takes two rate-limiting hits to eliminate a TSG with and without CIN. Therefore Knudson's two-hit hypothesis is compatible with the idea that the second hit occurs in a CIN gene.
- For a wide range of plausible parameter values, CIN can precede inactivation of the second allele of a TSG. In this case, the CIN mutation leads to the first phenotypic change on the way to cancer.
- The more CIN genes in the human genome, the more possibilities there are to commit CIN, and the more likely it is that CIN occurs early.
- One (or a few) neutral CIN genes are enough to ensure that CIN precedes inactivation of a TSG in an evolutionary pathway to cancer where one TSG must be inactivated in a rate limiting situation.
- One (or a few) costly CIN genes are enough to ensure that CIN precedes inactivation of the first TSG in an evolutionary pathway to cancer where at least two TSGs must be inactivated in rate-limiting situations.

LANGUAGE EVOLUTION 13

EVOLUTIONARY BIOLOGISTS have many reasons to be interested in language. Language is a truly human invention. It makes us special among animals. It is the fundamental tool of human interaction, the basis of human society, and the source of creativity. Moreover, language gives rise to a new mode of evolution.

Evolution requires transfer of information from one individual to another, from one generation to the next. During the first four billion years of life on earth, the primary means of information transfer was genetic. Language, however, has ignited cultural evolution. Some form of cultural evolution may be present in animals, but on a very limited scale. Human language offers a replication machinery for unlimited cultural evolution. Human language "makes infinite use of finite media." We spread our ideas rather than our genes. The enormous changes over the last five thousand years were caused not by genetic evolution but by cultural evolution. The emergence of language was the last of a series of major events in evolution, comparable in significance only to the origin of life, the first bacteria, the first higher cells, and the evolution of complex multicellularity. Language is the biggest invention of the last six hundred million years (Figure 13.1).

Language
is the most interesting invention
of the last 600 million years

?	Origin of life
3,500	Prokaryotes
1,500	Eukaryotes
600	Higher multicellularity
≈1?	Language

(millions of years ago)

Language gives rise to a
new mode of evolution

Figure 13.1 Language is the most interesting invention of the last 600 million years. The impact of language is comparable only to a few other major events in the history of life such as the origin of life and the emergence of the first bacterial cell, of the first higher cell, and of complex multicellular organisms. Prior to language, evolution used only genetic information. Human language is a tool for unlimited replication of cultural information, and led to an unprecedented explosion of cultural evolution. We spread our ideas and inventions, sometimes ignoring our genes. Bacteria invented a lot of genetics and all of biochemistry. Eukaryotes invented unlimited genetics and the ability to build complex multicellular organisms. Humans will be remembered for language.

Language is the one property that sets us apart from all other animal species. Some animals might have complicated and fascinating communication, but they cannot talk about events out of context. They can give warning calls for predators, but they cannot ask whether any lions were around yesterday. In contrast, we use language to share past experience, make plans, distribute information about others, and imagine stories of far greater extension in time and space than our own narrow existence. Everything we have discovered is due to language. All other cognitive specializations of humans may be derived from language. Mathematics and music, for example, are languages of a kind. The ability of our species to adapt to any environment is based on language. A discovery made by one person can benefit others. Our brains are problem solvers, searching for new insights. Language weaves these problem solvers

Dating the origin of language

- Humans and chimpanzees separated about 5 to 7 million years ago
- Some researchers provide evidence that the brain areas that are associated with human language began to increase in *australopithecus* about 3 million years ago
- Others argue that both *Homo ergaster* (1.8 million years ago) and *Homo neanderthalensis* (200,000 years ago) lacked important anatomical features of speech production
- Anatomically modern man appeared in Africa 100,000 years ago
- Some 50,000 years ago there was a wave out of Africa that populated the whole world
- Writing was invented around 5,000 years ago

Figure 13.2 There is no precise date for the origin of modern language, but these facts are of interest.

together into a worldwide web of human brains. A concept that originates in one brain spreads to others.

Language is a complex biological trait that requires the interplay of many anatomical and neuroanatomical features. Such a trait can only arise gradually and under the guidance of natural selection. Language is not the by-product of a big brain. If anything, it is the reverse: the cognitive demands of improved language might have provided selection pressure for increased brain size (Figure 13.2).

Evolution, however, always uses the same trick: structures that have evolved for one task are apportioned for something new. The language-processing machinery of the human brain is built of neuronal tissue that has evolved for some other purpose. There must be cognitive tasks in the brains of higher animals that are similar to learning the grammar of spoken language. Two possibilities are: (i) the interpretation of visual, auditory, and other information requires a parsing of the world into objects, groups of objects, and actions; many animals need to learn a "grammar" of how to interpret what they see, hear, smell, and so on; (ii) during development, different modules within the brain have to learn to "talk" to each other; very similar neural algorithms could allow language modules of different brains to learn to talk to each other.

This chapter is in three parts. First, I will explain fundamental aspects of formal language theory in order to analyze what language is. Second, I will discuss learning theory and the task of language acquisition. Finally, I will present a theory for the evolutionary dynamics of language. Here I will concentrate on the evolution of grammatical rules. Models for the evolution of simple communication systems, the lexical matrix, and the origin of syntax have been developed elsewhere (references can be found under Further Reading on page 308).

13.1 FORMAL LANGUAGE THEORY

Language is a mode of communication, a crucial part of human behavior, and a cultural object defining our social identity. There is also a fundamental aspect of language that makes it amenable to formal analysis: linguistic structures consist of smaller units that are grouped together according to certain rules. The combinatorial sequencing of small units into bigger structures occurs at several different levels. Phonemes form syllables and words. Words form phrases and sentences. The rules for such groupings are not arbitrary.

Individual languages have specific rules. Certain word orders are admissible in one language but not in another. In some languages word order is relatively free, but case marking is pronounced. There are always specific rules that generate valid or meaningful linguistic structures. Much of modern linguistic theory proceeds from this insight. The area of mathematics and computer science called formal language theory provides a mathematical machinery for dealing with such phenomena.

Formal language theory is a mathematical approach that attempts to describe the fundamental aspects of language. We start by defining an "alphabet" as a set containing a finite number of symbols. Possible alphabets for natural languages are the set of all phonemes or the set of all words of a language. For these two choices one obtains formal languages on different levels, but the mathematical principles are the same. Without loss of generality, we can consider the binary alphabet, {0, 1}, by enumerating the actual alphabet in binary code.

A "sentence" is a finite string of symbols. The set of all sentences over the binary alphabet is given by {0, 1, 00, 01, 10, 11, 000, . . .}. There are infinitely many sentences, as many as integers. This means that the set of all sentences is "countable."

A "language" is a set of sentences. Among all possible sentences, some are part of the language and some are not. A finite language contains a finite number of sentences. An infinite language contains an infinite number of sentences. There are infinitely many finite languages, as many as integers. The set of all finite languages is countable. There are infinitely many infinite languages, as many as real numbers. Therefore the set of all languages is not countable. This is important to remember, because we will find out that the set of all grammars *is* countable. Hence there are many more languages than grammars.

Here is an example of a finite language with three sentences:

$$L = \{0, 00, 111\} \tag{13.1}$$

Here is an example of an infinite language:

$$L = \{0, 01, 011, 0111, 01111, \ldots\} \tag{13.2}$$

This language contains all strings where a single 0 is followed by an arbitrary number of 1's. We can write this language in the form

$$L = 01^n. \tag{13.3}$$

Here 1^n denotes a string of n 1's. The number n can be any non-negative integer.

A "grammar" is a finite list of rules specifying a language. A grammar is expressed in terms of rewrite rules: a certain string can be rewritten as another string. Strings contain "terminals" and "nonterminals." Terminals are elements of the alphabet. Nonterminals are placeholders that can be replaced

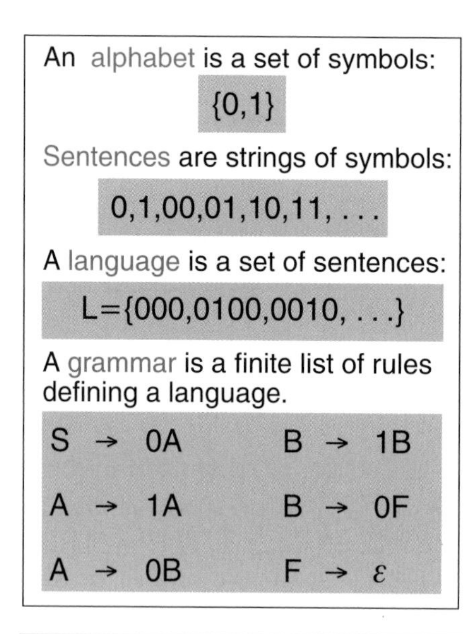

An alphabet is a set of symbols:
{0,1}
Sentences are strings of symbols:
0,1,00,01,10,11, . . .
A language is a set of sentences:
L={000,0100,0010, . . .}
A grammar is a finite list of rules defining a language.
S → 0A
B → 1B
A → 1A
B → 0F
A → 0B
F → ε

Figure 13.3 The basic objects of formal language theory are alphabets, sentences, languages, and grammars. Grammars consist of rewrite rules: a particular string can be rewritten as another string. Such rules contain symbols of the alphabet (here, 0 and 1), and so-called nonterminals (here, S, A, B, F), and a null element, ϵ. The grammar in this figure works as follows. Each sentence begins with the symbol S. S is rewritten as 0A. Now there are two choices: A can be rewritten as 1A or 0B. B can be rewritten as 1B or 0F. F always goes to ϵ. This grammar generates sentences of the form 01^m01^n0.

by other strings. After iterated application of the rewrite rules, the final string will contain only terminals (Figure 13.3).

Here is an example of a simple grammar

$$
\begin{aligned}
&S \to 0A \\
&A \to 1A \\
&A \to \epsilon
\end{aligned}
\tag{13.4}
$$

The nonterminals are S and A. The terminals are 0 and 1. There is a null element, ϵ, which stands for "nothing." Each sentence starts with S. The first rule states that S can be rewritten by the string $0A$. The remaining two rules state that A can be rewritten by $1A$ or by the null element ϵ. Let us use this grammar to generate some sentences:

$$
\begin{aligned}
&S \to 0A \to 0 \\
&S \to 0A \to 01A \to 01 \\
&S \to 0A \to 01A \to 011A \to 011
\end{aligned}
\tag{13.5}
$$

It is quite obvious that this grammar gives rise to the language

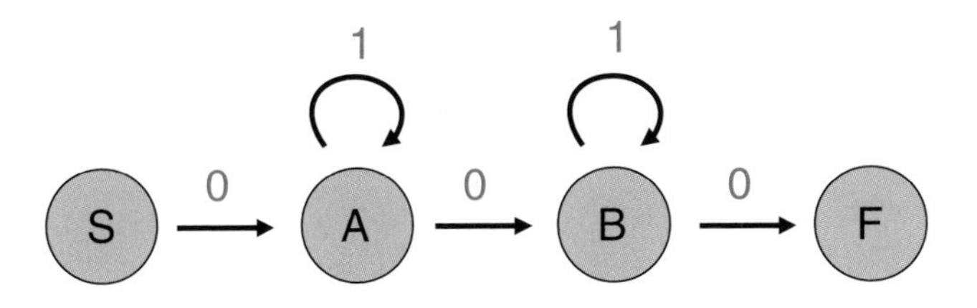
A finite-state automaton
1
1
S
0
A
0
B
0
F
This finite-state automaton generates the language
L=01^m01^n0

Figure 13.4 A finite-state automaton consists of a start, S, a finite number of intermediate states (here, A and B), and a finish, F. On each transition from one state to the next the automaton emits a symbol of the alphabet, 0 or 1, as indicated. If a finite-state automaton has at least one loop, then it can generate an infinite language. This finite-state automaton generates the infinite language $L = 01^m01^n0$. Some example sentences are: 000, 0100, 0010, 01010, 01100.

$$L = 01^n. \tag{13.6}$$

This language contains precisely those sentences where a single 0 is followed by a string of 1's.

13.1.1 Finite-State Grammars

Grammar 13.4 is an example of a finite-state grammar. Rewrite rules must be of the form: a single nonterminal (on the left) is rewritten as a single terminal, possibly followed by a nonterminal (on the right). Finite-state grammars can be represented by finite-state automata.

A finite-state automaton consists of a start, a finite number of states, and a finish. Whenever the machine moves from one state to another, it emits a single terminal. Any run from start to finish generates a particular sentence. There can be many different runs from start to finish, giving rise to many different sentences. As soon as a finite-state automaton has at least one loop, it will generate an infinite language.

Figure 13.4 shows a finite-state automaton that generates the same language as the grammar shown in Figure 13.3. Any transition corresponds to a rewrite rule. There are two loops, which correspond to the rules $A \to 1A$ and $B \to 1B$. These rules are recursive, because they allow the indefinite substitution of a nonterminal by itself. Recursion leads to infinite languages.

All the languages that can be generated by finite-state grammars or automata are called regular languages. Regular languages contain all finite languages and some infinite languages.

13.1.2 Context-Free Grammars

Consider the language

$$L = 0^n 1^n. \tag{13.7}$$

A sentence is part of this language if and only if a string of 0s is followed by a string of 1's, and both strings have the same length. It can be proved that there is no finite state grammar that can generate this language. Finite state grammars cannot count. They cannot ensure that 0^n is followed by 1^n for arbitrarily large n. But the following "context-free" grammar does the job:

$$\begin{aligned} S &\to 0S1 \\ S &\to \epsilon \end{aligned} \tag{13.8}$$

Let us generate some example sentences:

$$\begin{aligned} S &\to \epsilon \\ S &\to 0S1 \to 01 \\ S &\to 0S1 \to 00S11 \to 0011 \end{aligned} \tag{13.9}$$

A context-free grammar allows rewrite rules of the form: a single nonterminal (on the left) can be rewritten by an arbitrary string (on the right). Context-free grammars can be implemented by push-down automata. These are computers with a single memory stack: at any one time they only have access to the top register of their memory.

13.1.3 Context-Sensitive Grammars

There are languages that cannot be generated by context-free grammars (Figure 13.5). Consider the alphabet {0, 1, 2} and the language

$$L = 0^n 1^n 2^n. \tag{13.10}$$

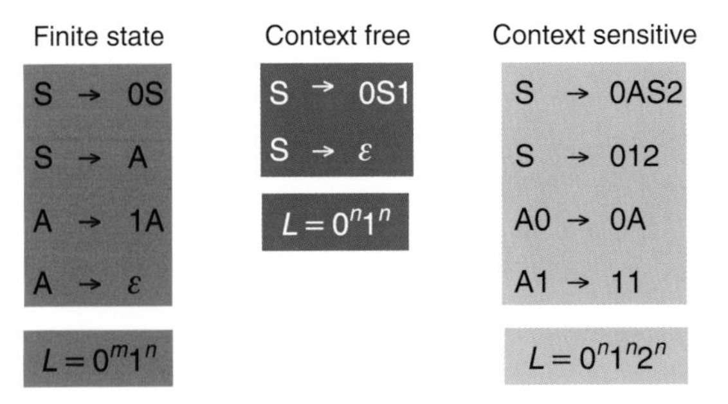

Three examples of grammars
Finite state
S → 0S
S → A
A → 1A
A → ε
L = 0^m1^n
Context free
S → 0S1
S → ε
L = 0^n1^n
Context sensitive
S → 0AS2
S → 012
A0 → 0A
A1 → 11
L = 0^n1^n2^n

Figure 13.5 Three grammars and their corresponding languages. Finite-state grammars have rewrite rules of the form: a single nonterminal (on the left) is rewritten as a single terminal possibly followed by a nonterminal (on the right). The finite-state grammar, in this figure, generates the regular language 0^m1^n; a valid sentence is any sequence of 0's followed by any sequence of 1's. A context-free grammar admits rewrite rules of the form: a single nonterminal is rewritten as an arbitrary string of terminals and nonterminals. The context-free grammar in this figure generates the language 0^n1^n; a valid sentence is a sequence of 0's followed by the same number of 1's. There is no finite-state grammar that could generate this language. A context-sensitive grammar admits rewrite rules of the form $\alpha A\beta \to \alpha\gamma\beta$. Here α, β, and γ are strings of terminals and nonterminals. While α and β may be empty, γ must be nonempty. The important restriction on the rewrite rules of context-sensitive grammars is that the complete string on the right must be at least as long as the complete string on the left. The context-sensitive grammar, in this figure, generates the language $0^n1^n2^n$. There is no context-free grammar that could generate this language.

The following grammar can generate this language

$$\begin{aligned} S &\to 0AS2 \\ S &\to 012 \\ A0 &\to 0A \\ A1 &\to 11 \end{aligned} \tag{13.11}$$

This grammar is not context free. The last two rewrite rules violate the constraints of context-free grammars, because they do not have a single nonterminal on the left. In fact, it is possible to prove that there exists no context-free grammar that can generate the language $0^n1^n2^n$. Grammar (13.11) is called context sensitive. Such grammars have rewrite rules of the form: in the context of some string, a nonterminal gets rewritten in a certain way. More precisely, context-sensitive grammars admit rewrite rules of the form

$$\alpha A \beta \to \alpha \gamma \beta. \tag{13.12}$$

Here A is a nonterminal and α, β, and γ are strings of terminals and nonterminals. Although α and β may be empty, γ must be nonempty. The complete string on the right must be at least as long as the complete string on the left.

Here are some derivations of sentences using the context-sensitive grammar (13.11):

$$\begin{aligned}
&S \to 012 \\
&S \to 0AS2 \to 0A0122 \to 00A122 \to 001122 \\
&S \to 0AS2 \to 0A0AS22 \to 0A0A01222 \to 00AA01222 \\
&\quad \to 00A0A1222 \to 000AA1222 \to 000A11222 \to 000111222
\end{aligned} \tag{13.13}$$

The set of all languages that are generated by context-sensitive grammars is a subset of so-called decidable languages. For each decidable language, there exists a Turing machine that takes as input any string of terminals and calculates whether or not this string is a valid sentence of this language. Hence the Turing machine can decide the grammaticality of any string.

A Turing machine is a general model of a computer. It has a read/write head and an infinite tape. In each position of the tape the entry 0 or 1 is stored. The head is a finite-state automaton. In any one elementary step of the computation, the head reads a position and then moves one increment either to the right or to the left. The head can modify the entry in each position. A Turing machine embodies the theoretical concept of a digital computer with infinite memory.

13.1.4 Phrase-Structure Grammars

Phrase-structure grammars have unrestricted rewrite rules,

$$\alpha \to \beta. \tag{13.14}$$

Both α and β can represent any finite string of terminals and nonterminals.

Phrase structure grammars generate the set of computable languages. For each computable language, there exists a Turing machine that can identify every sentence that is part of the language. If, however, the Turing machine receives as input a sentence that does not belong to the language, then it might compute forever. Hence the Turing machine cannot in principle decide whether a sentence is part of the language or not, because the computation may never stop.

Phrase-structure grammars are "Turing complete": for each Turing machine there exists a corresponding phrase-structure grammar and vice versa. Therefore phrase-structure grammars are as powerful as the set of all possible (digital) computers with infinite memory.

Let us construct an example for a computable language which is not decidable. Note that the set of all Turing machines can be enumerated. There are as many Turing machines as grammars. There are as many grammars as integers. Let M denote the integer representing a Turing machine. Let w denote a binary input string. Consider the language L which consists of all pairs (M, w) such that the Turing machine M accepts the input w:

$$L = (M, w). \tag{13.15}$$

There is no Turing machine that can compute whether or not M will accept w. Instead, we have to run the computation M using the input w and wait and see what happens. If the computation stops, then M has accepted w. If the computation does not stop, then we will never know. This is related to Turing's famous halting problem. But there obviously exists a computation that can list all pairs (M, w) that did come to a halt. Hence $L = (M, w)$ is a computable language. Computable languages that are not decidable are called semi-decidable: the grammatical sentences can be listed.

13.1.5 The Philosopher's Question

The example $L = (M, w)$, indicating a computable but undecidable language, is not very satisfying. It reminds me of the following story.

> A philosopher is visited by an angel who offers, "You can ask God a question." The philosopher is thrilled and wants to take his time. After a few days, the angel returns. The philosopher is ready: "I want to ask the following question: What is the pair of the best possible question that I could ask and its answer?" The angel

moves to God. God computes. The angel returns: "The best possible question that you could have asked is the question you did ask, and the answer is the answer I just gave you."

13.1.6 Chomsky and Gödel

There is a beautiful correspondence between languages, grammars, and machines. We have seen that the set of all grammars corresponds to the set of all digital computers with infinite memory and generates the set of computable languages. In this set, almost all languages are semi-decidable but not decidable. Decidable languages constitute only a small subset of all computable languages. Thus most languages that have a grammar cannot be decided by a Turing machine.

A subset of decidable languages is generated by context-sensitive grammars. A subset of those languages can be generated by context-free grammars. A subset of the context-free languages contains all regular languages. A subset of regular languages contains all finite languages.

This relationship among grammars is called the Chomsky hierarchy (Figure 13.6). It is named after the linguist Noam Chomsky, who introduced mathematical rigor in the study of human language. There are many different ways to represent rules and relationships in formal, discrete systems. But all rule systems are part of the Chomsky hierarchy. The Chomsky hierarchy is at the very foundation of computer science, mathematical logic, and formal language theory.

Any mathematical question is equivalent to asking whether a particular sentence is part of a language or not. A mathematical theorem is a sentence. The proof of a theorem is the derivation of this sentence via iterated application of the rewrite rules of the grammar. These rewrite rules are the axioms of a formal system.

There is an interesting link to Gödel's incompleteness theorem. You can envisage the language that contains all well-formed expressions of arithmetic. Here are three examples of well-formed expressions:

$$\begin{aligned} 3+5&=7 \\ 8/2&=3 \\ (1+4)*3&=15 \end{aligned} \tag{13.16}$$

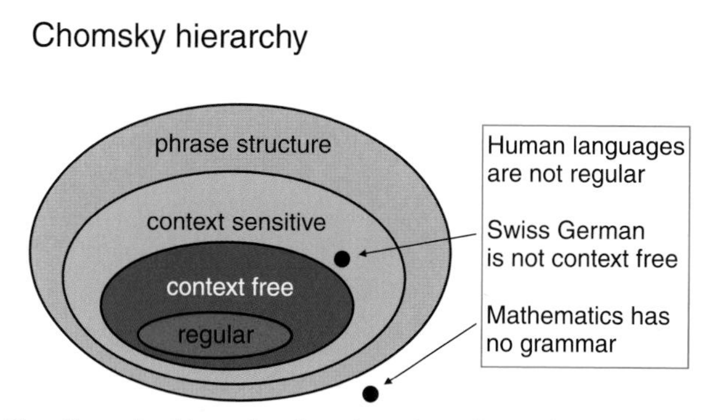

Figure 13.6 The Chomsky hierarchy describes the relationship among all possible grammars. Finite-state automata are equivalent to finite-state grammars and generate regular languages. Push-down automata are equivalent to context-free grammars and generate context-free languages. Linear-bounded automata are equivalent to context-sensitive grammars and are a subset of decidable languages. Turing machines are equivalent to phrase-structure grammars which generate semi-decidable languages. Linguists agree that human languages are not regular and that Swiss German is not context free. I argue that if Swiss German is not context free, then Austrian German is not entirely context free either.

All three expressions are well formed, but the first two are "wrong," while the third one is "correct." Here are examples of expressions that are not well formed:

$$
\begin{aligned}
&3+\ =4\\
&(2+6=)/3
\end{aligned}
\tag{13.17}
$$

It turns out that the language that contains all well-formed expressions of arithmetic can be generated by a context-free grammar. Gödel, however, proved that the language that contains exactly all correct expressions of arithmetic has no grammar. Thus Gödel's famous incompleteness theorem, which was first announced in a Viennese coffee house, implies that the language of mathematics is not computable.

Alas, the same God who talked to our philosopher must love practical jokes, to have ensured that mathematics, the most precise language of all, has no grammar.

13.1.7 A Place for Natural Languages

With the introduction of the Chomsky hierarchy, there was some interest in placing natural languages within this scheme. Natural languages are infinite: it is not possible to imagine a finite list that contains all English sentences. Furthermore, finite-state grammars are inadequate for natural languages. Such grammars are unable to represent long-range dependencies of the form "if . . . then." The string of words between "if" and "then" could be arbitrarily long and could itself contain more paired if-then constructions. Such pairings relate to rules that generate strings of the form $0^n 1^n$, which require context-free grammars.

There is a continuing debate whether context-free grammars are adequate for natural languages, or whether more complex grammars need to be evoked. Most linguists argue that human languages are mildly context sensitive: they need some context-sensitive rewrite rules, but they do not need the whole power of context-sensitive grammars.

The fundamental structures of natural languages are trees. The nodes represent phrases that can be composed of other phrases in a recursive manner. A tree is a "derivation" of a sentence within the rule system of a particular grammar. The interpretation of a sentence depends on the underlying tree structure. Ambiguity arises if more than one tree can be associated with a given sentence.

One can also define grammars that directly specify which trees are acceptable for a given language. There is a certain class of tree-adjoining grammars that is mildly context sensitive and that seems to be exactly as powerful as needed for the syntax of natural languages. Much of modern syntactic theory deals with grammars that specify rewrite rules of tree operations. All such grammars are, of course, part of the Chomsky hierarchy, and the results of learning theory, to be discussed next, apply to them.

13.2 LEARNING THEORY

Learning is inductive inference. The learner is presented with data and has to infer the rules that generate these data. The difference between "learning" and "memorization" is the ability to *generalize* beyond one's own experience to novel circumstances. In the context of language, the child will generalize to novel sentences never heard before. Any person can produce and understand sentences that are not part of his previous linguistic experience. Learning theory describes the mathematics of learning with the aim to outline conditions for successful generalization.

13.2.1 The Paradox of Language Acquisition

Children learn their native language by hearing grammatical sentences from their parents or others. From this "environmental input," children construct an internal representation of the underlying grammar. Children are not told the grammatical rules. Neither children nor adults are ever aware of the grammatical rules that specify their own language.

Chomsky pointed out that the environmental input available to the child does not uniquely specify the grammatical rules. This phenomenon is known as "poverty of stimulus." The "paradox of language acquisition" is that children of the same speech community reliably grow up to speak the same language. The proposed solution is that children learn the correct grammar by choosing from a restricted set of candidate grammars. The "theory" of this restricted set is "universal grammar" (UG). Formally, UG is not a grammar, but a theory of a collection of grammars.

The concept of an innate, genetically determined UG was controversial when introduced some fifty years ago and has remained so. The mathematical approach of learning theory, however, can explain in what sense UG is a logical necessity.

13.2.2 Learnability

Imagine a speaker-hearer pair. The speaker uses grammar, G, to construct sentences of language L. The hearer receives sentences and should after some time be able to use grammar G to construct other sentences of L. Mathematically speaking, the hearer is described by an algorithm (or more generally, a function), A, which takes a list of sentences as input and generates a language as output.

Let us introduce the notion of a "text" as a list of sentences. Specifically, text T of language L is an infinite list of sentences of L with each sentence of L

occurring at least once. Text T_N contains the first N sentences of T. We say that language L is learnable by algorithm A if for each text T of L we have

$$\lim_{N\to\infty} A(T_N) = L.$$

More precisely, for each text T there exists a number M such that for all $N > M$ we have $A(T_N) = L$. This means that given enough sentences as input, the algorithm will provide the correct language as output. This concept of learning is called "identification in the limit."

Furthermore, a set of languages is learnable by an algorithm, if each language of this set is learnable. We are interested in the question, What set of languages, $\mathcal{L} = \{L_1, L_2, \ldots\}$, can be learned by a given algorithm?

Gold's theorem was formulated in 1967 and represents a key result of learning theory. Consider a set of languages that contains all finite languages and at least one infinite language. Such a set is called "super-finite." Gold's theorem implies that there exists no algorithm that can learn a super-finite set of languages. Let us discuss the intuition behind the proof.

We can enumerate the set of all finite languages. Hence, we imagine an infinitely long list that contains all finite languages: $L_1, L_2, \ldots$. Let us add an infinite language that we call L_∞. The algorithm can consider all finite languages in the order of our listing, but at some stage it has to consider the infinite language, L_∞. The algorithm cannot consider L_∞ at the very end after having rejected all finite languages, because the list has no end. Moreover, L_∞ cannot be considered after all finite languages that are subsets of L_∞, because there are also infinitely many such languages. Thus the algorithm has to consider L_∞ at a particular time and before some of its finite subsets.

Suppose the algorithm considers languages in the following order:

$$L_1, L_2, L_3, \ldots, L_n, L_\infty, \ldots, L_k, \ldots \tag{13.18}$$

Imagine the teacher uses L_k, which is a finite subset of L_∞. The learner starts with the hypotheses L_1. As long as all sentences of the text are compatible with L_1, the learner stays there. If a sentence arrives that is not compatible with L_1, the learner moves to L_2. If all sentences are compatible with L_2, the learner stays there. Otherwise the learner moves to L_3 and so on. After having rejected the first n languages, the learner entertains the hypothesis L_∞. Since the teacher uses L_k, which is a subset of L_∞, no sentence will ever arrive that is incompatible with L_∞. Hence the learner will stay with this hypothesis forever and has thus converged to the wrong language.

The critical reader might be dissatisfied with this argument, because it relies on the subtle difference between a large finite language and its infinite superset. What could be less important for understanding natural language acquisition, one may think.

Adding the infinite language is, however, only an elegant trick to exclude the possibility of learning by memorization. Let us try to construct an algorithm that can learn the set of all finite languages. The learner simply memorizes all the sentences that are being presented by the teacher. At any one time the learner holds the hypothesis that the target language consists of exactly those sentences that are stored in her memory. This learner will identify all finite languages in the limit, but she will never guess an infinite language. Thus, by adding an infinite language, Gold excludes the possibility of learning by memorization.

Memorization is the wrong concept for natural language acquisition: the learner will identify the correct language only after having heard all sentences of this language. A learner who considers the set of all finite languages has no possibility of generalization: the learner can never extrapolate beyond the sentences she has already encountered. This is not the case for natural language acquisition: we can always say new sentences that we have never heard before. Hence Gold's clever theorem can also be interpreted in the following way: there exists no learner who can generalize on the set of all finite languages.

We can even extend the framework to consider a finite set of finite languages. Suppose there are only four sentences in the world: S_1, S_2, S_3, S_4. Hence there are sixteen possible languages. Learner A considers all sixteen languages as *a priori* possibilities, while learner B considers only two languages, for example, $L_1 = \{S_1, S_3\}$ and $L_2 = \{S_2, S_4\}$. If learner A receives sentence S_1, he has no information on whether sentences S_2 or S_3 will be part of the target language or not. He can only identify the target language after having heard all its sentences. If learner *B* receives sentence S_1, she knows that S_3 will be part of the language, while S_2 and S_4 will not. She can extrapolate beyond her

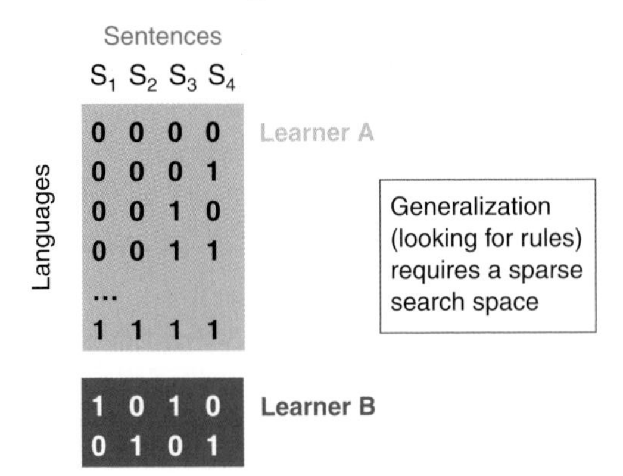

Figure 13.7 Imagine that there are only four sentences, S_1 to S_4. A language is a set of sentences. There are $2^4 = 16$ possible languages. One of these languages boasts all four sentences, while another (quiet) language has no sentence at all. Learner A entertains all sixteen languages as possible candidates in his effort to learn the target language. Any one sentence that is received by learner A does not provide information regarding the grammaticality of any other sentence. Learner A can converge to the target language by memorizing all the sentences that he has received. In contrast, learner B has only two candidate languages. If learner B notes that sentence S_4 is part of the target language, then she will know that S_2 is also part of this language, but not S_1 and S_3, because she knows that this language contains only S_2 and S_4. This simple example shows that a restricted search space is needed to generalize from one sentence to another. Generalization is an important part of human language acquisition. We can say sentences that we have not heard before. An unrestricted search space only allows learning by memorization. A restricted search space is required for learning by generalization.

experience. The ability to search for underlying rules requires a restricted search space (Figure 13.7).

Gold's framework is often formulated as identifying the correct language, but instead we can ask the learner to converge to a grammar that generates the correct language. The theorem holds also for this slightly modified situation: there is no algorithm that can learn a set of grammars that generates a super-finite set of languages. Note that finite-state grammars, context-free grammars, context-sensitive grammars, and phrase-structure grammars all generate super-finite sets of languages. Hence none of these sets can be learned by any algorithm.

Moreover, Gold's theorem holds in the greatest possible generality: "algorithm" here does not have to embody a computational device; it can be any function from text to language. Gold's theorem implies there is no mapping from text to language that can learn a super-finite set of languages.

13.2.3 Probably Almost Correct

Much criticism has been leveled against the Gold framework over the years: (i) the learner has to identify the target language exactly; (ii) the learner receives only positive examples; (iii) the learner has access to an arbitrarily large number of examples; (iv) the learner is not limited by any consideration of computational complexity. Assumptions (i) and (ii) are restrictive: relaxing these assumptions will enable particular learners to succeed on larger sets of languages. Assumptions (iii) and (iv) are unrestrictive: relaxing these assumptions will reduce the set of learnable languages. In fact, each assumption has been removed in various approaches to learning, but the essential conclusion remains the same: no algorithm can learn an unrestricted set of languages.

Perhaps the most significant extension of the classical framework is statistical learning theory. Here the learner is required to converge approximately to the right language with high probability. A deep result by Vladimir Vapnik and Alexey Chervonenkis states that a set of languages is learnable if and only if it has a finite VC dimension. The VC dimension is a combinatorial measure of the complexity of a set of languages. Thus if the set of possible languages is completely arbitrary (and therefore has infinite VC dimension), learning is not possible. It can be shown that the set of all regular languages (even the set of all finite languages) has infinite VC dimension and hence cannot be learned by any procedure in the framework of statistical learning theory. Subsets of regular languages that are generated by finite-state automata with n states, however, have finite VC dimension and one can estimate bounds on the number of sample sentences that are needed for learning.

Statistical learning theory in the VC framework removes assumptions (i), (ii), and (iii) of the Gold framework: it does not ask for convergence to exactly the right language, the learner receives positive and negative examples, and the learning process has to end after a certain number of examples. The theory provides bounds for how many example sentences are needed to converge approximately to the right language with high probability; this is the concept of informational complexity.

Les Valiant also added considerations of computational complexity, thereby removing assumption (iv) of the Gold framework: the learner is required to approximate the target grammar with high confidence using an efficient algorithm. Consequently, there are sets of languages that are learnable in principle (have finite VC dimension), but no algorithm can do this in polynomial time. Computer scientists consider a problem "intractable" if no algorithm exists that can solve this problem in polynomial time, which means in a number of time steps that is proportional to the size of the input raised to some power.

Some other models of learning deserve mention. For example, in one form of query-based learning, the learner is allowed to ask whether a particular sentence is in the target language or not. In this model, regular languages can be learned in polynomial time, but context-free languages cannot. Other query-based models of learning, with varying degrees of psychological plausibility, have been considered, and none permit all languages to be learnable. In summary, all extensions of learning theory underline the necessity of specific restrictions.

13.2.4 The Necessity of Innate Expectations

We can now state in what sense there has to be an innate UG. The human brain is equipped with a learning algorithm, A_H, which enables us to learn certain languages. This algorithm can learn each of the existing six thousand human languages and presumably many more, but it is impossible for A_H to learn every language (or grammar). Hence there is a restricted set of languages (grammars) that can be learned by A_H. UG is the theory of this restricted set (Figure 13.8).

Learning theory suggests that a restricted search space has to exist prior to data. By "data," we mean linguistic or other information the child uses to learn language or modify its language acquisition procedure. In our terminology, therefore, "prior to data" is equivalent to "innate." In this sense, learning theory shows there must be an innate UG, which is a consequence of the particular learning algorithm, A_H, used by humans. Discovering properties of A_H requires the empirical study of neurobiological and neurocognitive

Human language learning

- The human brain contains an algorithm, $\mathbf{A}_H$ that can learn language.
- The question is what is the set, $\mathbf{L}_H$, that can be learned by this algorithm?

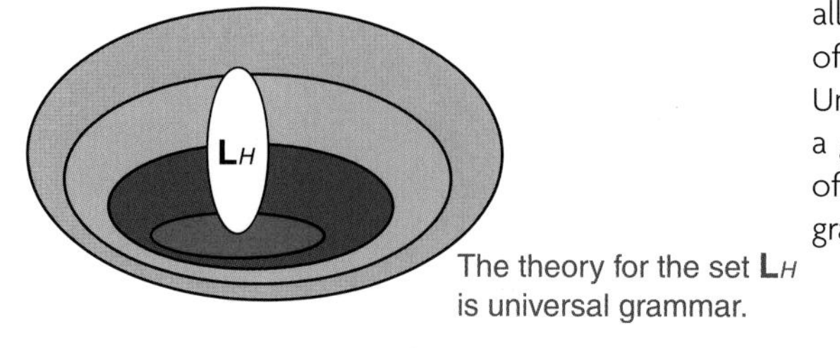

LH
The theory for the set LH is universal grammar.

Figure 13.8 The logical necessity of universal grammar. The human brain contains an algorithm that can learn grammar. There exists no algorithm that can learn an unrestricted class of grammars. Hence the human brain can only learn a certain subset of all possible grammars. The theory of this subset is universal grammar. Universal grammar is not necessarily a grammar, because a description of a set of grammars need not be a grammar.

functions of the human brain involved in language acquisition. Some aspects of UG, however, might be unveiled by studying common features of existing human languages. This has been a major goal of linguistic research during recent decades. A particular approach is the "principles and parameters theory," which assumes that the child comes equipped with innate principles and has to set parameters that are specific for individual languages. Another approach is "optimality theory," where learning a specific language is ordering innate constraints.

There is some discourse regarding whether the learning mechanism, A_H, is language specific or general purpose. Ultimately this is a question about the particular architecture of the brain and which neurons participate in which computations, but one cannot deny that there is a learning mechanism, A_H, that operates on linguistic input and enables the child to learn the rules of human language. This mechanism can learn a restricted set of languages; the theory of this set is UG. The continuing debate regarding an innate UG should not be whether there is one, but what form it takes. One can dispute individual linguistic universals, but one cannot generally deny their existence.

Neural networks are an important tool for modeling the neural mechanisms of language acquisition. The results of learning theory obviously apply to neural networks. No neural network can learn an unrestricted set of languages.

Sometimes it is claimed that the logical arguments for an innate UG rest on particular mathematical assumptions of generative grammars which deal only with syntax and not with semantics. Cognitive and functional linguistics are not based on formal language theory, but use psychological objects such as symbols, categories, schemas, and images. This does not remove the necessity of innate restrictions. The results of learning theory apply to any learning process where a "rule" has to be learned from some examples. Generalization is an inherent feature of any model of language acquisition and applies to semantics, syntax, and phonetics. Any procedure for successful generalization has to choose from a restricted range of hypotheses.

The results of learning theory also apply to learning mappings between linguistic form and meaning. If meaning is to be explicitly considered, then a language is not a set of sentences, but a set of sentence–meaning pairs. The task of language acquisition is then to learn which sentence–meaning pairs are "correct" (= part of the language) and which are not. Languages that describe form–meaning pairs are also part of the Chomsky hierarchy, and there exists no learning procedure that can succeed on an unrestricted set of such languages.

13.2.5 What Is Special about Language Acquisition?

Usually when we learn the grammar of generative systems, such as chess or arithmetic, somebody tells us the rules. We do not have to guess the moves of chess by looking at chess games. In contrast, the process of language acquisition occurs without being instructed about rules; neither teachers nor learners are aware of the rules. This is an important difference: if the learner is told the grammar of a language, then the set of all computable languages is learnable by an algorithm that memorizes the rules.

13.3 EVOLUTION OF GRAMMAR

We will now construct a theory that describes the deterministic evolutionary dynamics of grammar. We will calculate a condition for the evolution of grammatical coherence, which is similar to the error threshold of genetic evolution (Chapter 3).

Consider a learner with a search space that consists of n candidate grammars,

$$G_1, \ldots, G_n. \tag{13.19}$$

Each grammar, G_i, is a rule system that defines a set of valid sentences. Furthermore, a speaker who uses G_i will generate these sentences according to some probability distribution. Thus we can say that grammar G_i induces a measure, μ_i, on the set of all sentences. This measure has support (= is positive) on all sentences that are compatible with G_i. It is zero on all other sentences.

Consider two grammars G_i and G_j. The set of common sentences is given by the intersection $G_i \cap G_j$. The probability that a speaker who uses grammar G_i formulates a sentence that is compatible with grammar G_j is given by

$$a_{ij} = \mu_i(G_i \cap G_j). \tag{13.20}$$

This quantity denotes the weight of the probability distribution induced by G_i on the set of common sentences between G_i and G_j. In contrast, the probability that a speaker who uses grammar G_j formulates a sentence that is compatible with grammar G_i is given by

$$a_{ji} = \mu_j(G_i \cap G_j). \tag{13.21}$$

Note that a_{ij} is not necessarily identical to a_{ji}. The probability that a speaker of G_i formulates a sentence that is understood by another person who uses the same grammar, G_i, is given by

$$a_{ii} = \mu_i(G_i) = 1. \tag{13.22}$$

In this model, communication is perfect between any two individuals who use the same grammar. All grammars in the search space have this property. This assumption can be relaxed in more complicated models.

The matrix $A = [a_{ij}]$ describes the pairwise relationship among the n grammars. We have $0 \leq a_{ij} \leq 1$ and $a_{ii} = 1$. Matrix A is not necessarily symmetric.

We assume there is a reward for mutual understanding. The payoff for an individual using G_i communicating with an individual using G_j is given by

$$F(G_i, G_j) = \frac{1}{2}(a_{ij} + a_{ji}). \tag{13.23}$$

This is the average probability that G_i generates a sentence that is parsed by G_j and vice versa. Note that $F(G_i, G_i) = 1$. Thus all n grammars are equally powerful and allow the same level of communication.

13.3.1 The Replicator-Mutator Equation

We denote by x_i the frequency of individuals who use grammar G_i. The average payoff of each of these individuals is given by

$$f_i(\vec{x}) = \sum_{j=1}^{n} x_j F(G_i, G_j). \tag{13.24}$$

We assume that payoff translates into reproductive success: individuals with a higher payoff produce more offspring. Reproduction can be genetic or cultural.

The learning process can be subject to mistakes. Denote by Q_{ij} the probability that a child learning from a parent with grammar G_i will end up speaking grammar G_j. Obviously, Q is a stochastic matrix. With these assumptions the population dynamics are given by

$$\dot{x}_i = \sum_{j=1}^{n} x_j f_j(\vec{x}) Q_{ji} - \phi x_i \qquad i = 1, \ldots, n \tag{13.25}$$

Here $\phi = \sum_i x_i f_i$ is the average fitness or *grammatical coherence* of the population; it is the probability that a sentence said by one person is understood by another person. The total population size is constant; we have $\sum_i x_i = 1$.

Equation (13.25) is a "replicator-mutator equation." It contains both the quasispecies equation and the replicator equation as special cases. In the limit of perfect learning, where Q is the identity matrix, with all diagonal entries 1 and all off-diagonal entries 0, we obtain the replicator equation. In the limit of constant selection, where the fitness values f_i do not depend on the composition of the population, we obtain the quasispecies equation. The replicator-

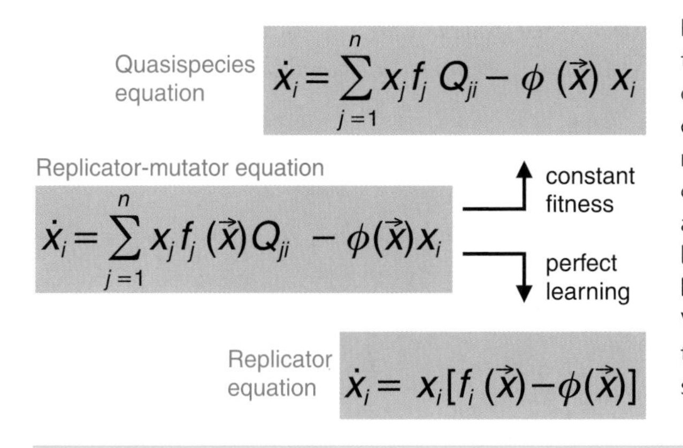

Quasispecies
equation
$\dot{x}_i = \sum_{j=1}^{n} x_j f_j Q_{ji} - \phi(\vec{x}) x_i$
Replicator-mutator equation
$\dot{x}_i = \sum_{j=1}^{n} x_j f_j(\vec{x}) Q_{ji} - \phi(\vec{x}) x_i$
constant
fitness
perfect
learning
Replicator
equation
$\dot{x}_i = x_i[f_i(\vec{x}) - \phi(\vec{x})]$

Figure 13.9 The language equation is a unifying description of deterministic evolutionary dynamics. It is also called the replicator-mutator equation. We obtain the replicator equation as the special case of perfect learning: the mutation matrix Q becomes the identity matrix I. We obtain the quasispecies equation as a special case of constant selection.

mutator equation is a general description of deterministic evolutionary dynamics, including frequency-dependent selection and mutation (Figure 13.9).

For an A matrix of the form $a_{ii} = 1$ and $0 \le a_{ij} < 1$, equation (13.25) can have multiple stable or unstable equilibria. For learning without mistakes, where Q is the identity matrix, there exists n asymmetric equilibria of the form $x_i = 1$ and $x_j = 0$ for all $j \neq i$. These equilibria are asymptotically stable. Such solutions correspond to situations where all individuals of a population have adopted the same grammar. In contrast, for high error rates the only stable equilibrium solution is one where all grammars occur at similar frequencies. We want to analyze the following question: How accurate does the learning process have to be for most individuals of the population to use the same grammar? In other words, when does a universal grammar induce coherent grammatical communication?

13.3.2 Super-Symmetry

Consider the special case where all grammars have the same distance from one another, hence

$$a_{ii} = 1 \quad \text{and} \quad a_{ij} = a \qquad \forall i \neq j \tag{13.26}$$

Here a is a number between 0 and 1. In accordance, we have

$$Q_{ii} = q \quad \text{and} \quad Q_{ij} = \frac{1-q}{n-1} \qquad \forall i \neq j \tag{13.27}$$

Here q is the accuracy of grammar acquisition: it denotes the probability of learning the correct grammar. The probability of learning an incorrect grammar is $u = (1 - q)/(n - 1)$. For this learning matrix, Q, the replicator-mutator equation can be written as

$$\dot{x}_i = x_i[f_i(q - u) - \phi] + u\phi. \tag{13.28}$$

The payoff for individuals using grammar G_i is given by

$$f_i = a + (1 - a)x_i. \tag{13.29}$$

For the average payoff (grammatical coherence) we obtain

$$\phi = \sum_i x_i f_i = a + (1 - a)\sum_i x_i^2. \tag{13.30}$$

One equilibrium of the super-symmetric replicator mutator equation is given by

$$x_1 = \cdots = x_n = 1/n \tag{13.31}$$

Here all grammars, $G_1, \ldots, G_n$, have the same frequency. This equilibrium always exists.

In addition there are asymmetric equilibria, where one grammar has frequency X and all other grammars have equal shares of the rest of the population. Hence these asymmetric equilibria are of the form

$$x_i = X \quad \text{and} \quad x_j = \frac{1 - X}{n - 1} \qquad \forall i \neq j \tag{13.32}$$

It is possible to show that such asymmetric equilibria exist and are stable, provided that the accuracy of grammar acquisition, q, exceeds a threshold value given by

$$q_1 = \frac{2\sqrt{a}}{1 + \sqrt{a}}. \tag{13.33}$$

The symmetric solution loses its stability when q exceeds the threshold

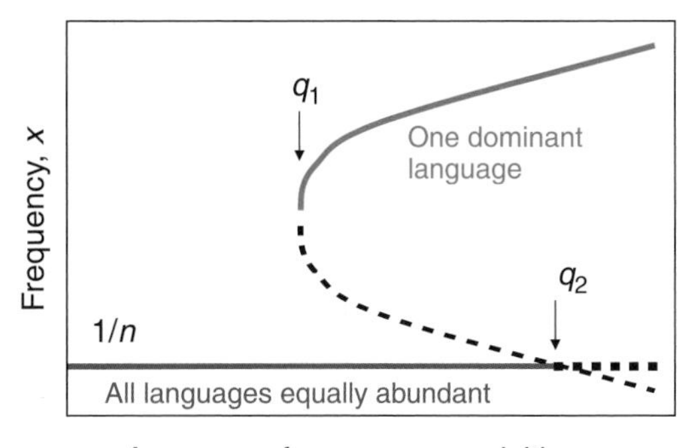

Figure 13.10 In the super-symmetric case, there is a simple bifurcation diagram. If the accuracy of grammar acquisition is less than a critical value, q_1, then the only stable equilibrium is the uniform distribution: all grammars are equally abundant in the population, and linguistic coherence is low. If the accuracy of grammar acquisition is greater than the critical value, q_1, then in addition to the uniform equilibrium there are n asymmetric equilibria, where one of the grammars is more abundant than all others. If the accuracy of grammar acquisition is greater than a second critical value, q_2, then only the asymmetric equilibria are stable. The broken lines indicate unstable equilibria.

$$q_2 = 1 - \frac{1-a}{na}. \tag{13.34}$$

These results hold for $n \gg 1/a$. If instead $1/a > n \gg 1$ then $q_1 = 2/\sqrt{n}$ and $q_2 = 1/2$.

Therefore if $q < q_1$, only the symmetric equilibrium is stable. If $q_1 < q < q_2$, then both the symmetric and the n asymmetric equilibria are stable; which of the $n+1$ equilibria will be adopted depends on the initial condition. Finally, for $q > q_2$, only the asymmetric equilibria are stable. Hence $q > q_1$ is a necessary condition for the population to converge to a coherent grammar, while $q > q_2$ is a sufficient condition (Figure 13.10).

These conditions specify a "coherence threshold" for universal grammar. In general, q will be a declining function of n. Therefore $q(n) > q_1$ is an implicit condition for the maximum size of the search space generated by universal grammar. The coherence threshold is a necessary condition for the evolution of complex language: only a universal grammar that satisfies the coherence threshold can lead to the emergence of grammatical communication.

13.3.3 Memoryless and Batch Learners

Let us stay with the super-symmetric language equation and calculate the coherence threshold for two specific learning procedures that determine how to evaluate the input sentences. We will consider a "memoryless learner" and a "batch learner." The first is the least powerful, the second the most powerful mechanism within a range of reasonable possibilities. Whatever the actual learning mechanism that is used by the human brain, it will have a performance better than the memoryless learner and worse than the batch learner.

The "memoryless learner" algorithm describes the interaction between a learner and a teacher. Suppose the teacher uses grammar G_k. The learner starts with a randomly chosen hypothesis, G_i. The teacher generates sentences consistent with G_k. As long as these sentences are also consistent with G_i, the learner maintains his hypothesis. If a sentence occurs that is not consistent with G_i, the learner picks at random a different hypothesis, G_j. After N sample sentences, the process stops, and the learner remains with his current hypothesis. This learning algorithm defines a Markov process. The transition probabilities depend on the teacher's grammar and on the a_{ij} values. For the special case where $a_{ij} = a$ for all $i \neq j$, and $n \gg 1$, the probability of learning the teacher's grammar after N sentences is

$$q = 1 - \left(1 - \frac{1-a}{n}\right)^N. \tag{13.35}$$

The threshold values, q_1 and q_2, can be restated in terms of the minimum number of sampling sentences per individual required for the population to converge to a coherent grammar. From $q > q_1$, we obtain

$$N > \frac{n}{1-a} \log \frac{1+\sqrt{a}}{1-\sqrt{a}}. \tag{13.36}$$

Hence for a memoryless learner, the number of sample sentences has to exceed a constant multiplied by the number of candidate grammars.

The memoryless learner makes the minimum demand on the cognitive ability of the individual. The other extreme is a "batch learner," who memorizes N sentences and then chooses the grammar that is most consistent with all memorized sentences. For the batch learner, we can show that the probability of learning the correct grammar, in a generic case, is given by

$$q = \frac{1 - (1 - a^N)^n}{na^N}. \tag{13.37}$$

Together with $q > q_1$, this leads to

$$N > \frac{\log n}{\log(1/a)}. \tag{13.38}$$

Hence a batch learner requires that the number of sample sentences exceeds a constant multiplied by the logarithm of the number of candidate grammars.

Since any realistic learning procedure has a performance somewhere between "memoryless learners" and "batch learners," equations (13.36) and (13.38) provide boundaries for the maximum size of the search space that is compatible with grammatical coherence within a population.

13.3.4 Breaking Super-Symmetry

Let us now consider the situation in which the candidate grammars, $G_1, \ldots, G_n$, have different distances from one another. Figure 13.11 shows the equilibrium solutions for a case with $n = 50$ candidate grammars, where the numbers a_{ij} are randomly chosen from a uniform distribution on (0, 1). For a small number of sample sentences, N, all grammars occur at roughly the same frequency and the grammatical coherence of the population is low. As N increases, equilibrium solutions become stable where the majority of the population uses a particular grammar. The critical transition occurs at an N value that is approximately given by equation (13.36) with $a = 1/2$. It can be shown that there are exactly n stable, one-grammar solutions if N is large enough and $a_{ij} < 1$ for all $i \neq j$.

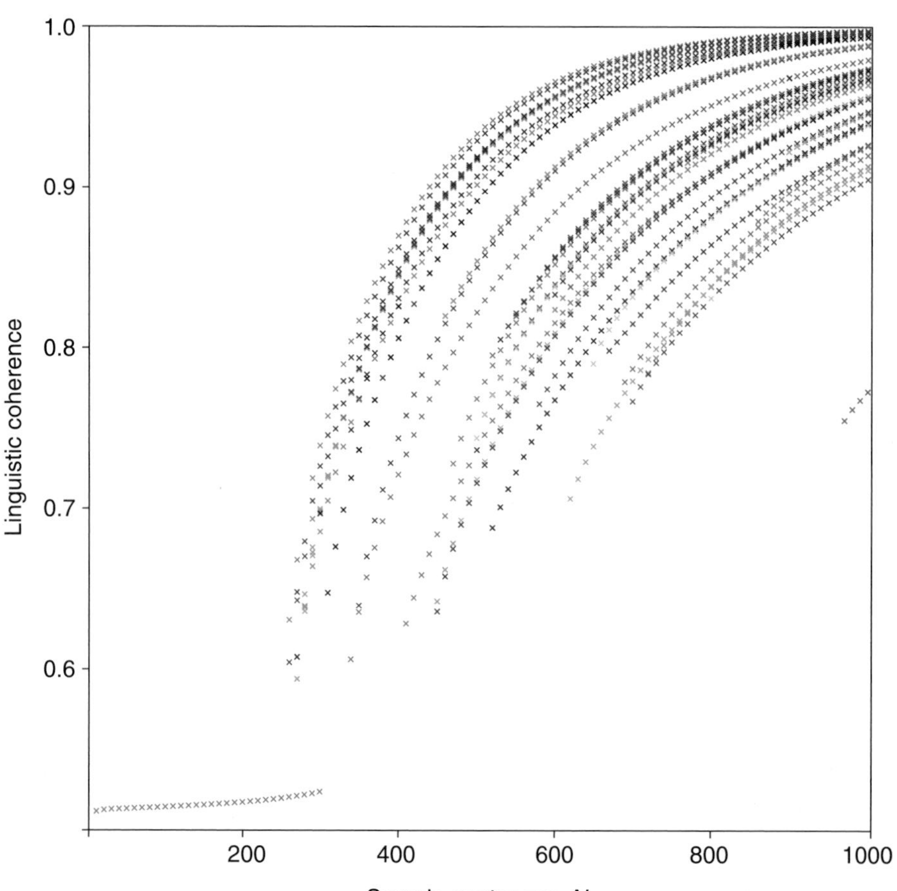

Figure 13.11 Grammatical coherence of a population versus the number of sample sentences, N, per individual for stable equilibrium solutions of equation (13.25). There are $n = 50$ grammars with randomly chosen pairwise distances; the a_{ij} values (for $i \neq j$) are taken from a uniform distribution on $[0, 1]$, and $a_{ii} = 1$. Children learn the grammar of their parents according to a memoryless learning algorithm. The grammatical coherence (or average fitness) of the population is given by $\phi = \sum_i x_i f_i$, where $f_i = (1/2) \sum_j x_j (a_{ij} + a_{ji})$. It is a measure of mutual understanding in the population. For small N all grammars occur at roughly similar frequency; the coherence is low (see red line in lower left corner). For larger values of N, stable equilibria appear with the majority of the population adopting the same grammar. Some grammars lead to stable equilibrium solutions only for large numbers of sample sentences. In the limit $N \to \infty$, there are n stable equilibria corresponding to all people using one of the n grammars.

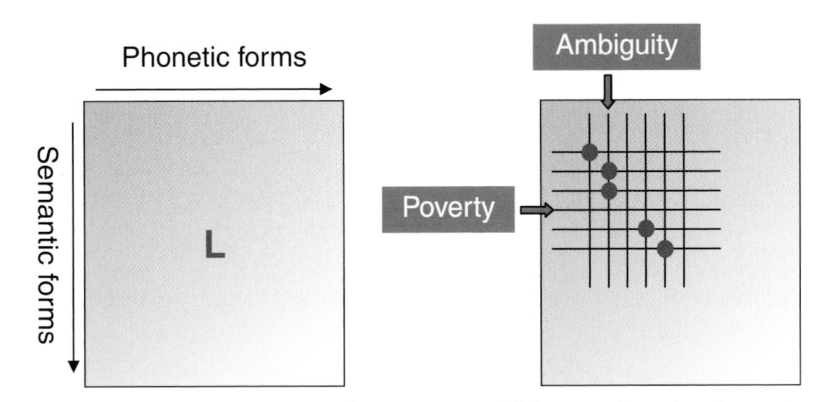

Figure 13.12 Language is a mapping between sound (phonetic forms) and meaning (semantic forms). A particular language may not encode all meanings and may not make use of all possible sounds. The ambiguity of a language is the loss of communicative capacity that arises if individual sounds are linked to more than one meaning. The poverty of a language is the fraction of meanings (measured by relevance) that cannot be expressed. The blue dots indicate sound–meaning pairs that make up a particular language.

13.3.5 Poverty and Ambiguity

Language is a mapping between sound and meaning (Figure 13.12). The mathematical formalism of language can be extended to include aspects of communicative behavior and performance. A language can be seen as an infinite matrix, L, that specifies mappings between phonetic forms and semantic forms that are "sound" and "meaning" insofar as they are linguistically determined. This matrix defines linguistic competence. For evaluating communicative success, we also need to describe linguistic performance. We assume the matrix, L, leads to a pair of matrices, P and Q, determining speaking and hearing. The element p_{ij}^I denotes the probability for a speaker of L_I to use sound j for encoding meaning i. The element q_{ij}^I denotes the probability for a hearer to decode sound j as meaning i. We have $\sum_j p_{ij}^I \leq 1$ and $\sum_i q_{ij}^I \leq 1$; a language may not encode all meanings and may not make use of all possible sounds.

Next we introduce a measure, σ, on the set of all meanings. Let us denote by σ_i the probability that communication is about meaning i. The measure σ

depends among other things on the environment, behavior, and phenotype of the individuals. It also defines which meanings are more relevant than others.

The probability that a speaker of L_I generates a sound that is understood by a hearer using L_J is given by $a_{IJ} = \sum_{ij} \sigma_i p_{ij}^I q_{ij}^J$. The communicative payoff between L_I and L_J can be defined as $F_{IJ} = \frac{1}{2}(a_{IJ} + a_{JI})$. The intrinsic communicative payoff of L_I is $F_{II} = a_{II}$.

In this framework, communicative payoff is a number between 0 and 1. A payoff of less than 1 arises as a consequence of ambiguity and poverty. Ambiguity, α_I, of language L_I is the loss of communicative capacity that arises if individual sounds are linked to more than one meaning. Poverty, β_I, is the fraction of meanings (measured by σ) that are not part of L_I. The communicative capacity of L_I can be written as $F_{II} = (1 - \alpha_I)(1 - \beta_I)$.

For language acquisition we need a measure for the similarity, s_{IJ}, between languages L_I and L_J. A possibility is $s_{IJ} = \sum_{ij} \sigma_i p_{ij}^I q_{ij}^J / \sum_{ij} \sigma_i p_{ij}^I$, denoting the probability that a sound generated by a speaker of L_I is correctly interpreted by a hearer using L_J. In this context, the similarity between L_I and L_J declines because of ambiguity, but not because of poverty. Ambiguity implies that a learner holding the correct hypothesis might think he is wrong and change his hypothesis. This has consequences for the notion of consistent learning; a consistent learner does not change his hypothesis if he already holds the correct hypothesis.

The candidate grammars could differ in their overall performance. Some grammars could describe a larger number of concepts or be less ambiguous than others. Hence candidate grammars can have different fitness values. In such a scenario, the one-grammar solutions assume different fitness values even for large N. Therefore we can imagine an evolutionary process where the population is searching for fitter candidate grammars. Suppose a population uses a particular grammar, G_1. Someone invents a modification that alters the grammar to G_2. A fluctuation could shift the whole population to adopt G_2. Such transitions are more likely to occur in a small population. They are favored if the two grammars are fairly similar and G_2 has a higher fitness than G_1. Hence the model provides a framework for studying the cultural, evolutionary adaptation of grammar within the same universal grammar.

13.4 EVOLVING A NEW RULE

Let us study the evolution of a new rule. Consider a population where everybody speaks the same language. A new linguistic feature arises, which confers a fitness advantage, s, whenever there is a communication between two individuals who use the new invention. This fitness advantage can be caused by increased linguistic performance or efficiency or simply by conferring an elevated status in the eyes of the observers. Others want to learn the new rule, which might require generalizing from examples. The probability of successful learning is given by q.

Denote by x the fraction of people who use the new feature. The remaining fraction of the population is given by y. Clearly, we have $x + y = 1$. The (cultural or biological) fitness of x and y individuals is respectively given by

$$f_x = 1 + sx \quad \text{and} \quad f_y = 1. \tag{13.39}$$

For the average fitness of the population, we obtain

$$\phi = xf_x + (1-x)f_y = 1 + sx^2. \tag{13.40}$$

Evolutionary dynamics are given by

$$\begin{aligned} \dot{x} &= f_x q x - \phi x \\ \dot{y} &= f_x(1-q)x + f_y y - \phi y \end{aligned} \tag{13.41}$$

An equilibrium of this differential equation is $x = 0$ and $y = 1$. This equilibrium always exists and is stable. A second equilibrium is given by the solution of the quadratic equation

$$sx^2 - sqx + 1 - q = 0. \tag{13.42}$$

Solutions in the real numbers exist if

$$q > 2(-1 + \sqrt{1+s})/s. \tag{13.43}$$

For small s, this inequality leads to

$$q > 1 - \frac{s}{4}. \tag{13.44}$$

This condition determines the minimum probability of successful learning that is needed for the new rule to be maintained in a large population.

Suppose the new rule has to be acquired by a memoryless learner. When generalizing on the data, the learner might entertain a total of n hypotheses. Denote by a the probability that a sentence demonstrating the new rule is consistent with any of the hypotheses. Thus the probability that the learner receives a sentence which is not compatible with his current hypothesis and jumps to the right hypothesis is simply given by $(1-a)/n$. The probability that this is not happening after N sample sentences is $[1-(1-a)/n]^N$. The probability that it did happen within N sentences is

$$q = 1 - \left(1 - \frac{1-a}{n}\right)^N. \qquad (13.45)$$

This quantity denotes the probability of successful generalization.

Combining equations (13.44) and (13.45), we obtain

$$N > \frac{n}{1-a} \log \frac{4}{s}. \qquad (13.46)$$

Again we obtain a linear relationship between the size of the search space, n, and the number of sample sentences, N, which are required for a new linguistic feature to be maintained in a population. The ability to hold on to advantageous traits is a necessary, but not a sufficient, condition for evolving them.

13.5 EVOLUTION OF UNIVERSAL GRAMMAR

To further illuminate the selective pressures that act on the design of universal grammar, we study the competition between different universal grammars (Figure 13.13). We state two specific results.

First, consider universal grammars with the same search space and the same learning procedure, the only difference being the number of input sentences, N. This quantity is proportional to the length of the learning period. We find that natural selection leads to intermediate values of N. For small N, the accuracy of learning the correct grammar is too low. For large N, the learning

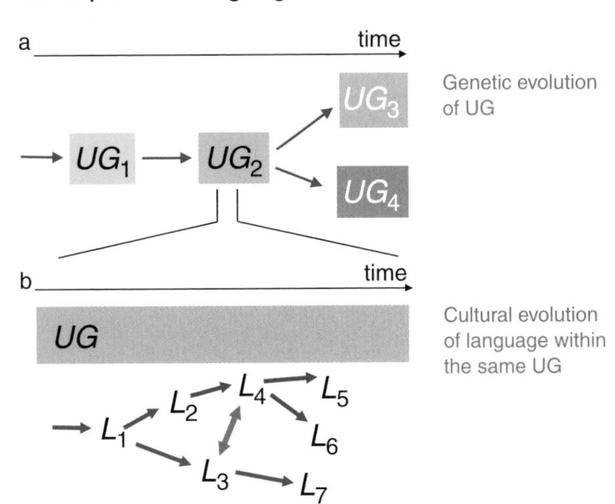

Figure 13.13 Two aspects of language evolution. (a) There is a biological evolution of universal grammar (UG) via genetic modifications that affect the architecture of the human brain and the class of languages it can learn. UG can change as a consequence of (i) random variation (neutral evolution), (ii) as a by-product of selection for other cognitive functions or (iii) under selection for language acquisition and communication. At some point in the evolutionary history of humans, a UG arose that allowed languages with infinite expressibility. (b) On a faster time scale, there is cultural evolution of language constrained by a constant UG. Languages change by (i) random variation, (ii) by contact with other languages (red arrow), (iii) by hitch-hiking on other cultural inventions, or (iv) by selection for increased learnability and communication. Although many language changes in historical linguistics may be neutral, a global picture of language evolution must include selection.

process takes too long and is too costly. This observation can explain why there is a limited language acquisition period in humans.

Second, consider universal grammars that differ in the size of their search space, n, but have the same learning mechanism and the same value of N. In general, there is selection pressure to reduce n. Only if n is below the coherence threshold can the universal grammar induce grammatical communication. In addition the smaller n is, the larger is the accuracy of grammar acquisition. There can, however, also be selection for larger n: suppose universal grammar

U_1 is larger than U_2 (that is, $n_1 > n_2$). If all individuals use a grammar, G_1, that is in both U_1 and U_2, then U_2 is selected. Now imagine that someone invents a new advantageous grammatical concept that leads to a modified grammar G_2 that is in U_1, but not in U_2. In this case, the larger universal grammar is favored. Hence there is selection both for reducing the size of the search space and for remaining open minded to be able to learn new concepts. For maximum flexibility, we expect search spaces to be as large as possible but still below the coherence threshold.

13.6 EVOLUTION OF RECURSION

Finally, we explore the conditions under which natural selection favors the emergence of a rule-based, recursive grammatical system with infinite expressibility. In contrast to such rule-based grammars, one might consider list-based grammars that consist only of a finite number of sentences. Such list-based grammars can be seen as very primitive evolutionary precursors (or alternatives) to rule-based grammars. Individuals would acquire their mental grammar not by searching for underlying rules, but by simply memorizing sentence types and their meaning (similar to memorizing the arbitrary meaning of words). List-based grammars do not allow for creativity on the level of syntax. Nevertheless, whether or not natural selection favors the more complicated rule-based grammars depends on circumstances that we need to explore.

Current human grammars can generate infinitely many sentence types, but for the purpose of transmitting information only a finite number of them can be relevant. Natural selection cannot directly reward the theoretical ability to construct infinitely long sentences. Let us therefore consider a group of individuals who use M different sentence types (or syntactic structures). Note that M specifies the number of sentence types that are relevant from the perspective of biological fitness.

Now imagine individuals who learn their mental grammar by memorizing lists of sentence types. We can ask how many sample sentences, N, a child must hear for the whole population to maintain M sentence types. If all sentence types occur equally often, we simply obtain $N > M$.

We can compare the performance of individuals using list-based versus rule-based grammars. Using the result for batch learners, who have memory requirements comparable to those of list learners, we obtain that the number of relevant sentence types, M, has to exceed a constant multiplied by the logarithm of the candidate grammars, n. We have

$$M > \frac{\log n}{\log(1/a)}. \tag{13.47}$$

If this condition holds, then rule-based grammars are more efficient than list-based grammars and will have a fitness advantage. Otherwise it would be more efficient to memorize sentence types associated with arbitrary meaning. In this case, language would have remained a rather dull communication system without any creative ability on the level of syntax. If, however, rule-based grammars are selected, then the potential for "making infinite use of finite means" comes as a by-product.

SUMMARY

- Human language is an unlimited replication device for cultural information.
- Human language gives rise to a new mode of evolution.
- Formal language theory offers a mathematical description of language and grammar.
- The Chomsky hierarchy shows the relationship between classes of languages, grammars, and machines (computational devices).
- Formal language theory is at the foundation of both mathematical logic and computer science. Phrase-structure grammars are equivalent to Turing machines. Gödel's theorem implies that the language of mathematics has no grammar.
- Learning theory describes conditions for successful generalization. The learner is presented with data (sentences or sentence–meaning pairs) and has to infer the rule system (grammar) that generates these data.
- Different approaches to learning theory show that no device can learn an unrestricted set of grammars. Generalization requires a restricted search space.

- Whatever the learning mechanism that is used by the human brain, it can only learn a restricted set of grammars. The description of this set is universal grammar.
- Because our brain evolved, universal grammar, too, is a product of evolution.
- Universal grammar is not a grammar, because a collection of grammars is not necessarily a grammar.
- Universal grammar is not universal, because what evolves must also be variable.
- The mathematical analysis of language evolution needs to combine three different fields: formal language theory, learning theory, and evolutionary dynamics.
- The underlying "replicator-mutator equation" contains as limiting cases the replicator equation (for perfect learning) and the quasispecies equation (for constant selection).
- The "coherence threshold" relates the maximum size of the search space, induced by universal grammar, to the performance of the learning procedure. Only a universal grammar that satisfies the coherence threshold allows the evolution of grammatical communication. This idea is similar to the error threshold of quasispecies theory.

CONCLUSION 14

WE BEGAN by studying selection and mutation. This led to the quasispecies equation,

$$\dot{x}_i = \sum_{j=0}^{n} x_j f_j q_{ji} - \phi x_i. \qquad (14.1)$$

The frequency of genome i is given by x_i and its fitness by f_i. The probability of mutating from j to i is denoted by q_{ji}, which is an element of the mutation matrix, Q. The average fitness of the population is $\phi = \sum_i x_i f_i$. The composition of the population, the quasispecies, is given by the vector $\vec{x} = (x_1, \ldots, x_n)$. Quasispecies live in sequence space and tend to climb uphill in the fitness landscape. An important result is the error threshold: adaptation is possible only if the mutation rate per base, u, is less than the inverse of the sequence length, $1/L$.

Evolutionary games represent situations in which fitness values are not constant, but frequency dependent. Deterministic evolutionary game dynamics of infinitely large populations are given by the replicator equation, which describes selection without mutation,

$$\dot{x}_i = x_i[f_i(\vec{x}) - \phi]. \tag{14.2}$$

Fitness values are often linear functions, $f_i = \sum_j x_j a_{ij}$. The coefficients a_{ij} are the entries of the payoff matrix, $A = [a_{ij}]$. If a strategy is a strict Nash equilibrium or an evolutionarily stable strategy (ESS), then it is a stable equilibrium of the replicator equation. For $n = 3$ there can be heteroclinic cycles. For $n \geq 4$ there can be limit cycles and chaos. The replicator equation is equivalent to the Lotka-Volterra equation of ecology.

A crucial question of evolutionary biology is: how can natural selection, which is based on fierce competition, lead to altruistic interactions? The essence of cooperation and defection is captured by the Prisoner's Dilemma:

$$\begin{array}{c} \\ C \\ D \end{array} \begin{array}{c} \begin{array}{cc} C & D \end{array} \\ \begin{pmatrix} R & S \\ T & P \end{pmatrix} \end{array} \tag{14.3}$$

The ranking $T > R > P > S$ implies that cooperation, C, is dominated by defection, D, which is a strict Nash equilibrium. Cooperation becomes an option if the game is repeated. This is the idea of direct reciprocity: I help you and you help me. Originally Tit-for-tat (TFT) emerged as the "best strategy," but subsequent work has challenged this perspective. In the presence of noise, Tit-for-tat is immediately replaced by Generous Tit-for-tat (GTFT), which sometimes forgives a defection. Both strategies, however, are vulnerable to drift to Always cooperate (ALLC), which in turn invites the invasion of Always defect (ALLD). In contrast, Win-stay, lose-shift (WSLS) can correct mistakes, is stable against invasion by ALLD, and dominates ALLC. Hence Win-stay, lose-shift is currently the "world champion" of strategies playing the repeated Prisoner's Dilemma.

Finite population size, N, requires a stochastic approach. In the Moran process, the fixation probability of a new mutant with relative fitness r is given by

$$\rho = \frac{1 - 1/r}{1 - 1/r^N}. \tag{14.4}$$

In the limit of neutrality, $r \to 1$, we have $\rho = 1/N$. The rate of evolution is given by the population size multiplied by the mutation rate multiplied by the fixation probability: $Nu\rho$. For the case of neutral evolution, N and ρ cancel out. Therefore the rate of neutral evolution is simply the mutation rate, u. This is the concept of the "molecular clock."

A frequency-dependent version of the Moran process describes evolutionary game dynamics in finite populations. In a game between two strategies A and B, if the fixation probability ρ_A is greater than $1/N$, then selection favors the replacement of B by A. This idea leads to a new understanding of evolutionary stability. The conditions for a strict Nash equilibrium or ESS are neither necessary nor sufficient to imply protection by selection in a finite population. Instead there is an interesting 1/3 law. Suppose A and B are best replies to themselves and x^* is the frequency of A at the unstable equilibrium. If

$$x^* < 1/3, \tag{14.5}$$

then $\rho_A > 1/N$ for weak selection and large N. This condition also holds in a frequency-dependent Wright-Fisher process.

Evolutionary graph theory is a new tool for studying stochastic evolutionary dynamics in structured populations. The individuals occupy the vertices of a graph, and the edges describe interactions. All graphs that are circulations have the same fixation probability as the Moran process. A graph is a circulation, if for each vertex the sum of all incoming edges equals the sum of all outgoing edges. Certain directed graphs, such as the star or superstar, are amplifiers of selection: the fixation probability of an advantageous mutant is greater than in the Moran process. Some graphs even guarantee the fixation of any advantageous mutant. Other graphs act as suppressors of selection: the fixation probability of advantageous mutants is less than in the Moran process. We have also encountered a particularly exciting result for games on graphs. The evolution of cooperation on a graph is favored by natural selection if

$$b/c > k. \tag{14.6}$$

The benefit of the altruistic act, b, divided by the cost, c, is greater than the average number of neighbors, k. In this case, cooperators behave like advantageous mutants. This mechanism for the evolution of cooperation can be called "network reciprocity."

Spatial evolutionary game theory can be formulated by deterministic cellular automata, leading to spatial chaos, evolutionary kaleidoscopes, and dynamic fractals. These studies lead to the concept of spatial reciprocity: help your neighbor. Unconditional cooperators survive by forming spatial clusters. Some clusters allow cooperators to invade a world of defectors.

Four specific applications of evolutionary dynamics were discussed. In HIV infection, evolutionary dynamics can explain the mechanism of disease progression. The basic model captures the essence of the interaction between HIV and the immune system:

$$\begin{aligned} \dot{v}_i &= v_i(r - px_i - qz) \\ \dot{x}_i &= cv_i - bx_i - uvx_i \\ \dot{z} &= kv - bz - uvz \end{aligned} \tag{14.7}$$

The abundance of virus strain i is given by v_i. The strain-specific immune response, x_i, is active against virus v_i. The cross-reactive immune response, z, is active against all viruses, $v = \sum_i v_i$. The index i runs from 1 to n, which is the number of antigenically different HIV strains present in a patient at a certain time. Each virus mutant can impair immune responses irrespective of its specificity. This antisymmetric interaction between HIV and the immune system leads to a "diversity threshold": if the antigenic diversity, n, of the virus exceeds a certain threshold, then HIV can no longer be controlled. At this point the asymptomatic period of the infection ends and the final phase, AIDS, begins. The idea of this model is that virus evolution in each individual patient leads to AIDS.

The evolution of the virulence of infectious agents can be studied by models of evolutionary dynamics. The basic equation of epidemiology is the following

$$\begin{aligned} \dot{x} &= k - ux - \beta xy \\ \dot{y} &= y(\beta x - u - v) \end{aligned} \tag{14.8}$$

Uninfected individuals, x, are born (immigrate) at a constant rate k, die at rate ux, and get infected at rate βxy. Infected individuals, y, die at rate $(u + v)y$, where v is the disease-induced mortality or virulence. In this model, natural selection between different parasite strains maximizes the basic reproductive ratio, R_0, which is defined as the number of secondary infections that arise from one infected individual if everybody else is still uninfected: $R_0 = \beta k/[u(u+v)]$. If virulence, v, is correlated with transmissibility, β, then selection can lead to intermediate levels of virulence. If we add superinfection of already-infected hosts, then selection causes higher levels of virulence than what would be optimum for the parasite. Moreover, superinfection leads to coexistence of parasite strains with different levels of virulence.

Many fundamental topics in cancer biology lead to questions of evolutionary dynamics. For example, how long does it take for a population of reproducing cells to activate an oncogene or inactivate a tumor suppressor gene (TSG)? What is the probability that genetic instability arises early in cancer progression? We have studied various stochastic processes to address such questions. One major conclusion was that it takes two rate-limiting hits to inactivate a tumor suppressor gene (in a small population of cells) with or without chromosomal instability (CIN).

$$\begin{array}{lll} A^{+/+} & \rightarrow A^{+/-} & \rightarrow A^{-/-} \\ \downarrow & \quad \downarrow & \\ A^{+/+}CIN & \rightarrow A^{+/-}CIN & \rightarrow A^{-/-}CIN \end{array} \tag{14.9}$$

Without CIN, the two hits are inactivating the first and second allele of the TSG. With CIN, the two rate-limiting hits are (i) inactivating the first allele of the TSG and (ii) mutating a gene that leads to CIN; subsequently the second allele of the TSG is inactivated rapidly (not rate limiting). Therefore, in order to evaluate whether CIN will precede inactivation of the first TSG on the way to cancer, we only have to compare the rate constants of the two hits, which essentially comes down to counting the number of available CIN genes.

But the best of all evolutionary games I have saved for last: the evolution of language. We have discussed what language is, how children learn language, and how a population can evolve language. A fundamental equation is

$$\dot{x}_i = \sum_{j=0}^{n} x_j f_j(\vec{x}) q_{ji} - \phi x_i. \tag{14.10}$$

In this replicator-mutator equation, a fraction x_i of people use grammar G_i. The fitness of a grammar depends on how many others use it. Thus $f_i(\vec{x})$ depends on the mixture of grammars in the population. The language-acquisition process is captured by the learning matrix $Q = [q_{ij}]$. For any given length of the learning period, there is a maximum size of the search space of possible grammars that is still compatible with linguistic coherence (language adaptation). This approach combines the quasispecies equation (14.1) with the replicator equation (14.2). Genetic reproduction is replaced by language learning. Structurally similar models of evolutionary dynamics span very different worlds.

Evolutionary thinking permeates all areas of biology. Evolution is the only theory that has explanatory power for the design and function of living systems. In principle, evolutionary biology can account for the amazing diversity and astonishing complexity of life.

The field of evolutionary dynamics explores the mathematical principles (natural laws) of evolutionary change. I do not know what the "ultimate understanding" of biology will look like, but one thing is clear: it will be based on precise mathematical descriptions of evolutionary processes. Mathematics is the proper language of evolution, because the basic evolutionary principles are of a mathematical nature. Though verbal in origin, the theory of evolution has become more and more mathematical over time, and all debates and new ideas regarding evolution should be formulated in mathematical equations. Verbal approaches are unsatisfactory if mathematical formulations exist. Mathematics allows a well-defined discourse and a clear description of natural phenomena. Mathematics is a way to think clearly.

In summary, the study of evolutionary dynamics leads to understanding the basic design features of biological systems from genes to people. Medical applications include the analysis of infectious diseases and cancer. By exploring cooperation and language, evolutionary dynamics can even lead to mathematical theories of human society. I have outlined the astonishing diversity of topics that can be discussed by evolutionary dynamics. I expect an ever-increasing diversity. In the future, the constructive power of evolutionary dynamics will be used to answer questions that are as yet unimaginable.

FURTHER READING

REFERENCES

INDEX

FURTHER READING

2. WHAT EVOLUTION IS

Evolutionary biology began with Lamarck (1809) and Darwin (1859, 1871). Some of my favorite books on evolution are Williams (1966, 1992), E. O. Wilson (1978, 2000), Mayr (1982, 2001), Kimura (1983), Dawkins (1982), Diamond (1992), Sigmund (1993), Maynard Smith and Szathmáry (1995), Dennett (1995), Hamilton (1996, 2001), Gould (2002), Trivers (2002), and Kirschner and Gerhart (2005). Schrödinger's (1944) "What is life?" brought many physicists to biology.

May (1973) is a brilliant text on modeling ecosystems. There is also a second edition with a new introduction: May (2001). May (1976) and May and Oster (1976) brought chaos to biology. Sugihara and May (1990) predict the unpredictable. A review on mathematical population biology is given by Levin et al. (1997).

As noted in the text, this book is not about population genetics. The origins of population genetics are R. A. Fisher (1930b), Wright (1931, 1932), and Haldane (1932). A classic is Wright (1968, 1969). Other great books on population genetics include Jacquard (1974), Lewontin (1974), Nei (1987), Gillespie

(1991), Hartl and Clark (1997), Bürger (2000), Ewens (2004), and Gavrilets (2004). Slatkin (1979) is a paper on frequency-dependent selection. Barton (2000) studies genetic hitchhiking. Barton (2001) explores the role of hybridization on evolution. The population genetics of speciation is discussed by Turelli, Barton, and Coyne (2001). A simple approach to evolutionary genetics is offered by Maynard Smith (1989).

Michod (1999) and Rice (2004) are books on evolutionary dynamics. Keller (1999) discusses evolution on multiple levels. Comprehensive overviews of many aspects of mathematical biology are offered by Murray (2002, 2003) and Edelstein-Keshet (2004). An excellent book on differential equations in biology is Taubes (2001). Strogatz (1994) is a great introduction to nonlinear dynamics.

May (2004) compares the Hardy-Weinberg law in biology to Newton's first law in physics. The latter states that bodies remain in their state of rest or uniform motion in a straight line, except insofar as acted upon by external forces. In comparison, the Hardy-Weinberg law states that gene frequencies in a population with random mating do not alter from generation to generation in the absence of selection, mutation, random drift, or migration.

3. FITNESS LANDSCAPES AND SEQUENCE SPACES

Quasispecies theory was invented by Eigen and Schuster (1979); see also Swetina and Schuster (1982), McCaskill (1984), Demetrius (1987), Leuthäusser (1987), Schuster and Swetina (1988), Eigen, McCaskill, and Schuster (1989), and Eigen (1992). Nowak and Schuster (1989) calculate the error threshold for finite population size. Important papers on quasispecies theory and RNA evolution are Fontana and Schuster (1987, 1998), Fontana et al. (1993), and Fontana (2002). Fontana and Buss (1994) discuss construction and other deep issues of evolutionary dynamics.

Kauffman and Levin (1987) study adaptive walks on fitness landscapes. A broad perspective of fitness landscapes and evolution on the edge of chaos is offered by Kauffman (1993). Stadler (1992) and Bonhoeffer and Stadler (1993) study error thresholds on correlated fitness landscapes. Krumhansl (1997) and Sherrington (1997) analyze landscapes in physics and biology. Stadler (1999) explores fitness landscapes that arise from sequence-structure maps.

Boerlijst, Bonhoeffer, and Nowak (1996) study quasispecies and recombination. Sasaki and Iwasa (1987) calculate optimum recombination rates in fluctuating environments. Schuster and Stadler (2002) discuss networks in evolution. Krakauer and Sasaki (2002) explore quasispecies on noisy fitness landscapes. Sasaki and Nowak (2003) introduce the concept of a mutation landscape: each sequence encodes for its own replication machinery and therefore has its own characteristic mutation rate. In this new framework, quasispecies move simultaneously over both fitness landscapes and mutation landscapes.

4. EVOLUTIONARY GAMES

The first applications of game theory to biology are Hamilton (1967), Trivers (1971), Maynard Smith and Price (1973), and Maynard Smith (1982, 1984). Taylor and Jonker (1978), Zeeman (1980), and Hofbauer, Schuster, and Sigmund (1979) introduce the replicator equation. The masterpiece on evolutionary game dynamics is Hofbauer and Sigmund (1998).

For the origins of the Lotka-Volterra equation, see Lotka (1925) and Volterra (1926). Kolmogorov (1936) describes the limit cycle theorem for predator-prey equations. A brilliant account of ecological equations is offered by May (1973). May and Leonard (1975) describe convergence to a heteroclinic cycle in a Lotka-Volterra system. The proof that the replicator equation is equivalent to the Lotka-Volterra equation can be found in Hofbauer and Sigmund (1998).

Taylor (1989) studies evolutionary stability under weak selection. Bomze (1983) and Stadler and Schuster (1990) provide complete classifications for replicator dynamics among three strategies. Stadler and Schuster (1992) study replicator equations with mutation. Fudenberg and Harris (1992) study evolutionary game dynamics with noise. Christiansen (1991) analyzes evolutionary stability in continuous strategy sets. An excellent paper on evolutionarily stable attractors is Rand, Wilson, and McGlade (1994).

Game theory was invented by von Neumann and Morgenstern (1944). Nash (1950) proposed the equilibrium that carries his name. Books on economic game theory include Fudenberg and Tirole (1991), Binmore (1994), Samuelson (1997), and Fudenberg and Levine (1998). Books on evolutionary game theory include Weibull (1995), Gintis (2000), and Vincent and Brown (2005). Evolutionary dynamics of extensive form games are described in the book by Cressman (2003).

Evolutionary game theory has been applied to many aspects of biology, including the evolution of sex (Trivers 1983), parent-offspring conflict (Godfray 1995), sibling rivalry (Mock and Parker 1997), genomic imprinting (Haig 2002), sex ratios (Hardy 2002), speciation (Hendry et al. 2000), animal behavior (Dugatkin and Reeve 1998; Houston and McNamara 1999), mate choice (Iwasa and Pomiankowski 1995), signaling (Johnstone 2002), cell organelles (Krakauer and Mira 2000), and interactions among plants (Falster and Westoby 2003). Sinervo and Lively (1996) study the Rock-Paper-Scissors game in lizards. Evolutionary dynamics of the Ultimatum game are studied by Nowak, Page, and Sigmund (2000). Experimental results of the Ultimatum game are reported in Henrich et al. (2001).

Iwasa and Sasaki (1987) study the evolution of the number of sexes. Iwasa and Pomiankowski (2001) explore evolutionary games of genomic imprinting. An attempt to bring population genetics into game theory is offered by Hammerstein (1996). Mylius and Diekmann (2001) study the complex consequences of invading an existing community under frequency-dependent selection. Metz, Nisbet, and Geritz (1992) discuss how to define fitness in various settings. Recent reviews of evolutionary game dynamics are Hofbauer and Sigmund (2003) and Nowak and Sigmund (2004).

5. PRISONERS OF THE DILEMMA

An early book on the Prisoner's Dilemma (PD) is Rapoport and Chammah (1965). Trivers (1971) brought the PD to biology. Selten (1975) studies equilibrium points in extensive games. Eshel (1977) is an early paper on the evolution of altruism. Maynard Smith (1979) points out that cooperation in hypercycles is threatened by defectors. Axelrod and Hamilton (1981) discuss the repeated PD in biology. Two iterated PD tournaments are described in Axelrod (1984). Selten and Hammerstein (1984) point out that Tit-for-tat is not an evolutionarily stable strategy (ESS). Molander (1985) calculates the optimum level of generosity. An important (nonmathematical) book with many ideas is Sugden (1986). See Fudenberg and Maskin (1986) for the "folk theorem" on repeated games. May (1987) calls for the evolution of cooperation with noise. Boyd and Lorberbaum (1987) argue that no pure strategy can be an ESS in the repeated PD. For a review of early work, see Axelrod and Dion (1988). Boyd (1989) explores the effect of noise. "Pavlov" is studied by Kraines and Kraines (1989). Reactive strategies in the repeated PD are analyzed by Nowak and Sigmund (1989a,b, 1990) and Nowak (1990b). "Perfect Tit-for-tat" is described by Fudenberg and Maskin (1990). "Pavlov" and "Perfect Tit-for-tat" are other names for Win-stay, lose-shift. Lindgren (1991) analyzes deterministic memory-3 strategies, which take into account the previous three moves and then choose to cooperate or defect, in the repeated Prisoner's Dilemma. Binmore and Samuelson (1992) study finite-state automata in repeated games.

TFT is a catalyst for the initiation of cooperation but is not the ultimate goal; Generous Tit-for-tat replaces TFT in a noisy world (Nowak and Sigmund 1992). Win-stay, lose-shift outperforms Generous Tit-for-tat in the simultaneous PD (Nowak and Sigmund 1993). The alternating PD is another game (Nowak and Sigmund 1994; Frean 1994). Boerlijst, Nowak, and Sigmund (1997) study Contrite Tit-for-tat. The continuous PD is studied by Roberts and Sherratt (1998), Wahl and Nowak (1999a,b), and Killingback and Doebeli (2002). Punishment and cooperation are studied by Fehr and Gächter (2000, 2002), Sigmund, Hauert, and Nowak (2001), and Boyd et al. (2003). The option not to play can favor cooperation (Hauert et al. 2002; Szabó and Vukov 2004). Hammerstein (2003) contains a collection of papers presented at a Dahlem conference on the evolution of cooperation.

Experimental studies of the evolution of cooperation in biological systems include Wilkinson (1984), Lombardo (1985), Dugatkin (1988, 1997), Bull and Molineux (1992), Heinsohn and Packer (1995), and Velicer and Yu (2003). Milinski (1987) reports TFT in sticklebacks. Turner and Chao (1999) describe a PD among RNA viruses. Cooperation of coviruses is analyzed by Nee (2000). Pfeiffer, Schuster, and Bonhoeffer (2001) study cooperation in ATP metabolism. Cooperation among humans is studied by Milinski and Wedekind (1998), Wedekind and Milinski (1996), Semmann, Krambeck, and Milinski (2003). The experimental economics of human altruism is reviewed by Fehr and Fischbacher (2003).

Adaptive dynamics were introduced by Nowak and Sigmund (1990) to study the evolution of cooperation in the repeated Prisoner's Dilemma. For the further development of adaptive dynamics, see Hofbauer and Sigmund (1990), Metz et al. (1996), Geritz et al. (1997), and Dieckmann (1997). Applications of adaptive dynamics can be found in Dieckmann and Doebeli (1999), Dieckmann, Law, and Metz (2000), Page and Nowak (2000), Le Galliard, Ferrière, and Dieckmann (2003), Doebeli and Dieckmann (2004), and Doebeli, Hauert, and Killingback (2004). An ESS need not be an attractor of adaptive dynamics (Nowak 1990a). That an ESS can be unreachable was first shown by Eshel (1983); see also Eshel (1996).

The concept of kin selection was proposed by Hamilton (1964a,b); see also Hamilton (1971) and Frank (1998). Wilson (1980), Szathmáry and Demeter (1987), Wilson, Pollock, and Dugatkin (1992), and Sober and Wilson (1998) study the evolution of cooperation by group selection. Levin (1999) gives an ecologist's perspective of the tragedy of the commons.

Indirect reciprocity is the idea that individuals help those who have helped others (Nowak and Sigmund 1998). Experimental evidence is given by Wedekind and Milinski (2000) and Milinski, Semmann, and Krambeck (2002). Important recent developments include Leimar and Hammerstein (2001), Fishman (2003), Panchanathan and Boyd (2003, 2004), Ohtsuki and Iwasa (2004), and Brandt and Sigmund (2005). For a review, see Nowak and Sigmund (2005).

6. FINITE POPULATIONS

The Moran process is described by Moran (1958) and Moran (1962). The standard book on stochastic processes is Karlin and Taylor (1975). For stochastic processes in population genetics, see Ewens (2004). For the neutral theory of evolution, see Kimura (1968, 1983). For the selected papers of this great evolutionary biologist, see Kimura (1994). Neutral evolution in spatially continuous populations is studied by Barton, Depaulis, and Etheridge (2002).

7. GAMES IN FINITE POPULATIONS

Evolutionary game dynamics in finite populations are studied by Riley (1979), Schaffer (1988), Fogel, Fogel, and Andrews (1998), Ficici and Pollack (2000), Schreiber (2001), and Alos-Ferrer (2003). The approach that is described here was introduced by Nowak et al. (2004b) and Taylor et al. (2004). The latter is a complete classification of all selective scenarios among two strategies. Imhof et al. (2005) study evolutionary game dynamics in finite populations consisting of the three strategies ALLD, TFT, and ALLC. They find that the equilibrium distribution of a mutation-selection process is almost entirely centered on TFT, even if ALLD is the only strict Nash equilibrium. Fudenberg et al. (2006) calculate limiting distributions for the mutation-selection process concerning games in finite populations.

The frequency-dependent Moran process is a natural analog to the replicator equation, but one can envisage many stochastic processes that describe game dynamics in finite populations. An interesting possibility is the following. Pick two players at random. One is chosen for reproduction, the other for elimination. Hence only mixed pairs can change the population. Suppose that player A is chosen for reproduction with probability $f_i/(f_i + g_i)$ and player B with probability $g_i/(f_i + g_i)$. In this case, we obtain the same process that we have analyzed up to rescaling time. If instead the fitter player is always chosen for reproduction, then the resulting process is stochastic in speed, but deterministic in direction: it will always follow the gradient of selection. If, on the other hand, player A is chosen for reproduction with probability $1/(1 + \exp[-(f_i - g_i)/\tau])$, then parameter w cancels. There is, however, a new parameter, τ, which plays a similar role. If $\tau \to 0$ then the fitter player is always chosen; selection is strong. If $\tau \to \infty$ then selection is weak, and the process is dominated by random drift. In the limit of large τ, we obtain exactly the same results as presented in this chapter (see Nowak et al. 2004b).

Another possibility is studying a frequency-dependent Wright-Fisher process. In this case, we do not have explicit expressions for the fixation probabilities, but it is still possible to show that the 1/3 law also holds (Imhof and Nowak 2006).

8. EVOLUTIONARY GRAPH THEORY

The balance between drift and selection is discussed by Wright (1931, 1932) and Fisher and Ford (1950). Maruyama (1970), and Slatkin (1981) show that the fixation probability is unaffected by symmetric spatial structure. See also

Nagylaki and Lucier (1980). Pulliam (1988) studies "sources" and "sinks" in population dynamics. Barton (1993) and Whitlock (2003) study fixation probability and time in subdivided populations.

Mathematical properties of random graphs are analyzed by Erdös and Rényi (1960). Watts and Strogatz (1998) and Watts (1999) introduce small-world networks. Barabasi and Albert (1999) introduce scale-free networks. Strogatz (2001) offers a nice review on networks. Boyd, Diaconis, and Xiao (2004) study random walks on graphs.

Games on networks are analyzed by Ellison (1993), Nakamaru, Matsuda, and Iwasa (1997), Nakamaru, Nogami, and Iwasa (1998), Abramson and Kuperman (2001), Ebel and Bornholdt (2002), Szabó and Hauert (2002), Nakamaru and Iwasa (2005), Santos and Pacheco (2005), and Santos, Pacheco, and Leanerts (2006). Bala and Goyal (2000) and Skyrms and Pemantle (2000) study games of network formation. Newman (2001) evaluates collaboration networks among scientists. Flack, Krakauer, and de Waal (2005) and Flack et al. (2006) study networks among primates. Liggett (1999) is a book on voting models.

Evolutionary graph theory, as presented here, was introduced by Lieberman, Hauert, and Nowak (2005). For evolution of cooperation on graphs and the rule $b/c > k$, see Ohtsuki et al. (2006).

9. SPATIAL GAMES

Cellular automata were invented by Stanislav Ulam. John von Neumann used them in his work on self-reproducing automata. The "game of life" was invented by the mathematician John Conway in 1970. See Berlekamp, Conway, and Guy (2001, 2003), as well as Poundstone (1985) and Sigmund (1993). Wolfram (1984, 1994, 2002) and Toffoli and Margolus (1987) are important contributions to cellular automata theory. Langton (1986) uses cellular automata to study artificial life.

Spatial models in ecology are studied by Levin and Paine (1974), Durrett (1988, 1999), Levin (1992), Hassel, Comins, and May (1991, 1994), Durrett and Levin (1994a,b, 1998), Pacala and Tilman (1994), Tilman and Kareiva (1997), and Neuhauser (2001). Lloyd and Jansen (2004) describe the spatial dynamics of epidemics. Boerlijst and Hogeweg (1991a,b) show that hypercycles lead to spiral waves in spatial models, which confer some protection against parasites.

For spatial evolutionary game theory, see Nowak and May (1992, 1993). Analytic results in a simplified setting are obtained by Herz (1994). Nowak, Bonhoeffer, and May (1994a,b) study asynchronous updating and other extensions. More evolution of cooperation in spatial models is described by Lindgren and Nordahl (1994), Ferrière and Michod (1996), Epstein (1998), van Baalen and Rand (1998), Mitteldorf and Wilson (2000), Irwin and Taylor (2001), and Ifti, Killingback, and Doebeli (2004). Hauert and Doebeli (2004) show that spatial structure can harm cooperators in the Snowdrift game. Killingback and Doebeli (1996) study the spatial Hawk-Dove game. Killingback and Doebeli (1998) analyze self-organized criticality in spatial games. Sasaki, Hamilton, and Ubeda (2002) is an insightful paper on pacemakers in spatial models. An experimental study of a spatial Rock-Paper-Scissors game among bacteria is Kerr et al. (2002).

10. HIV INFECTION

The molecular biology of retroviruses is described in Coffin, Hughes, and Varmus (1997). Levine (1992) is an introduction to viruses. Nowak and May (2000) is a book on virus dynamics.

Nowak, May, and Anderson (1990) and Nowak et al. (1991) proposed that HIV disease progression is the consequence of virus evolution in individual hosts. Experimental evidence that HIV escapes from CTL (cytotoxic T lymphocyte) responses is provided by Phillips et al. (1991), McAdam et al. (1995), McMichael et al. (1995), Nowak et al. (1995b), Borrow et al. (1997), Price et al. (1997), and Goulder et al. (1997). A early paper on the genetic variation of HIV-1 during single infections is Saag et al. (1988). Wei et al. (2003) describe escape from neutralizing antibodies.

McLean and Nowak (1992) study the interaction between HIV and other pathogens. Bonhoeffer and Nowak (1994a) model viral strategies of immune function impairment. De Boer and Boerlijst (1994) analyze diversity and virulence thresholds. Sasaki (1994) and Sasaki and Haraguchi (2000) analyze the evolutionary dynamics of antigenic drift and shift. A theory for antigenic variation in multiple epitopes has been developed by Nowak et al. (1995b) and Nowak, May, and Sigmund (1995). Antia, Nowak, and Anderson (1996) study antigenic variation of parasites. Krakauer and Komarova (2003) explore levels of selection in virus dynamics. Bonhoeffer et al. (2004) study epistasis in drug-treatment–resistant mutations in HIV.

Ho et al. (1995), Wei et al. (1995), Nowak et al. (1995a), Perelson et al. (1996), Perelson, Essunger, and Ho (1997), and Bonhoeffer et al. (1997) analyze data on HIV decline during drug therapy and the emergence of resistance. Wodarz and Nowak (1999) show that specific drug therapy regimes or the combination of drug treatment with immunotherapy can lead to control of infection. Lifson et al. (1997, 2000, 2001) study virus dynamics and early treatment in SIV infection.

11. EVOLUTION OF VIRULENCE

For the beginnings of mathematical epidemiology, see Bernoulli (1760), Farr (1840), Hamer (1906), Ross (1908), and Kermack and McKendrick (1933). Anderson and May (1979) and May and Anderson (1979), emphasizing the need for mathematical simplicity and close connection to data, are crucial papers that have shaped much of the modern approach to theoretical epidemiology. For a comprehensive treatment, see Anderson and May (1991). Equally excellent but more mathematical, are the approaches of Bailey (1975), Dietz (1975), Diekmann, Heesterbeek, and Metz (1990), and Diekmann and Heesterbeek (2000).

The evolution of virulence is examined by May and Anderson (1979, 1983), Anderson and May (1981), S. A. Levin and Pimentel (1981), B. R. Levin (1982), Bremermann and Pickering (1983), Stewart and B. R. Levin (1984), Seger (1988), Seger and Hamilton (1988), Bremermann and Thieme (1989), Knolle (1989), Frank (1992), Ewald (1993), Read and Harvey (1993), Lenski and May (1994), and Haraguchi and Sasaki (2000).

Host-parasite coevolution is studied by May and Anderson (1990) and Sasaki (2000). Yamamura (1993) analyzes a host-parasite coevolution model, where vertical transmission leads to a reduction of parasite virulence. Antia, B. R. Levin, and May (1994) analyze the effect of within-host population dynamics on evolution. B. R. Levin and Bull (1994) develop the "short-sighted evolution of virulence." An excellent review on the population biology of infectious agents is offered by B. R. Levin, Lipsitch, and Bonhoeffer (1999).

Stewart and B. R. Levin (1984) discuss different conditions for the evolution and maintainance of temperate and virulent phages in a model where phage reproduction can occur via the infection of new cells or (vertically) via cell division. Nowak (1991) shows that vertical transmission can lead to complicated selective dynamics even for very simple models. Here selection need not optimize R_0.

Nowak and May (1994) and May and Nowak (1994) explore superinfection. Bonhoeffer and Nowak (1994b) analyze the effect of mutation on the evolution of virulence. May and Nowak (1995) study coinfection. Lipsitch, Herre, and Nowak (1995) introduce a law of diminishing returns. Lipsitch and Nowak (1995) study evolution of virulence in sexually transmitted HIV. Vertically and horizontally transmitted parasites are studied by Lipsitch et al. (1995) and Lipsitch, Siller, and Nowak (1996).

The classical empirical study of myxomatosis in Australian rabbits is Fenner and Ratcliffe (1965). Dieter Ebert is the world's leading expert on gut infections of water fleas (Ebert 1994). Bull, Molineux, and Rice (1991) describe experimental evolution of avirulence. Herre (1993) has studied nematodes of fig wasps. Busenberg and Cooke (1993) is a book on the epidemiology of vertically transmitted infectious diseases. Frank (2002) gives an evolutionary approach to immunology and infection.

Superinfection is related to metapopulation dynamics in ecology (Tilman et al. 1994). For metapopulation models see Sabelis, Diekmann, and Jansen (1991), Nee and May (1992), Doebeli and Ruxton (1997), Parvinen (1999), Gyllenberg, Parvinen, and Diekmann (2002), and Wakeley (2004).

12. EVOLUTIONARY DYNAMICS OF CANCER

An excellent introduction to cancer genetics is Vogelstein and Kinzler (1998). A classic is Boveri (1914). Muller (1927) showed that ionizing radiation, which was known to be carcinogenic, was also mutagenic. Knudson (1971) is the crucial study of retinoblastoma; see also Knudson (1993). The discovery of the retinoblastoma tumor suppressor gene is reported in Friend et al. (1986). The discovery of the APC locus in colorectal cancer is described by Kinzler et al. (1991). Somatic mutation rates of humans are measured by Grist et al.

(1992). Weinberg (1991) and Levine (1993) provide reviews of tumor suppressor genes. APC inactivation is studied by Nagase and Nakamura (1993) and Lamlum et al. (1999). Boyer et al. (1995) study microsatellite instability. The terms "gatekeepers" and "caretakers" are introduced by Kinzler and Vogelstein (1997). The concept of genetic instability in human cancers is developed by Lengauer, Kinzler, and Vogelstein (1997, 1998). Strauss (1998) offers a review of genetic instability. Wheeler et al. (1999) study microsatellite instability (MIN) in colon cancer. For a review on the fidelity of DNA replication, see Kunkel and Bebenek (2000). Shonn, McCarroll, and Murray (2000) study chromosome segregation (and chromosomal instability, CIN) in yeast. Somatic stem cells and cancer stem cells are studied by Reya et al. (2001). Knudson (2001) provides an illuminating perspective on cancer genetics. CIN in murine cancer models is studied by Chang, Khoo, and DePinho (2001) and Chang et al. (2003). Bach, Renehan, and Potten (2000) describe intestinal stem cells. Bardelli et al. (2001) show that different carcinogens can select for different types of genetic instability. Bissell and Radisky (2001) offer review articles on tumor biology. Colon stem cells are studied by Yatabe, Tavare, and Shibata (2001). Adenomas without genetic instability are characterized by Haigis et al. (2002). Hermsen et al. (2002) find CIN during progression from colorectal adenoma to carcinoma. Genomic instability in yeast is reviewed by Kolodner, Putnam, and Myung (2002). Maser and DePinho (2002) review CIN in cancer. Nasmyth (2002) describes the molecular biology of chromosome separation. Mutations that interfere with this process can lead to CIN. Sieber et al. (2002) do not observe CIN in colorectal adenomas, but their experimental method may not be sensitive enough. Pihan et al. (2003) find CIN in carcinomas. Different perspectives on CIN are offered by Sieber, Heinimann, and Tomlinson (2003) and Rajagopalan et al. (2003). Inactivation of hCDC4 leads to CIN (Rajagopalan et al. 2004).

Nordling (1953), Armitage and Doll (1954, 1957), and J. C. Fisher (1959) explain the age-incidence curves of human cancers with multistage probabilistic models. The idea of a mutator phenotype was introduced by Loeb, Springgate, and Battula (1974); see also Loeb (1991, 2001). Cairns (1975) offers some penetrating ideas on mutation, selection, and cancer. Mathematical models of chemotherapy are developed by Goldie and Coldman (1979, 1983). Moolgavkar and Knudson (1981) study a two-stage model of cancer initiation. Sherratt and Nowak (1992) explore spatial models of cancer progression. Tomlinson, Novelli, Bodmer (1996) argue that an increased mutation rate is not beneficial for cancer. Taddei et al. (1997) study the effect of increased mutation rates in normal evolution (not in somatic evolution). Anderson and Chaplain (1998) describe a mathematical model for angiogenesis. Nunney (1999) analyzes lineage selection in multistage carcinogenesis. Owen and Sherratt (1999) model immune responses against tumors. Wodarz and Krakauer (2001) study genetic instability and angiogenesis. Cairns (2002) describes the role of somatic stem cells in carcinogenesis. Tomlinson, Sasieni, and Bodmer (2002) estimate the number of mutations in a cancer. Luebeck and Moolgavkar (2002) fit a model of multistage carcinogenesis to colon cancer incidence data. Nowak et al. (2002) calculate the probability that a CIN mutation represents the first phenotypic change on the way to colon cancer. Komarova et al. (2002) and Komarova, Sengupta, and Nowak (2003) study the dynamics of genetic instability in sporadic and familial colorectal cancer. Plotkin and Nowak (2002) analyze the affect of apoptosis on tumorigenesis. Frank and Nowak (2003) and Frank, Iwasa, and Nowak (2003) consider at what stage of development cancer mutations are likely to occur. Little and Wright (2003) present a stochastic model of colon cancer. Gatenby and Vincent (2003) provide a mathematical model of carcinogenesis. Gatenby and Maini (2003) give a perspective on mathematical oncology. Michor et al. (2003a,b,c) show that small compartments protect against cancer initiation via mutations in oncogenes and tumor suppressor genes, but favor genetic instability. The concept that epithelial tissues are organized into small compartments goes back to Mintz (1971). The linear process of somatic evolution is described by Nowak et al. (2003). The theory of stochastic tunneling is analyzed by Iwasa, Michor, and Nowak (2004). Nowak et al. (2004a) and Iwasa et al. (2005) describe the evolutionary dynamics of inactivating tumor suppressor genes. For a related problem in classical population genetics, see Robertson (1978) and Karlin and Tavare (1983). Michor, Iwasa, and Nowak (2004) is a review of mathematical models of cancer progression. Michor et al. (2004) offer a simple mechanical model of colon cancer initiation. Michor et al. (2005a) show that CIN initiates tumorigenesis if two tumor suppression genes are inactivated in rate-limiting situations. Zheng, Wise, and Cristini (2005) present a computer simulation of spatial tumor growth. Frank (2005) offers a new perspective on age-incidence curves. Michor et al. (2005b) provide the first quantitative analysis of a human cancer in vivo by studying imatinib therapy of chronic myeloid leukemia (CML).

13. LANGUAGE EVOLUTION

Modern linguistics was shaped by Chomsky (1956, 1957, 1965). The principle and parameter theory was introduced by Chomsky (1981), as was government and binding (Chomsky 1993). Jackendoff (1997) explores the architecture of the language faculty. Jackendoff (2002) describes the foundations of language. Robins (1979) offers a history of linguistics. Two highly informative books on language are Miller (1991) and Pinker (1994).

Formal language theory is closely related to the foundations of computer science (Turing 1936, 1950). Bar-Hillel (1953) gives an arithmetic approach to syntax. Introductions to formal language theory are Harrison (1978) and Partee, ter Meulen, and Wall (1990). Tree adjunct grammars were invented by Joshi et al. (1975). Pullum and Gazdar (1982) explore whether natural languages are context free. Shieber (1985) argues that Swiss German is not context free. Sadock (1991) explores autolexical syntax. For a book on certain phrase-structure grammars, see Pollard and Sag (1994). Stabler (2004) studies mildly context-sensitive grammars. Optimality theory is the idea that language acquisition is based on the ordering of constraints (Prince and Smolensky 1997, 2004; Tesar and Smolensky 2000).

Gold (1967, 1978) is one of the originators of learning theory. Statistical learning theory is due to Vapnik and Chervonenkis (1971, 1981) and Vapnik (1998). Valiant (1984) introduced the important concept of "probably almost correct." Statistical learning theory is related to inductive inference (Pitt 1989). Sakakibara (1988, 1990) provide a model for learning context-free grammars from structural data; see also Sakakibara (1997). The trigger learning algorithm is introduced in Gibson and Wexler (1994). Angluin (1987), Gasarch and Smith (1992), and Angluin and Kharitonov (1995) study models of learning in which the student can ask questions. Siskind (1996) studies the learning of mappings from word to meaning. Niyogi (1998) studies the informational complexity of learning. Osherson, Stob, and Weinstein (1986) and Jain et al.

(1999) are important introductions to learning theory. An excellent review is Pinker (1979). Saffran, Aslin, and Newport (1996) provide an empirical study of learning in infants. Wexler and Culicover (1980) study formal principles of language acquisition; see also Yang (2002). Goldsmith (2001) studies unsupervised learning of morphology. An interesting class of learnable languages is described by Stabler (1998).

Language universals are studied by Greenberg et al. (1978) and Comrie (1981). Baker (2001) offers a noteworthy attempt to explain the diversity of human language by considering a small number of parameters. For parameter setting, refer to Manzini and Wexler (1987).

Cognitive aspects of language are discussed by Lakoff (1987). Bates and MacWhinney (1982) develop functionalist grammars. Langacker (1987) gives the foundations of cognitive grammars. Batali (1994) describes innate biases in language acquisition. Elman et al. (1996) study development and innateness of language learning. Bresnan (2000) gives a lexical-functional approach to syntax. Domjan and Burkhard (1986) is a book on learning and behavior. Brent (1997) and Bertolo (2001) focus on language acquisition.

Gopnik and Crago (1991) identify a language deficiency that follows a Mendelian inheritance pattern. A neurological analysis is found in Vargha-Khadem et al. (1998). The genetic mutation was identified by Lai et al. (2001).

For ideas and models of language evolution, see Brandon and Hornstein (1986), Aoki and Feldman (1987), Hurford (1989), Pinker and Bloom (1990), Hashimoto and Ikegami (1996), Kirby and Hurford (1997), Niyogi and Berwick (1997), Steels (1997), Hazlehurst and Hutchins (1998), Wang (1998), Jackendoff (1999), Fitch (2000), Cangelosi and Parisi (2002), and Christiansen et al. (2002).

Evolutionary dynamics of simple communication systems are studied by Nowak and Krakauer (1999), Nowak, Krakauer, and Dress (1999), and Trapa and Nowak (2000). These papers make a connection between evolutionary game theory and language evolution. An information-theoretic approach to language evolution is discussed by Nowak, Plotkin, and Krakauer (1999) and Plotkin and Nowak (2000). Natural selection of syntactic communication is explored by Nowak, Plotkin, and Jansen (2000). The evolution of a private sign system is studied by Krakauer (2001).

Language change in the perspective of historical linguistics is studied by Kroch (1989), Lightfoot (1991, 1999), Ringe, Warnow, and Taylor (2002), Warnow et al. (2005), and Nakhleh et al. (2005).

Smith (1977) is an important book on the behavior of communication. Books on language evolution include Lieberman (1984, 1991), Bickerton (1990), Newmeyer (1991), Hawkins and Gell-Mann (1992), Aitchinson (1996), Dunbar (1996), Hauser (1996), Deacon (1997), Hurford, Studdert-Kennedy, and Knight (1998), Knight, Studdert-Kennedy, and Hurford (2000), and Sampson (2005). Human evolution is examined by Boyd and Silk (1997). A book on cultural evolution is Cavalli-Sforza and Feldman (1981). Cultural evolution and cognition are studied by Tomasello (1999).

Nowak, Komarova, and Niyogi (2001) and Komarova, Niyogi, and Nowak (2001) propose models for the evolution of grammar. The evolutionary dynamics of lexical items is explored by Nowak (2000) and Komarova and Nowak (2001a). Natural selection of a critical period of language acquisition is analyzed by Komarova and Nowak (2001b). Komarova and Rivin (2001) and Rivin (2001) analyze the perfomance of memoryless and other learners on random matrices describing the similarity between candidate grammars. Komarova and Nowak (2003) study language evolution in finite population models. Nowak and Komarova (2001) and Nowak, Komarova, and Niyogi (2002) are reviews of the evolutionary dynamics of language. Page and Nowak (2002) relate the replicator-mutator equation to the Price equation. Mitchener and Nowak (2003) study natural selection of different universal grammars. Chaotic switching among strict Nash equilibria is described by Mitchener and Nowak (2004).

REFERENCES

Abrams, P. A., and H. Matsuda. 1997. "Fitness minimization and dynamic instability as a consequence of predator-prey coevolution." *Evol. Ecol.* 11: 1–20.

Abramson, G., and M. Kuperman. 2001. "Social games in a social network." *Phys. Rev. E* 63: 030901R.

Aitchinson, J. 1996. *The seeds of speech*. Cambridge: Cambridge University Press.

Alos-Ferrer, C. 2003. "Finite population dynamics and mixed equilibria." *Int. Game Theory Review* 5: 263–290.

Anderson, A. R., and M. A. Chaplain. 1998. "Continuous and discrete mathematical models of tumor-induced angiogenesis." *B. Math. Biol.* 60: 857–899.

Anderson, R. M., and R. M. May. 1979. "Population biology of infectious diseases: Part I." *Nature* 280: 361–367.

——— 1981. "The population dynamics of microparasites and their invertebrate hosts." *Philos. T. Roy. Soc. B* 291: 451–524.

——— 1991. *Infectious diseases of humans*. Oxford: Oxford University Press.

Angluin, D. 1987. "Learning regular sets from queries and counterexamples." *Inform. Comput.* 75: 87–106.

Angluin, D., and M. Kharitonov. 1995. "When won't membership queries help?" *J. Comput. Syst. Sci.* 50: 336–355.

Antia, R., B. R. Levin, and R. M. May. 1994. "Within-host population dynamics and the evolution and maintenance of microparasite virulence." *Am. Nat.* 144: 457–472.

Antia, R., M. A. Nowak, and R. M. Anderson. 1996. "Antigenic variation and the within-host dynamics of parasites." *P. Natl. Acad. Sci. USA* 93: 985–989.

Aoki, K., and M. W. Feldman. 1987. "Toward a theory for the evolution of cultural communication: Coevolution of signal transmission and reception." *P. Natl. Acad. Sci. USA* 84: 7164–7168.

Armitage, P., and R. Doll. 1954. "The age distribution of cancer and a multi-stage theory of carcinogenesis." *Brit. J. Cancer* 8: 1–12.

——— 1957. "A two-stage theory of carcinogenesis in relation to the age distribution of human cancer." *Brit. J. Cancer* 11: 161–169.

Asavathiratham, C., S. Roy, B. Lesieutre, and G. Verghese. 2001. "The influence model." *IEEE Contr. Syst. Mag.* 21: 52–64.

Axelrod, R. 1984. *The evolution of cooperation.* New York: Basic Books. (Reprinted 1989, Harmondsworth, UK: Penguin.)

——— 1987. "The evolution of strategies in the iterated prisoner's dilemma." In L. Davis, ed., *Genetic algorithms and simulated annealing*, 32–41. London: Pitman.

Axelrod, R., and D. Dion. 1988. "The further evolution of cooperation." *Science* 242: 1385–1390.

Axelrod, R., and W. D. Hamilton. 1981. "The evolution of cooperation." *Science* 211: 1390–1396.

Bach, S. P., A. G. Renehan, and C. S. Potten. 2000. "Stem cells: The intestinal stem cell as a paradigm." *Carcinogenesis* 21: 469–476.

Bailey, N. J. T. 1975. *The mathematical theory of infectious diseases and its application.* London: Griffin.

Baker, M. C. 2001. *The atoms of language: The mind's hidden rules of grammar*. New York: Basic Books.

Bala, V., and S. Goyal. 2000. "A noncooperative model of network formation." *Econometrica* 68: 1181–1229.

Balfe, P., P. Simmonds, C. A. Ludlam, J. O. Bishop, and A. J. L. Brown. 1990. "Concurrent evolution of human-immunodeficiency-virus type-1 in patients infected from the same source—rate of sequence change and low-frequency of inactivating mutations." *J. Virol.* 64: 6221–6233.

Barabasi, A., and R. Albert. 1999. "Emergence of scaling in random networks." *Science* 286: 509–512.

Bardelli, A., D. P. Cahill, G. Lederer, M. R. Speicher, K. W. Kinzler, B. Vogelstein, and C. Lengauer. 2001. "Carcinogen-specific induction of genetic instability." *P. Natl. Acad. Sci. USA* 98: 5770–5775.

Bar-Hillel, Y. 1953. "A quasi-arithmetical notation for syntactic description." *Language* 29: 47–58.

Barton, N. 1993. "The probability of fixation of a favoured allele in a subdivided population." *Genet. Res.* 62: 149–158.

——— 2000. "Genetic hitchhiking." *Philos. T. Roy. Soc. B* 355: 1553–1562.

——— 2001. "The role of hybridization in evolution." *Mol. Ecol.* 10: 551–568.

Barton, N., F. Depaulis, and A. M. Etheridge. 2002. "Neutral evolution in spatially continuous populations." *Theor. Popul. Biol.* 61: 31–48.

Batali, J. 1994. "Innate biases and critical periods: Combining evolution and learning in the acquisition of syntax." In R. A. Brooks and P. Maes, eds., *Artificial life IV: Proceedings of the fourth international workshop on the synthesis and simulation of living systems, MIT*, 160–171. Cambridge: MIT Press.

Bates, E., and B. MacWhinney. 1982. "Functionalist approaches to grammar." In E. Wanner and L. R. Gleitman, eds., *Language acquisition: The state of the art*, 173–218. Cambridge: Cambridge University Press.

Berlekamp, E. R., J. H. Conway, and R. K. Guy. 1982a. *Winning ways for your mathematical plays*. Vol. 1: *Games in general*. New York: Academic Press.

——— 1982b. *Winning ways for your mathematical plays*. Vol. 2: *Games in particular*. New York: Academic Press.

——— 2001. *Winning ways for your mathematical plays*. Vol. 1. 2nd ed. Natick, MA: A K Peters, Ltd.

——— 2001. *Winning ways for your mathematical plays*. Vol. 2. 2nd ed. Natick, MA: A K Peters, Ltd.

Bernoulli, D. 1760. "Essai d'une nouvelle analyse de la mortalité causée par la petite vérole et des advantages de l'inoculation pour la prévenir." *Mém. Math. Phys. Acad. Roy. Sci., Paris*, 1–45.

Bertolo, S., ed. 2001. *Language acquisition and learnability*. Cambridge: Cambridge University Press.

Bickerton, D. 1990. *Language and species*. Chicago: University of Chicago Press.

Binmore, K. 1994. *Game theory and the social contract*. Cambridge: MIT Press.

Binmore, K., and L. Samuelson. 1992. "Evolutionary stability in repeated games played by finite automata." *J. Econ. Theory* 57: 278–305.

Bissell, M. J., and D. Radisky. 2001. "Putting tumors in context." *Nat. Rev. Cancer* 1: 46–54.

Boerlijst, M. C., S. Bonhoeffer, and M. A. Nowak. 1996. "Viral quasi-species and recombination." *P. Roy. Soc. Lond. B Bio.* 263: 1577–1584.

Boerlijst, M., and P. Hogeweg. 1991a. "Self-structuring and selection: Spiral waves as a substrate for prebiotic evolution." In C. G. Langton, C. Taylor, J. D. Farmer, and S. Rasmussen, eds., *Artificial Life II*, SFI studies in the sciences of complexity, vol. 10, 255–276. Boston: Addison-Wesley.

——— 1991b. "Spiral wave structure in pre-biotic evolution: Hypercycles stable against parasites." *Physica D* 48: 17–28.

Boerlijst, M. C., M. A. Nowak, and K. Sigmund. 1997. "The logic of contrition." *J. Theor. Biol.* 185: 281–293.

Bomze, I. M. 1983. "Lotka-Volterra equations and replicator dynamics: A two-dimensional classification." *Biol. Cybern.* 48: 201–211.

Bonhoeffer, S., C. Chappey, N. T. Parkin, J. M. Whitcomb, and C. J. Petropoulos. 2004. "Evidence for positive epistasis in HIV-1." *Science* 306: 1547–1550.

Bonhoeffer, S., R. M. May, G. M. Shaw, and M. A. Nowak. 1997. "Virus dynamics and drug therapy." *P. Natl. Acad. Sci. USA* 94: 6971–6976.

Bonhoeffer, S., and M. A. Nowak. 1994a. "Intra-host versus inter-host selection: Viral strategies of immune function impairment." *P. Natl. Acad. Sci. USA* 91: 8062–8066.

——— 1994b. "Mutation and the evolution of virulence." *P. Roy. Soc. Lond. B Bio.* 258: 133–140.

Bonhoeffer, S., and P. F. Stadler. 1993. "Error thresholds on correlated fitness landscapes." *J. Theor. Biol.* 164: 359–372.

Borrow, P., H. Lewicki, X.-P. Wei, M. S. Horwitz, N. Peffer, H. Meyers, J. A. Nelson, J. E. Gairin, B. H. Hahn, M. B. A. Oldstone, and G. M. Shaw. 1997. "Antiviral pressure exerted by HIV-1 specific cytotoxic T lymphocytes (CTLs) during primary infection demonstrated by rapid selection of CTL escape virus." *Nat. Med.* 3: 205–211.

Boveri, T. 1914. *Zur Frage der Entstehung maligner Tumoren*. Jena, Germany: Gustav Fischer. (English translation, 1929. *The origin of malignant tumors*. Trans. M. Boveri. Baltimore: Williams and Wilkins.)

Boyd, R. 1989. "Mistakes allow evolutionary stability in the repeated prisoner's dilemma game." *J. Theor. Biol.* 136: 47–56.

Boyd, R., H. Gintis, S. Bowles, and P. J. Richerson. 2003. "The evolution of altruistic punishment." *P. Natl. Acad. Sci. USA* 100: 3531–3535.

Boyd, R., and J. P. Lorberbaum. 1987. "No pure strategy is evolutionarily stable in the repeated prisoner's dilemma game." *Nature* 327: 58–59.

Boyd, R., and J. B. Silk. 1997. *How humans evolved.* 1st ed. New York: W. W. Norton. (4th ed. 2005).

Boyd, S., P. Diaconis, and L. Xiao. 2004. "Fastest mixing Markov chain on a graph." *SIAM Rev.* 46: 667–689.

Boyer, J. C., A. Umar, J. I. Risinger, J. R. Lipford, M. Kane, S. Yin, J. C. Barrett, R. D. Kolodner, and T. A. Kunkel. 1995. "Microsatellite instability, mismatch repair deficiency, and genetic defects in human cancer cell lines." *Cancer Res.* 55: 6063–6070.

Brandon, R. N., and N. Hornstein. 1986. "From icons to symbols: Some speculations on the origins of language." *Biol. Philos.* 1: 169–189.

Brandt, H., and K. Sigmund. 2005. "Indirect reciprocity, image scoring, and moral hazard." *P. Natl. Acad. Sci. USA* 102: 2666–2670.

Bremermann, H. J., and J. Pickering. 1983. "A game-theoretical model of parasite virulence." *J. Theor. Biol.* 100: 411–426.

Bremermann, H. J., and H. R. Thieme. 1989. "A competitive-exclusion principle for pathogen virulence." *J. Math. Biol.* 27: 179–190.

Brent, M., ed. 1997. *Computational approaches to language acquisition.* Cambridge: MIT Press.

Bresnan, J. 2000. *Lexical-functional syntax.* Oxford: Blackwell.

Bull, J. J., and I. J. Molineux. 1992. "Molecular genetics of adaptation in an experimental model of cooperation." *Evolution* 46: 882–895.

Bull, J. J., I. J. Molineux, and W. R. Rice. 1991. "Selection of benevolence in a host-parasite system." *Evolution* 45: 875–882.

Bürger, R. 2000. *The mathematical theory of selection, recombination, and mutation.* Chichester, UK: Wiley.

——— 2002. "On a genetic model of intraspecific competition and stabilizing selection." *Am. Nat.* 160: 661–682.

Busenberg, S., and K. Cooke. 1993. *Vertically transmitted diseases.* Berlin: Springer Verlag.

Cairns, J. 1975. "Mutation selection and the natural history of cancer." *Nature* 255: 197–200.

——— 2002. "Somatic stem cells and the kinetics of mutagenesis and carcinogenesis." *P. Natl. Acad. Sci. USA* 99: 10567–10570.

Cangelosi, A., and D. Parisi, eds. 2002. *Simulating the evolution of language*. London: Springer.

Cavalli-Sforza, L. L., and M. W. Feldman. 1981. *Cultural transmission and evolution: A quantitative approach*. Princeton, NJ: Princeton University Press.

Chang, S., C. Khoo, and R. A. DePinho. 2001. "Modeling chromosomal instability and epithelial carcinogenesis in the telomerase-deficient mouse." *Semin. Cancer Biol.* 11: 227–239.

Chang, S., C. Khoo, M. L. Naylor, R. S. Maser, and R. A. DePinho. 2003. "Telomere-based crisis: Functional differences between telomerase activation and ALT in tumor progression." *Gene. Dev.* 17: 88–100.

Chomsky, N. A. 1956. "Three models for the description of language." *IRE T. Inform. Theor.* 2: 113–124.

——— 1957. *Syntactic structures*. Berlin: Mouton.

——— 1965. *Aspects of the theory of syntax*. Cambridge: MIT Press.

——— 1972. *Language and mind*. New York: Harcourt Brace Jovanovich.

——— 1981. "Principles and parameters in syntactic theory." In N. Hornstein and D. Lightfoot, eds., *Explanation in linguistics: The logical problem of language acquisition*, 123–146. London: Longman.

——— 1993. *Lectures on government and binding: The Pisa lectures*. 7th ed. Berlin: Mouton de Gruyter. (First published 1981. Dordrecht: Foris Publications.)

Christiansen, F. B. 1991. "On conditions for evolutionary stability for a continuously varying character." *Am. Nat.* 138: 37–50.

Christiansen, M. H., R. A. C. Dale, M. R. Ellefson, and C. M. Conway. 2002. "The role of sequential learning in language evolution: Computational and experimental studies." In A. Cangelosi and D. Parisi, eds., *Simulating the evolution of language*, 165–187. London: Springer.

Coffin, J. M., S. H. Hughes, and H. E. Varmus, eds. 1997. *Retroviruses*. Cold Spring Harbor, NY: Cold Spring Harbor Laboratory Press.

Comrie, B. 1981. *Language universals and linguistic typology*. Chicago: University of Chicago Press.

Cressman, R. 2003. *Evolutionary dynamics and extensive form games*. Cambridge: MIT Press.

Darwin, C. 1859. *On the origin of species*. London: J. Murray.

——— 1871. *The descent of man*. London: J. Murray.

Davidson, J. M., K. L. Gorringe, S.-F. Chin, B. Orsetti, C. Besret, C. Courtay-Cahen, I. Roberts, C. Theillet, C. Caldas, and P. A. W. Edwards. 2000. "Molecular cytogenetic analysis of breast cancer cell lines." *Brit. J. Cancer* 83: 1309–1317.

Dawkins, R. 1982. *The extended phenotype*. Oxford: W. H. Freeman.

Deacon, T. 1997. *The symbolic species*. London: Penguin Books.

De Boer, R. J., and M. C. B. Boerlijst. 1994. "Diversity and virulence thresholds in AIDS." *P. Nat. Acad. Sci. USA* 91: 544–548.

Demetrius, L. 1987. "Random spin models and chemical kinetics." *J. Chem. Phys.* 87: 6939–6946.

Dennett, D. C. 1995. *Darwin's dangerous idea: Evolution and the meanings of life*. New York: Simon & Schuster.

Diamond, J. M. 1992. *The third chimpanzee: The evolution and future of the human animal*. New York: HarperCollins.

Dieckmann, U. 1997. "Can adaptive dynamics invade?" *Trends Ecol. Evol.* 12: 128–131.

Dieckmann, U., and M. Doebeli. 1999. "On the origin of species by sympatric speciation." *Nature* 400: 354–357.

Dieckmann, U., and R. Law. 1996. "The dynamical theory of coevolution: A derivation from stochastic ecological processes." *J. Math. Biol.* 34: 579–612.

Dieckmann, U., R. Law, and J. A. J. Metz, eds. 2000. *The geometry of ecological interactions: Simplifying spatial complexity*. Cambridge: Cambridge University Press.

Dieckmann, U., P. Marrow, and R. Law. 1995. "Evolutionary cycling in predator-prey interactions: Population dynamics and the red queen." *J. Theor. Biol.* 176: 91–102.

Diekmann, O., and J. A. P. Heesterbeek. 2000. *Mathematical epidemiology of infectious diseases: Model building, analysis, and interpretation*. Chichester, UK: Wiley.

Diekmann, O., J. A. P. Heesterbeek, and J. A. J. Metz. 1990. "On the definition and the computation of the basic reproductive ratio R_0 in models for infectious diseases in heterogeneous populations." *J. Math. Biol.* 28: 365–382.

Dietz, K. 1975. "Transmission and control of arbovirus diseases." In D. Ludwig and K. L. Cooke, eds., *Epidemiology*, 104–121. Philadelphia: SIAM.

Doebeli, M., and U. Dieckmann. 2004. "Adaptive dynamics of speciation: Spatial structure." In U. Dieckmann, J. A. J. Metz, M. Doebeli, and D. Tautz, eds., *Adaptive Speciation*, 140–167. Cambridge: Cambridge University Press.

Doebeli, M., C. Hauert, and T. Killingback. 2004. "The evolutionary origin of cooperators and defectors." *Science* 306: 859–862.

Doebeli, M., and G. D. Ruxton. 1997. "Evolution of dispersal rates in metapopulation models: Branching and cyclic dynamics in phenotype space." *Evolution* 51: 1730–1741.

Domjan, M., and B. Burkhard. 1986. *The principles of learning and behavior*. Monterey, CA: Brooks/Cole.

Drake, J. W. 1991. A constant rate of spontaneous mutation in DNA-based microbes. *P. Natl. Acad. Sci. USA* 88: 7160–7164.

——— 1993. Rates of spontaneous mutation among RNA viruses. *P. Natl. Acad. Sci. USA* 90: 4171–4175.

Drake, J. W., B. Charlesworth, D. Charlesworth, and J. F. Crow. 1998. Rates of spontaneous mutation. *Genetics* 148: 1667–1686.

Dugatkin, L. A. 1988. "Do guppies play tit for tat during predator inspection visits?" *Behav. Ecol. Sociobiol.* 23: 395–399.

——— 1997. *Cooperation among animals*. Oxford: Oxford University Press.

Dugatkin, L. A., and H. K. Reeve, eds. 1998. *Game theory and animal behaviour*. Oxford: Oxford University Press.

Dunbar, R. 1996. *Grooming, gossip, and the evolution of language*. Cambridge: Cambridge University Press.

Durrett, R. 1988. *Lecture notes on particle systems and percolation*. Stamford, CT: Wadsworth and Brooks/Cole Advanced Books and Software.

——— 1999. "Stochastic spatial models." *SIAM Rev.* 41: 677–718.

Durrett, R., and S. A. Levin. 1994a. "The importance of being discrete (and spatial)." *Theor. Popul. Biol.* 46: 363–394.

——— 1994b. "Stochastic spatial models: A user's guide to ecological applications." *Philos. T. Roy. Soc. B* 343: 329–350.

——— 1998. "Spatial aspects of interspecific competition." *Theor. Popul. Biol.* 53: 30–43.

Ebel, H., and S. Bornholdt. 2002. "Coevolutionary games on networks." *Phys. Rev. E* 66: 056118.

Ebert, D. 1994. "Virulence and local adaptation of a horizontally transmitted parasite." *Science* 265: 1084–1086.

Edelstein-Keshet, L. 2004. *Mathematical models in biology*. Philadelphia: SIAM.

Eigen, M. 1992. *Steps towards life: A perspective on evolution*. Oxford: Oxford University Press.

Eigen, M., J. McCaskill, and P. Schuster. 1989. "The molecular quasi-species." *Adv. Chem. Phys.* 75: 149–263.

Eigen, M., and P. Schuster. 1979. *The hypercycle*. New York: Springer.

Ellison, G. 1993. "Learning, local interaction, and coordination." *Econometrica* 61: 1047–1071.

Elman, J. L., E. A. Bates, M. H. Johnson, A. Karmiloff-Smith, D. Parisi, and K. Plunkett. 1996. *Rethinking innateness: A connectionist perspective on development*. Cambridge: MIT Press.

Epstein, J. M. 1998. "Zones of cooperation in demographic prisoner's dilemma." *Complexity* 4: 36–48.

Erdös, P., and A. Rényi. 1960. "On the evolution of random graphs." *Acta Math. Acad. Sci. H.* 5: 17–61.

Eshel, I. 1977. "On the founder effect and the evolution of altruistic traits: An ecogenetic approach." *Theor. Popul. Biol.* 11: 410–424.

——— 1983. "Evolutionary and continuous stability." *J. Theor. Biol.* 103: 99–111.

——— 1996. "On the changing concept of evolutionary population stability as a reflection of a changing point of view in the quantitative theory of evolution." *J. Math. Biol.* 34: 485–510.

Ewald, P. W. 1993. "The evolution of virulence." *Sci. Am.* 268: 86–93.

Ewens, W. J. 2004. *Mathematical population genetics*. 2nd ed. Berlin: Springer.

Falster, D. S., and M. Westoby. 2003. "Plant height and evolutionary games." *Trends Ecol. Evol.* 18: 337–343.

Farr, W. 1840. "Progress of epidemics." *Second report of the Registrar General of England*, 91–98.

Fehr, E., and U. Fischbacher. 2003. "The nature of human altruism." *Nature* 425: 785–791.

Fehr, E., and S. Gächter. 2000. "Cooperation and punishment in public goods experiments." *Am. Econ. Rev.* 90: 980–994.

——— 2002. "Altruistic punishment in humans." *Nature* 415: 137–140.

Fenner, F., and F. N. Ratcliffe. 1965. *Myxomatosis*. Cambridge: Cambridge University Press.

Ferrière, R., and R. E. Michod. 1996. "The evolution of cooperation in spatially heterogeneous populations." *Am. Nat.* 147: 692–717.

Ficici, S. G., and J. B. Pollack. 2000. "Effects of finite populations on evolutionary stable strategies." In D. Whitley, D. Goldberg, E. Cantú-Paz, L. Spector, I. Parmee, and H.-G. Beyer, eds., *Gecco 2000: Proceedings of the genetic and evolutionary computation conference, July 10–12, Las Vegas, NV*, 927–934. San Francisco: Morgan-Kaufmann.

Fisher, J. C. 1959. "Multiple-mutation theory of carcinogenesis." *Nature* 181: 651–652.

Fisher, R. A. 1930a. "The distribution of gene ratios for rare mutations." *P. Roy. Soc. Edinb. B* 50: 205–220.

——— 1930b. *The genetical theory of natural selection*. Oxford: Oxford University Press.

Fisher, R. A., and E. B. Ford. 1950. "The Sewall Wright effect." *Heredity* 4: 117–119.

Fishman, M. A. 2003. "Indirect reciprocity among imperfect individuals." *J. Theor. Biol.* 225: 285–292.

Fitch, W. T. 2000. "The evolution of speech: A comparative review." *Trends Cogn. Sci.* 4: 258–267.

Flack, J. C., M. C. Girvan, F. B. M. de Waal, and D. C. Krakauer. 2006. "Policing stabilizes construction of social niches in primates." *Nature* 439: 426–429.

Flack, J. C., D. C. Krakauer, and F. B. M. de Waal. 2005. "Robustness mechanisms in primate societies: A perturbation study." *P. Roy. Soc. Lond. B Bio.* 272: 1091–1099.

Fogel, D., G. Fogel, and P. Andrews. 1998. "On the instability of evolutionary stable strategies in small populations." *Ecol. Model.* 109: 283–294.

Fontana, W. 2002. "Modelling 'evo-devo' with RNA." *BioEssays* 24: 1164–1177.

Fontana, W., and L. W. Buss. 1994. "What would be conserved if 'the tape were played twice'?" *P. Natl. Acad. Sci. USA* 91: 757–761.

Fontana, W., and P. Schuster. 1987. "A computer model of evolutionary optimization." *Biophys. Chem.* 26: 123–147.

——— 1998. "Continuity in evolution: On the nature of transitions." *Science* 280: 1451–1455.

Fontana, W., P. F. Stadler, E. G. Bornberg-Bauer, T. Griesmacher, I. L. Hofacker, M. Tacker, P. Tarazona, E. D. Weinberger, and P. Schuster. 1993. "RNA folding and combinatory landscapes." *Phys. Rev. E* 47: 2083–2099.

Frank, S. A. 1992. "A kin selection model for the evolution of virulence." *P. Roy. Soc. Lond. B Bio.* 250: 195–197.

——— 1998. *Foundations of social evolution*. Princeton, NJ: Princeton University Press.

——— 2002. *Immunology and evolution of infectious disease*. Princeton, NJ: Princeton University Press.

——— 2005. "Age-specific incidence of inherited versus sporadic cancers: A test of the multistage theory of carcinogenesis." *P. Natl. Acad. Sci. USA* 102: 1071–1075.

Frank, S. A., Y. Iwasa, and M. A. Nowak. 2003. "Patterns of cell division and the risk of cancer." *Genetics* 163: 1527–1532.

Frank, S. A., and M. A. Nowak. 2003. "Cell biology: Developmental predisposition to cancer." *Nature* 422: 494.

Frauenfelder, H., A. R. Bishop, A. Garcia, A. Perelson, P. Schuster, D. Sherrington, and P. J. Swart, eds. 1997. "Sixteenth annual international conference of the Center for Nonlinear Studies." *Physica D* 107: 117–439.

Frean, M. R. 1994. "The prisoner's dilemma without synchrony." *P. Roy. Soc. Lond. B Bio.* 257: 75–79.

Friend, S. H., R. Bernards, S. Rogelj, R. A. Weinberg, J. M. Rapaport, D. M. Albert, and T. P. Dryja. 1986. "A human DNA segment with properties of the gene that predisposes to retinoblastoma and osteosarcoma." *Nature* 323: 643–646.

Fudenberg, D., and C. Harris. 1992. "Evolutionary dynamics with aggregate shocks." *J. Econ. Theory* 57: 420–441.

Fudenberg, D., L. A. Imhof, M. A. Nowak, and C. Taylor. 2006. "Stochastic evolution as a generalized Moran process." Preprint.

Fudenberg, D., and D. K. Levine. 1998. *The theory of learning in games.* Cambridge: MIT Press.

Fudenberg, D., and E. Maskin. 1986. "The folk theorem in repeated games with discounting or with incomplete information." *Econometrica* 50: 533–554.

——— 1990. "Evolution and cooperation in noisy repeated games." *Am. Econ. Rev.* 80: 274–279.

Fudenberg, D., and J. Tirole. 1991. *Game theory*. Cambridge: MIT Press.

Gasarch, W. I., and C. H. Smith. 1992. "Learning via queries." *J. Assoc. Comput. Mach.* 39: 649–674.

Gatenby, R. A., and P. K. Maini. 2003. "Mathematical oncology: Cancer summed up." *Nature* 421: 321.

Gatenby, R. A., and T. L. Vincent. 2003. "An evolutionary model of carcinogenesis." *Cancer Res.* 63: 6212–6220.

Gavrilets, S. 2004. *Fitness landscapes and the origin of species*. Princeton, NJ: Princeton University Press.

Geritz, S. A. H., J. A. J. Metz, É. Kisdi, and G. Meszéna. 1997. "Dynamics of adaptation and evolutionary branching." *Phys. Rev. Lett.* 78: 2024–2027.

Gibson, E., and K. Wexler. 1994. "Triggers." *Linguist. Inq.* 25: 407–454.

Gillespie, J. H. 1991. *The causes of molecular evolution.* Oxford: Oxford University Press.

Gintis, H. 2000. *Game theory evolving: A problem-centered introduction to modeling strategic interaction.* Princeton, NJ: Princeton University Press.

Godfray, H. C. J. 1995. "Evolutionary theory of parent-offspring conflict." *Nature* 376: 133–138.

Gold, E. M. 1967. "Language identification in the limit." *Inform. Control* 10: 447–474.

——— 1978. "Complexity of automaton identification from given data." *Inform. Control* 37: 303–320.

Goldie, J. H., and A. J. Coldman. 1979. "A mathematic model for relating the drug sensitivity of tumors to their spontaneous mutation rate." *Cancer Treat. Rep.* 63: 1727–1733.

——— 1983. "Quantitative model for multiple levels of drug resistance in clinical tumors." *Cancer Treat. Rep.* 67: 923–931.

Goldsmith, J. 2001. "Unsupervised learning of the morphology of a natural language." *Comput. Linguist.* 27: 153–198.

Gopnik, M., and M. Crago. 1991. "Familial aggregation of a developmental language disorder." *Cognition* 39: 1–50.

Gould, S. J. 2002. *The structure of evolutionary theory*. Cambridge, MA: Belknap Press of Harvard University Press.

Goulder, P. J. R., R. E. Phillips, R. A. Colbert et al. 1997. "Late escape from an immunodominant cytotoxic T-lymphocyte response associated with progression to AIDS." *Nature Medicine*, 3: 212–217.

Greenberg, J. H., C. A. Ferguson, and E. A. Moravcsik, eds. 1978. *Universals of human language*. Stanford: Stanford University Press.

Grist, S. A., M. McCarron, A. Kutlaca, D. R. Turner, and A. A. Morley. 1992. "In vivo human somatic mutation: Frequency and spectrum with age." *Mutat. Res.* 266: 189–196.

Gyllenberg, M., K. Parvinen, and U. Dieckmann. 2002. "Evolutionary suicide and evolution of dispersal in structured metapopulations." *J. Math. Biol.* 45: 79–105.

Haig, D. 2002. *Genomic imprinting and kinship*. Piscataway, NJ: Rutgers University Press.

Haigis, K. M., J. G. Caya, M. Reichelderfer, and W. F. Dove. 2002. "Intestinal adenomas can develop with a stable karyotype and stable microsatellites." *P. Natl. Acad. Sci. USA* 99: 8927–8931.

Haldane, J. B. S. 1932. *The causes of evolution*. London: Longmans, Green.

Hamer, W. H. 1906. "Epidemic disease in England." *The Lancet*, i: 733–739.

Hamilton, W. D. 1964a. "The genetical evolution of social behaviour I." *J. Theor. Biol.* 7: 1–16.

——— 1964b. "The genetical evolution of social behaviour II." *J. Theor. Biol.* 7: 17–52.

——— 1967. "Extraordinary sex ratios." *Science* 156: 477–488.

——— 1971. "Selection of selfish and altruistic behavior in some extreme models." In J. F. Eisenberg and W. S. Dillon, eds., *Man and beast: Comparative social behavior*, 57–91. Washington, DC: Smithsonian Press.

——— 1996. *Narrow roads of gene land 1: Evolution of social behaviour*. New York: W. H. Freeman.

——— 2001. *Narrow roads of gene land 2: Evolution of sex*. Oxford: Oxford University Press.

Hammerstein, P. 1996. "Darwinian adaptation, population genetics, and the streetcar theory of evolution." *J. Math. Biol.* 34: 511–532.

———, ed. 2003. *Genetic and cultural evolution of cooperation*. Cambridge: MIT Press.

Haraguchi, Y., and A. Sasaki. 2000. "The evolution of parasite virulence and transmission rate in a spatially structured population." *J. Theor. Biol.* 203: 85–96.

Hardy, I. C. W., ed. 2002. *Mother knows best. Sons or daughters? A review of sex ratios. Concepts and methods*. Cambridge: Cambridge University Press.

Harrison, M. A. 1978. *Introduction to formal language theory*. Boston: Addison-Wesley.

Hartl, D. L., and A. G. Clark. 1997. *Principles of population genetics*. 3rd ed. Sunderland, MA: Sinauer.

Hashimoto, T., and T. Ikegami. 1996. "Emergence of net-grammar in communicating agents." *BioSystems* 38: 1–14.

Hassell, M. P., H. N. Comins, and R. M. May. 1991. "Spatial structure and chaos in insect population dynamics." *Nature* 353: 255–258.

——— 1994. "Species coexistence and self-organizing spatial dynamics." *Nature* 370: 290–292.

Hauert, C., S. De Monte, J. Hofbauer, and K. Sigmund. 2002. "Volunteering as red queen mechanism for cooperation in public goods games." *Science* 296: 1129–1132.

Hauert, C., and M. Doebeli. 2004. "Spatial structure often inhibits the evolution of cooperation in the snowdrift game." *Nature* 428: 643–646.

Hauser, M. D. 1996. *The evolution of communication*. Cambridge: Harvard University Press.

Hawkins, J. A., and M. Gell-Mann. 1992. *The evolution of human languages*. Reading, MA: Addison-Wesley.

Hazlehurst, B., and E. Hutchins. 1998. "The emergence of propositions from the coordination of talk and action in a shared world." *Lang. Cognitive Proc.* 13: 373–424.

Heinsohn, R., and C. Packer. 1995. "Complex cooperative strategies in group-territorial African lions." *Science* 269: 1260–1262.

Hendry, A. P., J. K. Wenburg, P. Bentzen, E. C. Volk, and T. P. Quinn. 2000. "Rapid evolution of reproductive isolation in the wild: Evidence from introduced salmon." *Science* 290: 516–518.

Henrich, J., R. Boyd, S. Bowles, C. Camerer, E. Fehr, H. Gintis, and R. McElreath. 2001. "In search of *Homo economicus*: Behavioral experiments in 15 small-scale societies." *Am. Econ. Rev.* 91(2): 73–78.

Hermsen, M., C. Postma, J. Baak, M. Weiss, A. Rapallo, A. Sciutto, G. Roemen, J.-W. Arends, R. Williams, W. Giaretti, A. de Goeij, and G. Meijer. 2002. "Colorectal adenoma to carcinoma progression follows multiple pathways of chromosomal instability." *Gastroenterology* 123: 1109–1119.

Herre, E. A. 1993. "Population structure and the evolution of virulence in nematode parasites of fig wasps." *Science* 259: 1442–1445.

Herz, A. V. M. 1994. "Collective phenomena in spatially extended evolutionary games." *J. Theor. Biol.* 169: 65–87.

Ho, D. D., A. U. Neumann, A. S. Perelson, W. Chen, J. M. Leonard, and M. Markowitz. 1995. "Rapid turnover of plasma virions and CD4 lymphocytes in HIV-1 infection." *Nature* 373: 123–126.

Hofbauer, J., P. Schuster, and K. Sigmund. 1979. "A note on evolutionary stable strategies and game dynamics." *J. Theor. Biol.* 81: 609–612.

Hofbauer, J., and K. Sigmund. 1990. "Adaptive dynamics and evolutionary stability." *Appl. Math. Lett.* 3: 75–79.

——— 1998. *Evolutionary games and population dynamics*. Cambridge: Cambridge University Press.

——— 2003. "Evolutionary game dynamics." *B. Am. Math. Soc.* 40: 479–519.

Houston, A. I., and J. M. McNamara. 1999. *Models of adaptive behaviour: An approach based on state*. Cambridge: Cambridge University Press.

Hurford, J. R. 1989. "Biological evolution of the Saussurean sign as a component of the language acquisition device." *Lingua* 77: 187–222.

Hurford, J. R., M. Studdert-Kennedy, and C. Knight, eds. 1998. *Approaches to the evolution of language*. Cambridge: Cambridge University Press.

Ifti, M., T. Killingback, and M. Doebeli. 2004. "Effects of neighbourhood size and connectivity on the spatial continuous prisoner's dilemma." *J. Theor. Biol.* 231: 97–106.

Imhof, L. A., D. Fudenberg, and M. A. Nowak. 2005. "Evolutionary cycles of cooperation and defection." *P. Natl. Acad. Sci. USA* 102: 10797–10800.

Imhof, L. A., and M. A. Nowak. 2006. "Evolutionary game dynamics in a Wright-Fisher process" *J. Math. Biol.* Submitted.

Irwin, A. J., and P. D. Taylor. 2001. "Evolution of altruism in stepping-stone populations with overlapping generations." *Theor. Popul. Biol.* 60: 315–325.

Iwasa, Y., F. Michor, N. L. Komarova, and M. A. Nowak. 2005. "Population genetics of tumor suppressor genes." *J. Theor. Biol.* 233: 15–23.

Iwasa, Y., F. Michor, and M. A. Nowak. 2004. "Stochastic tunnels in evolutionary dynamics." *Genetics* 166: 1571–1579.

Iwasa, Y., and A. Pomiankowski. 1995. "Continual change in mate preferences." *Nature* 377: 420–422.

——— 2001. "The evolution of X-linked genomic imprinting." *Genetics* 158: 1801–1809.

Iwasa, Y., and A. Sasaki. 1987. "Evolution of the number of sexes." *Evolution* 41: 49–65.

Jackendoff, R. 1997. *The architecture of the language faculty*. Cambridge: MIT Press.

——— 1999. "Possible stages in the evolution of the language capacity." *Trends Cogn. Sci.* 3: 272–279.

——— 2002. *Foundations of language: Brain, meaning, grammar, evolution*. Oxford: Oxford University Press.

Jacquard, A. 1974. *The genetic structure of populations*. New York: Springer Verlag.

Jain, S., D. Osherson, J. S. Royer, and A. Sharma. 1999. *Systems that learn: An introduction to learning theory*. 2nd ed. Cambridge: MIT Press.

Johnstone, R. A. 2002. "Signalling theory." In M. D. Pagel, ed., *Encyclopedia of evolution*, Vol. 2, 1059–1062. Oxford: Oxford University Press.

Joshi, A. K., L. Levy, and M. Takahashi. 1975. "Tree adjunct grammars." *J. Comput. Syst. Sci.* 10: 136–163.

Karlin, S., and S. Tavare. 1983. "A class of diffusion processes with killing arising in population genetics." *SIAM J. Appl. Math.* 43: 31–41.

Karlin, S., and H. E. Taylor. 1975. *A first course in stochastic processes*. 2nd ed. London: Academic Press.

Kauffman, S. 1993. *The origins of order: Self-organization and selection in evolution*. Oxford: Oxford University Press.

Kauffman, S., and S. Levin. 1987. "Towards a general theory of adaptive walks on rugged landscapes." *J. Theor. Biol.* 128: 11–45.

Keller, L., ed. 1999. *Levels of selection in evolution*. Princeton, NJ: Princeton University Press.

Kermack, W. O., and A. G. McKendrick. 1933. "Contributions to the mathematical theory of epidemics III—further studies of the problem of endemicity." *P. Roy. Soc. Lond. A Mat.* 141: 94–122.

Kerr, B., M. A. Riley, M. W. Feldman, and B. J. M. Bohannan. 2002. "Local dispersal promotes biodiversity in a real-life game of rock-paper-scissors." *Nature* 418: 171–174.

Killingback, T., and M. Doebeli. 1996. "Spatial evolutionary game theory: Hawks and Doves revisited." *P. Roy. Soc. Lond. B Bio.* 263: 1135–1144.

——— 1998. "Self-organized criticality in spatial evolutionary game theory." *J. Theor. Biol.* 191: 335–340.

——— 2002. "The continuous prisoner's dilemma and the evolution of cooperation through reciprocal altruism with variable investment." *Am. Nat.* 160: 421–438.

Kimura, M. 1968. "Evolutionary rate at the molecular level." *Nature* 217: 624–626.

——— 1983. *The neutral theory of molecular evolution*. Cambridge: Cambridge University Press.

——— 1985. "The role of compensatory neutral mutations in molecular evolution." *J. Genet.* 64: 7–19.

——— 1994. *Population genetics, molecular evolution, and the neutral theory: Selected papers*. Chicago: University of Chicago Press.

Kinzler, K. W., M. C. Nilbert, B. Vogelstein, T. M. Bryan, D. B. Levy, K. J. Smith, A. C. Preisinger, S. R. Hamilton, P. Hedge, A. Markham, M. Carlson, G. Joslyn, J. Groden, R. White, Y. Miki, Y. Miyoshi, I. Nishisho, and Y. Nakamura. 1991. "Identification of a gene located at chromosome 5q21 that is mutated in colorectal cancers." *Science* 251: 1366–1370.

Kinzler, K. W., and B. Vogelstein. 1997. "Gatekeepers and caretakers." *Nature* 386: 761–763.

Kirby, S., and J. Hurford. 1997. "Learning, culture, and evolution in the origin of linguistic constraints." In P. Husbands and I. Harvey, eds., *Fourth European conference on artificial life*, 493–502. Cambridge: MIT Press.

Kirschner, M. W., and J. C. Gerhart. 2005. *The plausibility of life: Resolving Darwin's dilemma*. New Haven, CT: Yale University Press.

Knight, C., M. Studdert-Kennedy, and J. Hurford. 2000. *The evolutionary emergence of language: Social function and the origins of linguistic form*. Cambridge: Cambridge University Press.

Knolle, H. 1989. "Host density and the evolution of parasite virulence." *J. Theor. Biol.* 136: 199–207.

Knudson, A. G. 1971. "Mutation and cancer: Statistical study of retinoblastoma." *P. Natl. Acad. Sci. USA* 68: 820–823.

——— 1993. "Antioncogenes and human cancer." *P. Natl. Acad. Sci. USA* 90: 10914–10921.

——— 2001. "Two genetic hits (more or less) to cancer." *Nat. Rev. Cancer* 1: 157–162.

Kolmogorov, A. N. 1936. "Sulla teoria di Volterra della lotta per l'esistenza." *Giorn. Instituto Ital. Attuari* 7: 74–80.

Kolodner, R. D., C. D. Putnam, and K. Myung. 2002. "Maintenance of genome stability in *Saccharomyces cerevisiae*." *Science* 297: 552–557.

Komarova, N. L., C. Lengauer, B. Vogelstein, and M. A. Nowak. 2002. "Dynamics of genetic instability in sporadic and familial colorectal cancer." *Cancer Biotherapy* 1: 685–692.

Komarova, N. L., P. Niyogi, and M. A. Nowak. 2001. "The evolutionary dynamics of grammar acquisition." *J. Theor. Biol.* 209: 43–59.

Komarova, N. L., and M. A. Nowak. 2001a. "The evolutionary dynamics of the lexical matrix." *B. Math. Biol.* 63: 451–484.

——— 2001b. "Natural selection of the critical period for language acquisition." *P. Roy. Soc. Lond. B Bio.* 268: 1189–1196.

——— 2003. "Language dynamics in finite populations." *J. Theor. Biol.* 221: 445–457.

Komarova, N. L., and I. Rivin. 2001. "Mathematics of learning." arXiv.org preprint math.PR/0105235

Komarova, N. L., A. Sengupta, and M. A. Nowak. 2003. "Mutation-selection networks of cancer initiation: Tumor suppressor genes and chromosomal instability." *J. Theor. Biol.* 223: 433–450.

Kraines, D., and V. Kraines. 1989. "Pavlov and the prisoner's dilemma." *Theor. Decis.* 26: 47–79.

Krakauer, D. C. 2001. "Kin imitation for a private sign system." *J. Theor. Biol.* 213: 145–157.

Krakauer, D. C., and N. L. Komarova. 2003. "Levels of selection in positive strand virus dynamics." *J. Evolution. Biol.* 16: 64–73.

Krakauer, D. C., and A. Mira. 2000. "Mitochondria and the death of oocytes." *Nature* 403: 501.

Krakauer, D. C., and A. Sasaki. 2002. "Noisy clues to the origin of life." *P. Roy. Soc. Lond. B Bio.* 269: 2423–2428.

Kroch, A. 1989. "Reflexes of grammar in patterns of language change." *Language Variation and Change* 1: 199–244.

Krumhansl, J. A. 1997. "Landscapes in physics and biology: A tourist's impression." *Physica D* 107: 430–435.

Kunkel, T. A., and K. Bebenek. 2000. "DNA replication fidelity." *Annu. Rev. Biochem.* 69: 497–529.

Lai, C. S. L., S. E. Fisher, J. A. Hurst, F. Vargha-Khadem, and A. P. Monaco. 2001. "A forkhead-domain gene is mutated in a severe speech and language disorder." *Nature* 413: 519–523.

Lakoff, G. 1987. *Women, fire, and dangerous things: What categories reveal about the mind*. Chicago: University of Chicago Press.

Lamarck, J. B. P. A. 1809. *Philosophie zoologique: ou exposition des considérations relative à l'histoire naturelle des animaux*. Paris: Dentu et l'Auteur.

——— 2004. *Zoological philosophy: An exposition with regard to the natural history of animals*. Gold Beach, OR: High Sierra Books. (Originally published 1809.)

Lamlum, H., M. Ilyas, A. Rowan, S. Clark, V. Johnson, J. Bell, I. Frayling, J. Efstathiou, K. Pack, S. Payne, R. Roylance, P. Gorman, D. Sheer, K. Neale, R. Phillips, I. Talbot, W. Bodmer, and I. Tomlinson. 1999. "The type of somatic mutation at APC in familial adenomatous polyposis is determined by the site of the germline mutation: A new facet to Knudson's 'two-hit' hypothesis." *Nat. Med.* 5: 1071–1075.

Langacker, R. W. 1987. *Foundations of cognitive grammar.* Vol. 1: *Theoretical prerequisites*. Stanford: Stanford University Press.

Langton, C. G. 1986. "Studying artificial life with cellular automata." *Physica D* 22: 120–149.

Le Galliard, J.-F., R. Ferrière, and U. Dieckmann. 2003. "The adaptive dynamics of altruism in spatially heterogeneous populations." *Evolution* 57: 1–17.

Leimar, O., and P. Hammerstein. 2001. "Evolution of cooperation through indirect reciprocation." *P. Roy. Soc. Lond. B Bio.* 268: 745–753.

Lengauer, C., K. W. Kinzler, and B. Vogelstein. 1997. "Genetic instability in colorectal cancers." *Nature* 386: 623–627.

——— 1998. "Genetic instabilities in human cancers." *Nature* 396: 623–649.

Lenski, R. E., and R. M. May. 1994. "The evolution of virulence in parasites and pathogens: Reconciliation between two competing hypotheses." *J. Theor. Biol.* 169: 253–265.

Leuthäusser, I. 1987. "Statistical mechanics of Eigen's evolution model." *J. Stat. Phys.* 48: 343–360.

Levin, B. R. 1982. "Evolution of parasites and hosts." In R. M. Anderson and R. M. May, eds., *Population biology of infectious diseases*, 212–243. New York: Springer Verlag.

Levin, B. R., and J. J. Bull. 1994. "Short-sighted evolution, and the virulence of pathogenic microorganisms." *Trends Microbiol.* 2: 76–81.

Levin, B. R., M. Lipsitch, and S. Bonhoeffer. 1999. "Population biology, evolution, and infectious disease: Convergence and synthesis." *Science* 283: 806–809.

Levin, S. A. 1992. "The problem of pattern and scale in ecology." *Ecology* 73: 1943–1967.

——— 1999. *Fragile dominion: Complexity and the commons*. Reading, MA: Perseus Books.

Levin, S. A., and R. T. Paine. 1974. "Disturbance, patch formation, and community structure." *P. Natl. Acad. Sci. USA* 71: 2744–2747.

Levin, S. A., B. Grenfell, A. Hastings, and A. S. Perelson. 1997. "Mathematical and computational challenges in population biology and ecosystems science." *Science* 275: 334–343.

Levin, S. A., and D. Pimentel. 1981. "Selection of intermediate rates of increase in parasite-host systems." *Am. Nat.* 117: 308–315.

Levine, A. J. 1992. *Viruses*. New York: Scientific American Library.

——— 1993. "The tumor suppressor genes." *Annu. Rev. Biochem.* 62: 623–651.

Lewontin, R. C. 1974. *The genetic basis of evolutionary change*. Columbia Biological Series. New York: Columbia University Press.

Lieberman, E., C. Hauert, and M. A. Nowak. 2005. "Evolutionary dynamics on graphs." *Nature* 433: 312–316.

Lieberman, P. 1984. *The biology and evolution of language*. Cambridge, MA: Harvard University Press.

——— 1991. *Uniquely human: The evolution of speech, thought, and selfless behavior*. Cambridge, MA: Harvard University Press.

Lifson, J. D., M. A. Nowak, S. Goldstein, J. L. Rossio, A. Kinter, G. Vasquez, T. A. Wiltrout, C. Brown, D. Schneider, L. Wahl, A. L. Lloyd, J. Williams, W. R. Elkins, A. S. Fauci, and V. M. Hirsch. 1997. "The extent of early viral replication is a critical determinant of the natural history of simian immunodeficiency virus infection." *J. Virol.* 71: 9508–9514.

Lifson, J. D., J. L. Rossio, R. Arnaout, L. Li, T. L. Parks, D. K. Schneider, R. F. Kiser, V. J. Coalter, G. Walsh, R. J. Imming, B. Fisher, B. M. Flynn, N. Bischofberger, M. Piatak, V. M. Hirsch, M. A. Nowak, and D. Wodarz. 2000. "Containment of simian immunodeficiency virus infection: Cellular immune responses and protection from rechallenge following transient postinoculation antiretroviral treatment." *J. Virol.* 74: 2584–2593.

Lifson, J. D., J. L. Rossio, M. Piatak, T. Parks, L. Li, R. Kiser, V. Coalter, B. Fisher, B. M. Flynn, S. Czajak, V. M. Hirsch, K. A. Reimann, J. E. Schmitz, J. Ghrayeb, N. Bischofberger, M. A. Nowak, R. C. Desrosiers, and D. Wodarz. 2001. "Role of $CD8^+$ lymphocytes in control of simian immunodeficiency virus infection and resistance to rechallenge after transient early antiretroviral treatment." *J. Virol.* 75: 10187–10199.

Liggett, T. M. 1999. *Stochastic interacting systems: Contact, voter, and exclusion processes*. Berlin: Springer.

Lightfoot, D. 1991. *How to set parameters: Arguments from language change*. Cambridge: MIT Press.

——— 1999. *The development of language: Acquisition, change, and evolution*. Maryland Lectures in Language and Cognition. Oxford: Blackwell Publishers.

Lindgren, K. 1991. "Evolutionary phenomena in simple dynamics." In C. G. Langton, C. Taylor, J. D. Farmer, and S. Rasmussen, eds., *Artificial life II*, SFI studies in the sciences of complexity, vol. 10, 295–312. Boston: Addison-Wesley.

Lindgren, K., and M. G. Nordahl. 1994. "Evolutionary dynamics of spatial games." *Physica D* 75: 292–309.

Lipsitch, M., E. A. Herre, and M. A. Nowak. 1995. "Host population structure and the evolution of virulence: A law of diminishing returns." *Evolution* 49: 743–748.

Lipsitch, M., and M. A. Nowak. 1995. "The evolution of virulence in sexually transmitted HIV/AIDS." *J. Theor. Biol.* 174: 427–440.

Lipsitch, M., M. A. Nowak, D. Ebert, and R. M. May. 1995. "The population dynamics of vertically and horizontally transmitted parasites." *P. Roy. Soc. Lond. B Bio.* 260: 321–327.

Lipsitch, M., S. Siller, and M. A. Nowak. 1996. "The evolution of virulence in pathogens with vertical and horizontal transmission." *Evolution* 50: 1729–1741.

Little, M. P., and E. G. Wright. 2003. "A stochastic carcinogenesis model incorporating genomic instability fitted to colon cancer data." *Math. Biosci.* 183: 111–134.

Lloyd, A. L., and V. A. A. Jansen. 2004. "Spatiotemporal dynamics of epidemics: Synchrony in metapopulation models." *Math. Biosci.* 188: 1–16.

Loeb, L. A. 1991. "Mutator phenotype may be required for multistage carcinogenesis." *Cancer Res.* 51: 3075–3079.

——— 2001. "A mutator phenotype in cancer." *Cancer Res.* 61: 3230–3239.

Loeb, L. A., C. F. Springgate, and N. Battula. 1974. "Errors in DNA replication as a basis of malignant changes." *Cancer Res.* 34: 2311–2321.

Lombardo, M. P. 1985. "Mutual restraint in tree swallows: A test of the tit for tat model of reciprocity." *Science* 227: 1363–1365.

Lotka, A. J. 1925. *Elements of physical biology*. Baltimore: Williams and Wilkins. (Reissued as *Elements of mathematical biology* by Dover, 1956.)

Luebeck, E. G., and S. H. Moolgavkar. 2002. "Multistage carcinogenesis and the incidence of colorectal cancer." *P. Natl. Acad. Sci. USA* 99: 15095–15100.

Manzini, M. R., and K. Wexler. 1987. "Parameters, binding theory, and learnability." *Linguist. Inq.* 18: 413–444.

Maruyama, T. 1970. "Effective number of alleles in a subdivided population." *Theor. Popul. Biol.* 1: 273–306.

Maser, R. S., and R. A. DePinho. 2002. "Connecting chromosomes, crisis, and cancer." *Science* 297: 565–569.

Matessi, C., A. Gimelfarb, and S. Gavrilets. 2001. "Long-term buildup of reproductive isolation promoted by disruptive selection: How far does it go?" *Selection* 2: 41–64.

Matsuda, H., N. Ogita, A. Sasaki, and K. Satō. 1992. "Statistical mechanics of population: The lattice Lotka-Volterra model." *Prog. Theor. Phys.* 88: 1035–1049.

May, R. M. 1973. *Stability and complexity in model ecosystems*. Princeton, NJ: Princeton University Press. (Second edition with new introduction, 2001.)

——— 1976. "Simple mathematical models with very complicated dynamics." *Nature* 261: 459–467.

——— 1987. "More evolution of cooperation." *Nature* 327: 15–17.

——— 2004. "Uses and abuses of mathematics in biology." *Science* 303: 790–793.

May, R. M., and R. M. Anderson. 1979. "Population biology of infectious diseases: Part II." *Nature* 280: 455–461.

May, R. M., and R. M. Anderson. 1983. "Epidemiology and genetics in the coevolution of parasites and hosts." *P. Roy. Soc. Lond. B Bio.* 219: 281–313.

——— 1990. "Parasite-host coevolution." *Parasitology* 100: S89–S101.

May, R. M., and W. Leonard. 1975. "Nonlinear aspects of competition between three species." *SIAM J. Appl. Math.* 29: 243–252.

May, R. M., and M. A. Nowak. 1994. "Superinfection, metapopulation dynamics, and the evolution of diversity." *J. Theor. Biol.* 170: 95–114.

May, R. M., and M. A. Nowak. 1995. "Coinfection and the evolution of parasite virulence." *P. Roy. Soc. Lond. B Bio.* 261: 209–215.

May, R. M., and G. F. Oster. 1976. "Bifurcations and dynamic complexity in simple ecological models." *Am. Nat.* 110: 573–599.

Maynard Smith, J. 1979. "Hypercycles and the origin of life." *Nature* 280: 445–446.

——— 1982. *Evolution and the theory of games*. Cambridge: Cambridge University Press.

——— 1984. "Game-theory and the evolution of behavior." *Behav. Brain Sci.* 7: 95–101.

——— 1989. *Evolutionary genetics*. Oxford: Oxford University Press.

Maynard Smith, J., and G. R. Price. 1973. "Logic of animal conflict." *Nature* 246: 15–18.

Maynard Smith, J., and E. Szathmáry. 1995. *The major transitions in evolution*. Oxford: W. H. Freeman.

Mayr, E. E. 1982. *The growth of biological thought: Diversity, evolution, and inheritance*. Cambridge, MA: Harvard University Press.

——— 2001. *What evolution is*. New York: Basic Books.

McAdam, S. N., P. Klenerman, L. G. Tussey, S. Rowland-Jones, D. Lalloo, A. L. Brown, et al. 1995. "Immunogenic HIV variant peptides that bind to HLA-B8 but fail to stimulate cytotoxic T lymphocyte responses." *J. Immunol.* 155: 2729–2736.

McCaskill, J. S. 1984. "A stochastic theory of macromolecular evolution." *Biol. Cybern.* 50: 63–73.

McLean, A. R., and M. A. Nowak. 1992. "Models of interactions between HIV and other pathogens." *J. Theor. Biol.* 155: 69–86.

McMichael, A., S. Rowland-Jones, P. Klenerman, S. McAdam, F. Gotch, R. Phillips, and M. Nowak. 1995. "Epitope variation and T-cell recognition." *J. Cell. Biochem.* 59 (S21A): 60.

Metz, J. A. J., S. A. H. Geritz, G. Meszéna, F. J. A. Jacobs, and J. S. van Heerwarden. 1996. "Adaptive dynamics, a geometrical study of the consequences of nearly faithful reproduction." In S. J. van Strien and S. M. Verduyn Lunel, eds., *Stochastic and spatial structures of dynamical systems, K. Ned. Akad. Van Wet. B* 45: 183–231. Amsterdam: North-Holland Publishing Company.

Metz, J. A. J., R. M. Nisbet, and S. A. H. Geritz. 1992. "How should we define fitness for general ecological scenarios?" *Trends Ecol. Evol.* 7: 198–202.

Michod, R. E. 1999. *Darwinian dynamics: Evolutionary transitions in fitness and individuality*. Princeton, NJ: Princeton University Press.

Michor, F., S. A. Frank, R. M. May, Y. Iwasa, and M. A. Nowak. 2003a. "Somatic selection for and against cancer." *J. Theor. Biol.* 225: 377–382.

Michor, F., Y. Iwasa, N. L. Komarova, and M. A. Nowak. 2003b. "Local regulation of homeostasis favors chromosomal instability." *Curr. Biol.* 13: 581–584.

Michor, F., Y. Iwasa, and M. A. Nowak. 2004. "Dynamics of cancer progression." *Nat. Rev. Cancer* 4: 197–205.

Michor, F., Y. Iwasa, H. Rajagopalan, C. Lengauer, and M. A. Nowak. 2004. "Linear model of colon cancer initiation." *Cell Cycle* 3: 358–362.

Michor, F., Y. Iwasa, B. Vogelstein, C. Lengauer, and M. A. Nowak. 2005a. "Can chromosomal instability initiate tumorigenesis?" *Semin. Cancer Biol.* 15: 43–49.

Michor, F., T. Hughes, Y. Iwasa, S. Branford, N. P. Shah, C. L. Sawyers, and M. A. Nowak. 2005b. "Dynamics of chronic myeloid leukemia." *Nature* 435: 1267–1270.

Michor, F., M. A. Nowak, S. A. Frank, and Y. Iwasa. 2003c. "Stochastic elimination of cancer cells." *P. Roy. Soc. Lond. B Bio.* 270: 2017–2024.

Milinski, M. 1987. "Tit for tat in sticklebacks and the evolution of cooperation." *Nature* 325: 433–435.

Milinski, M., D. Semmann, and H.-J. Krambeck. 2002. "Reputation helps solve the 'tragedy of the commons.'" *Nature* 415: 424–426.

Milinski, M., and C. Wedekind. 1998. "Working memory constrains human cooperation in the prisoner's dilemma." *P. Natl. Acad. Sci. USA* 95: 13755–13758.

Miller, G. A. 1991. *The science of words. Scientific American* Library Series. New York: W. H. Freeman.

Mintz, B. 1971. "Clonal basis of mammalian differentiation." In D. D. Davies and M. Balls, eds., *Control mechanisms of growth and differentiation, Sym. Soc. Exp. Biol.* 25: 345–370. Cambridge: Cambridge University Press.

Mitchener, W. G., and M. A. Nowak. 2003. "Competitive exclusion and coexistence of universal grammars." *B. Math. Biol.* 65: 67–93.

——— 2004. "Chaos and language." *P. Roy. Soc. Lond. B Bio.* 271: 701–704.

Mitteldorf, J., and D. S. Wilson. 2000. "Population viscosity and the evolution of altruism." *J. Theor. Biol.* 204: 481–496.

Mock, D. W., and G. A. Parker. 1997. *The evolution of sibling rivalry*. Oxford: Oxford University Press.

Molander, P. 1985. "The optimal level of generosity in a selfish, uncertain environment." *J. Conflict Resolut.* 29: 611–618.

Moolgavkar, S. H., and A. G. Knudson. 1981. "Mutation and cancer: A model for human carcinogenesis." *J. Natl. Cancer I.* 66: 1037–1052.

Moran, P. A. P. 1958. "Random processes in genetics." *P. Camb. Philos. Soc.* 54: 60–71.

——— 1962. *The statistical processes of evolutionary theory*. Oxford: Clarendon Press.

Muller, H. J. 1927. "Artificial transmutation of the gene." *Science* 46: 84–87.

Murray, J. D. 2002. *Mathematical biology I: An introduction.* 3rd ed. New York: Springer-Verlag.

——— 2003. *Mathematical biology II: Spatial models and biomedical applications.* 3rd ed. New York: Springer-Verlag.

Mylius, S. D., and O. Diekmann. 2001. "The resident strikes back: Invader-induced switching of resident attractor." *J. Theor. Biol.* 211: 297–311.

Nagase, H., and Y. Nakamura. 1993. "Mutations of the APC (adenomatous polyposis-coli) gene." *Hum. Mutat.* 2: 425–434.

Nagylaki, T., and B. Lucier. 1980. "Numerical analysis of random drift in a cline." *Genetics* 94: 497–517.

Nakamaru, M., and Y. Iwasa. 2005. "The evolution of altruism by costly punishment in lattice structured populations: Score-dependent viability versus score-dependent fertility." *Evol. Ecol. Res.* 7: 853–870.

Nakamaru, M., H. Matsuda, and Y. Iwasa. 1997. "The evolution of cooperation in a lattice-structured population." *J. Theor. Biol.* 184: 65–81.

Nakamaru, M., H. Nogami, and Y. Iwasa. 1998. "Score-dependent fertility model for the evolution of cooperation in a lattice." *J. Theor. Biol.* 194: 101–124.

Nakhleh, L., D. Ringe, and T. Warnow. 2005. "Perfect phylogenetic networks: A new methodology for reconstructing the evolutionary history of natural languages." *Language* 81: 382–420.

Nash, J. F. 1950. "Equilibrium points in n-person games." *P. Natl. Acad. Sci. USA* 36: 48–49.

Nasmyth, K. 2002. "Segregating sister genomes: The molecular biology of chromosome separation." *Science* 297: 559–565.

Nee, S. 2000. "Mutualism, parasitism, and competition in the evolution of coviruses." *Philos. T. Roy. Soc. B* 355: 1607–1613.

Nee, S., and R. M. May. 1992. "Dynamics of metapopulations: Habitat destruction and competitive coexistence." *J. Anim. Ecol.* 61: 37–40.

Nei, M. 1987. *Molecular evolutionary genetics.* New York: Columbia University Press.

Neuhauser, C. 2001. "Mathematical challenges in spatial ecology." *Not. Am. Math. Soc.* 48: 1304–1314.

Newman, M. E. J. 2001. "The structure of scientific collaboration networks." *P. Natl. Acad. Sci. USA* 98: 404–409.

Newmeyer, F. J. 1991. "Functional explanation in linguistics and the origins of language." *Lang. Commun.* 11: 3–28.

Niyogi, P. 1998. *The informational complexity of learning: Perspectives on neural networks and generative grammar*. Dordrecht: Kluwer Academic Publishers.

Niyogi, P., and R. C. Berwick. 1997. "Evolutionary consequences of language learning." *Linguist. Philos.* 20: 697–719.

Nordling, C. O. 1953. "A new theory on cancer–inducing mechanism." *Brit. J. Cancer* 7: 68–72.

Nowak, M. 1990a. "An evolutionarily stable strategy may be inaccessible." *J. Theor. Biol.* 142: 237–241.

——— 1990b. "Stochastic strategies in the prisoner's dilemma." *Theor. Popul. Biol.* 38: 93–112.

——— 1991. "The evolution of viruses: Competition between horizontal and vertical transmission of mobile genes." *J. Theor. Biol.* 150: 339–347.

——— 2000. "The basic reproductive ratio of a word, the maximum size of a lexicon." *J. Theor. Biol.* 204: 179–189.

Nowak, M. A., R. M. Anderson, A. R. McLean, T. F. W. Wolfs, J. Goudsmit, and R. M. May. 1991. "Antigenic diversity thresholds and the development of AIDS." *Science* 254: 963–969.

Nowak, M. A., S. Bonhoeffer, and R. M. May. 1994a. "More spatial games." *Int. J. Bifurcat. Chaos* 4: 33–56.

——— 1994b. "Spatial games and the maintenance of cooperation." *P. Natl. Acad. Sci. USA* 91: 4877–4881.

Nowak, M. A., S. Bonhoeffer, C. Loveday et al. 1995c. "HIV dynamics: Results confirmed." *Nature* 375: 193.

Nowak, M. A., and N. L. Komarova. 2001. "Towards an evolutionary theory of language." *Trends Cogn. Sci.* 5: 288–295.

Nowak, M. A., N. L. Komarova, and P. Niyogi. 2001. "Evolution of universal grammar." *Science* 291: 114–118.

——— 2002. "Computational and evolutionary aspects of language." *Nature* 417: 611–617.

Nowak, M. A., N. L. Komarova, A. Sengupta, P. V. Jallepalli, I.-M. Shih, B. Vogelstein, and C. Lengauer. 2002. "The role of chromosomal instability in tumor initiation." *P. Natl. Acad. Sci. USA* 99: 16226–16231.

Nowak, M. A., and D. C. Krakauer. 1999. "The evolution of language." *P. Natl. Acad. Sci. USA* 96: 8028–8033.

Nowak, M. A., D. C. Krakauer, and A. Dress. 1999. "An error limit for the evolution of language." *P. Roy. Soc. Lond. B Bio.* 266: 2131–2136.

Nowak, M. A., and R. M. May. 1992. "Evolutionary games and spatial chaos." *Nature* 359: 826–829.

——— 1993. "The spatial dilemmas of evolution." *Int. J. Bifurcat. Chaos* 3: 35–78.

——— 1994. "Superinfection and the evolution of parasite virulence." *P. Roy. Soc. Lond. B Bio.* 255: 81–89.

——— 2000. *Virus dynamics*. Oxford: Oxford University Press.

Nowak, M. A., R. M. May, and R. M. Anderson. 1990. "The evolutionary dynamics of HIV-1 quasispecies and the development of immunodeficiency disease." *AIDS* 4: 95–103.

Nowak, M. A., R. M. May, and K. Sigmund. 1995. "Immune-responses against multiple epitopes." *J. Theor. Biol.* 175: 325–353.

Nowak, M. A., R. M. May, R. E. Phillips, S. Rowland-Jones, D. G. Lalloo, S. McAdam, P. Klenerman, B. Köppe, K. Sigmund, C. R. M. Bangham, and A. J. McMichael. 1995b. "Antigenic oscillations and shifting immunodominance in HIV-1 infections." *Nature* 375: 606–611.

Nowak, M. A., F. Michor, and Y. Iwasa. 2003. "The linear process of somatic evolution." *P. Natl. Acad. Sci. USA* 100: 14966–14969.

Nowak, M. A., F. Michor, N. L. Komarova, and Y. Iwasa. 2004a. "Evolutionary dynamics of tumor suppressor gene inactivation." *P. Natl. Acad. Sci. USA* 101: 10635–10638.

Nowak, M. A., K. M. Page, and K. Sigmund. 2000. "Fairness versus reason in the ultimatum game." *Science* 289: 1773–1775.

Nowak, M. A., J. B. Plotkin, and V. A. A. Jansen. 2000. "The evolution of syntactic communication." *Nature* 404: 495–498.

Nowak, M. A., J. B. Plotkin, and D. C. Krakauer. 1999. "The evolutionary language game." *J. Theor. Biol.* 200: 147–162.

Nowak, M. A., A. Sasaki, C. Taylor, and D. Fudenberg. 2004b. "Emergence of cooperation and evolutionary stability in finite populations." *Nature* 428: 646–650.

Nowak, M., and P. Schuster. 1989. "Error thresholds of replication in finite populations: Mutation frequencies and the onset of Muller's ratchet." *J. Theor. Biol.* 137: 375–395.

Nowak, M., and K. Sigmund. 1989a. "Game-dynamical aspects of the prisoner's dilemma." *Appl. Math. Comput.* 30: 191–213.

——— 1989b. "Oscillations in the evolution of reciprocity." *J. Theor. Biol.* 137: 21–26.

——— 1990. "The evolution of stochastic strategies in the prisoner's dilemma." *Acta Appl. Math.* 20: 247–265.

——— 1992. "Tit for tat in heterogeneous populations." *Nature* 355: 250–253.

——— 1993. "A strategy of win-stay, lose-shift that outperforms tit-for-tat in the prisoner's dilemma game." *Nature* 364: 56–58.

——— 1994. "The alternating prisoner's dilemma." *J. Theor. Biol.* 168: 219–226.

——— 1998. "Evolution of indirect reciprocity by image scoring." *Nature* 393: 573–577.

——— 2004. "Evolutionary dynamics of biological games." *Science* 303: 793–799.

——— 2005. "Evolution of indirect reciprocity." *Nature* 437: 1291–1298.

Nunney, L. 1999. "Lineage selection and the evolution of multistage carcinogenesis." *P. Roy. Soc. Lond. B Bio.* 266: 493–498.

Ohtsuki, H., C. Hauert, E. Lieberman, and M. A. Nowak. 2006. "A simple rule for evolution of cooperation on graphs and social networks." *Nature*. In press.

Ohtsuki, H., and Y. Iwasa. 2004. "How should we define goodness? Reputation dynamics in indirect reciprocity." *J. Theor. Biol.* 231: 107–120.

Osherson, D. N., M. Stob, and S. Weinstein. 1986. *Systems that learn: An introduction to learning theory for cognitive and computer scientists*. Cambridge: MIT Press.

Otsuka, K., T. Suzuki, H. Shibata, S. Kato, M. Sakayori, H. Shimodaira, R. Kanamaru, and C. Ishioka. 2003. "Analysis of the human APC mutation spectrum in a *Saccharomyces cerevisiae* strain with a mismatch repair defect." *Int. J. Cancer* 103: 624–630.

Owen, M. R., and J. A. Sherratt. 1999. "Mathematical modelling of macrophage dynamics in tumours." *Math. Mod. Meth. Appl. S.* 9: 513–539.

Pacala, S. W., and D. Tilman. 1994. "Limiting similarity in mechanistic and spatial models of plant competition in heterogeneous environments." *Am. Nat.* 143: 222–257.

Page, K. M., and M. A. Nowak. 2000. "A generalized adaptive dynamics framework can describe the evolutionary ultimatum game." *J. Theor. Biol.* 209: 173–179.

——— 2002. "Unifying evolutionary dynamics." *J. Theor. Biol.* 219: 93–98.

Panchanathan, K., and R. Boyd. 2003. "A tale of two defectors: The importance of standing for evolution of indirect reciprocity." *J. Theor. Biol.* 224: 115–126.

——— "Indirect reciprocity can stabilize cooperation without the second-order free-rider problem." *Nature* 432: 499–502.

Partee, B. H., A. ter Meulen, and R. E. Wall. 1990. *Mathematical methods in linguistics*. Dordrecht: Kluwer Academic Publishers.

Parvinen, K. 1999. "Evolution of migration in a metapopulation." *B. Math. Biol.* 61: 531–550.

Perelson, A. S., P. Essunger, and D. D. Ho. 1997. "Dynamics of HIV-1 and CD4+ lymphocytes *in vivo*." *AIDS* 11: S17–S24.

Perelson, A. S., A. U. Neumann, M. Markowitz, J. M. Leonard, and D. D. Ho. 1996. "HIV-1 dynamics *in vivo*: Virion clearance rate, infected cell life-span, and viral generation time." *Science* 271: 1582–1586.

Pfeiffer, T., S. Schuster, and S. Bonhoeffer. 2001. "Cooperation and competition in the evolution of ATP-producing pathways." *Science* 292: 504–507.

Phillips, R. E., S. Rowland-Jones, D. F. Nixon, F. M. Gotch, J. P. Edwards, A. O. Ogunlesi, J. G. Elvin, J. A. Rothbard, C. R. M. Bangham, C. R. Rizza, and A. J. McMichael. 1991. "Human immunodeficiency virus genetic variation that can escape cytotoxic T cell recognition." *Nature* 354: 453–459.

Pihan, G. A., J. Wallace, Y. Zhou, and S. J. Doxsey. 2003. "Centrosome abnormalities and chromosome instability occur together in pre-invasive carcinomas." *Cancer Res.* 63: 1398–1404.

Pinker, S. 1979. "Formal models of language learning." *Cognition* 7: 217–283.

——— 1994. *The language instinct*. New York: William Morrow and Company.

Pinker, S., and P. Bloom. 1990. "Natural language and natural selection." *Behav. Brain Sci.* 13: 707–784.

Pitt, L. 1989. "Probabilistic inductive inference." *J. ACM* 36: 383–433.

Plotkin, J. B., and M. A. Nowak. 2000. "Language evolution and information theory." *J. Theor. Biol.* 205: 147–159.

——— 2002. "The different effects of apoptosis and DNA repair on tumorigenesis." *J. Theor. Biol.* 214: 453–467.

Pollard, C. J., and I. A. Sag. 1994. *Head-driven phrase structure grammar*. Chicago: University of Chicago Press.

Poundstone, W. 1985. *The recursive universe*. Oxford: Oxford University Press.

Price, D. A., P. J. R. Goulder, P. Klenerman, A. K. Sewell, P. J. Easterbrook, M. Troop, et al. 1997. "Positive selection of HIV-1 cytotoxic T lymphocyte escape variants during primary infection." *Proc. Natl. Acad. Sci. USA* 94: 1890–1895.

Prince, A., and P. Smolensky. 1997. "Optimality: From neural networks to universal grammar." *Science* 275: 1604–1610.

——— 2004. *Optimality theory: Constraint interaction in generative grammar*. Oxford: Blackwell Publishing. (First circulated 1993 as Rutgers University Center for Cognitive Science Technical Report 2, http://roa.rutgers.edu)

Pulliam, H. R. 1988. "Sources, sinks, and population regulation." *Am. Nat.* 132: 652–661.

Pullum, G. K., and G. Gazdar, 1982. "Natural languages and context free languages." *Linguist. Philos.* 4: 471–504.

Rajagopalan, H., P. V. Jallepalli, C. Rago, V. E. Velculescu, K. W. Kinzler, B. Vogelstein, and C. Lengauer. 2004. "Inactivation of hCDC4 can cause chromosomal instability." *Nature* 428: 77–81.

Rajagopalan, H., M. A. Nowak, B. Vogelstein, and C. Lengauer. 2003. "The significance of unstable chromosomes in colorectal cancer." *Nat. Rev. Cancer* 3: 695–701.

Rand, D. A., H. B. Wilson, and J. M. McGlade. 1994. "Dynamics and evolution: Evolutionarily stable attractors, invasion exponents, and phenotype dynamics." *Philos. T. Roy. Soc. B* 343: 261–283.

Rapoport, A., and A. M. Chammah. 1965. *Prisoner's dilemma*. Ann Arbor: University of Michigan Press.

Read, A. F., and P. H. Harvey. 1993. "Parasitology: The evolution of virulence." *Nature* 362: 500–501.

Reya, T., S. J. Morrison, M. Clarke, and I. L. Weissman. 2001. "Stem cells, cancer, and cancer stem cells." *Nature* 414: 105–111.

Rice, S. H. 2004. *Evolutionary theory: Mathematical and conceptual foundations*. Sunderland, MA: Sinauer.

Riley, J. G. 1979. "Evolutionary equilibrium strategies." *J. Theor. Biol.* 76: 109–123.

Ringe, D., T. Warnow, and A. Taylor. 2002. "Indo-European and computational cladistics." *T. Philol. Soc.* 100: 59–129.

Rivin, I. 2001. "Yet another zeta function and learning." arXiv.org preprint cs.LG/0107033.

Roberts, G., and T. N. Sherratt. 1998. Development of cooperative relationships through increasing investment. *Nature* 394: 175–179.

Robertson, A. 1978. "Time of detection of recessive visible genes in small populations." *Genet. Res.* 31: 255–264.

Robins, R. H. 1979. *A short history of linguistics*. 2nd ed. London: Longman.

Ross, R. 1908. *Report on the prevention of malaria in Mauritius*. London.

Saag, M. S., B. H. Hahn, J. Gibbons, Y. X. Li, E. S. Parks, W. P. Parks, and G. M. Shaw. 1988. "Extensive variation of human immunodeficiency virus type-1 in vivo." *Nature* 334: 440–444.

Sabelis, M. W., O. Diekmann, and V. A. A. Jansen. 1991. "Metapopulation persistence despite local extinction: Predator-prey patch models of the Lotka-Volterra type." *Biol. J. Linn. Soc.* 42: 267–283.

Sadock, J. M. 1991. *Autolexical syntax: A theory of parallel grammatical representations*. Studies in Contemporary Linguistics. Chicago: University of Chicago Press.

Saffran, J. R., R. N. Aslin, and E. L. Newport. 1996. "Statistical learning by 8-month-old infants." *Science* 274: 1926–1928.

Sakakibara, Y. 1988. "Learning context-free grammars from structural data in polynomial time." In D. Haussler and L. Pitt, eds., *Proceedings of the first annual workshop on computational learning theory, MIT, Cambridge, MA*, 330–344. San Francisco: Morgan Kaufmann Publishers.

——— 1990. "Learning context-free grammars from structural data in polynomial time." *Theor. Comput. Sci.* 76: 223–242. (First presented at COLT 1988; see Sakakibara 1988.)

——— 1997. "Recent advances of grammatical inference." *Theor. Comput. Sci.* 185: 15–45.

Sampson, G. 2005. *Educating Eve: The 'language instinct' debate*. London: Continuum International Publishing Group. (First published 1997, London: Cassell Academic.)

Samuelson, L. 1997. *Evolutionary games and equilibrium selection*. Cambridge: MIT Press.

Santos, F. C., and J. M. Pacheco. 2005. "Scale-free networks provide a unifying framework for the emergence of cooperation." *Phys. Rev. Lett.* 95: 098104.

Santos, F. C., J. M. Pacheco, and Tom Leanerts. 2006. "Evolutionary dynamics of social dilemmas in structured heterogeneous populations." *Proc. Nat. Acad. Sci. (USA)* 103: 3490–3494.

Sasaki, A. 1994. "Evolution of antigen drift/switching: Continuously evading pathogens." *J. Theor. Biol.* 168: 291–308.

——— 2000. "Host-parasite coevolution in a multilocus gene-for-gene system." *P. Roy. Soc. Lond. B Bio.* 267: 2183–2188.

Sasaki, A., W. D. Hamilton, and F. Ubeda. 2002. "Clone mixtures and a pacemaker: New facets of red-queen theory and ecology." *P. Roy. Soc. Lond. B Bio.* 269: 761–772.

Sasaki, A., and Y. Haraguchi. 2000. "Antigenic drift of viruses within a host: A finite site model with demographic stochasticity." *J. Mol. Evol.* 51: 245–255.

Sasaki, A., and Y. Iwasa. 1987. "Optimal recombination rate in fluctuating environments." *Genetics* 115: 377–388.

Sasaki, A., and M. A. Nowak. 2003. "Mutation landscapes." *J. Theor. Biol.* 224: 241–247.

Satō, K., H. Matsuda, and A. Sasaki. 1994. "Pathogen invasion and host extinction in lattice structured populations." *J. Math. Biol.* 32: 251–268.

Schaffer, M. 1988. "Evolutionary stable strategies for a finite population and a variable contest size." *J. Theor. Biol.* 132: 469–478.

Schreiber, S. 2001. "Urn models, replicator processes, and random genetic drift." *SIAM J. Appl. Math.* 61: 2148–2167.

Schrödinger, E. 1944. *What is life? The physical aspect of the living cell.* Cambridge: Cambridge University Press.

——— 1992. *What is life? The physical aspect of the living cell: With mind and matter and autobiographical sketches.* Cambridge: Cambridge University Press.

Schuster, P., and P. F. Stadler. 2002. "Networks in molecular evolution." *Complexity* 8(1): 34–42.

Schuster, P., and J. Swetina. 1988. "Stationary mutant distributions and evolutionary optimization." *B. Math. Biol.* 50: 635–660.

Seger, J. 1988. "Dynamics of simple host parasite models with more than two genotypes in each species." *Philos. T. Roy. Soc. B* 319: 541–555.

Seger, J., and W. D. Hamilton. 1988. "Parasite and sex." In R. E. Michod and B. R. Levin, eds., *The evolution of sex*, 176–193. Sunderland, MA: Sinauer.

Selten, R. 1975. "Reexamination of the perfectness concept for equilibrium points in extensive games." *Int. J. Game Theory* 4: 25–55.

Selten, R., and P. Hammerstein. 1984. "Gaps in Harley argument on evolutionarily stable learning rules and in the logic of tit for tat." *Behav. Brain Sci.* 7: 115–116.

Semmann, D., H.-J. Krambeck, and M. Milinski. 2003. "Volunteering leads to rock-paper-scissors dynamics in a public goods game." *Nature* 425: 390–393.

Shahshahani, S. 1979. "A new mathematical framework for the study of linkage and selection." *Mem. Am. Math. Soc.*, no. 211. Providence, RI: American Mathematical Society.

Sherratt, J. A., and M. A. Nowak. 1992. "Oncogenes, anti-oncogenes and the immune response to cancer: A mathematical model." *P. Roy. Soc. Lond. B Bio.* 248: 261–271.

Sherrington, D. 1997. "Landscape paradigms in physics and biology: Introduction and overview." *Physica D* 107: 117–121.

Shieber, S. M. 1985. "Evidence against the context-freeness of natural language." *Linguist. Philos.* 8: 333–343.

Shih, I. M., W. Zhou, S. N. Goodman, C. Lengauer, K. W. Kinzler, and B. Vogelstein. 2001. "Evidence that genetic instability occurs at an early stage of colorectal tumorigenesis." *Cancer Research* 61: 818–822.

Shonn, M. A., R. McCarroll, and A. W. Murray. 2000. "Requirement of the spindle checkpoint for proper chromosome segregation in budding yeast meiosis." *Science* 289: 300–303.

Sieber, O. M., K. Heinimann, P. Gorman, H. Lamlum, M. Crabtree, C. A. Simpson, D. Davies, K. Neale, S. V. Hodgson, R. R. Roylance, R. K. S. Phillips, W. F. Bodmer, and I. P. M. Tomlinson. 2002. "Analysis of chromosomal instability in human colorectal adenomas with two mutational hits at APC." *P. Natl. Acad. Sci. USA* 99: 16910–16915.

Sieber, O. M., K. Heinimann, and I. P. Tomlinson. 2003. "Genomic instability—the engine of tumorigenesis?" *Nat. Rev. Cancer* 3: 701–708.

Sigmund, K. 1993. *Games of life: Explorations in ecology, evolution and behaviour*. Oxford: Oxford University Press.

Sigmund, K., C. Hauert, and M. A. Nowak. 2001. "Reward and punishment." *P. Natl. Acad. Sci. USA* 98: 10757–10762.

Sinervo, B., and C. M. Lively. 1996. "The rock-paper-scissors game and the evolution of alternative male strategies." *Nature* 380: 240–243.

Siskind, J. M. 1996. "A computational study of cross-situational techniques for learning word-to-meaning mappings." *Cognition* 61: 39–91.

Skyrms, B., and R. Pemantle. 2000. "A dynamic model of social network formation." *P. Natl. Acad. Sci. USA* 97: 9340–9346.

Slatkin, M. 1979. "Frequency- and density-dependent selection on a quantitative character." *Genetics* 93: 755–771.

——— 1981. "Fixation probabilities and fixation times in a subdivided population." *Evolution* 35: 477–488.

Smith, W. J. 1977. *The behavior of communicating*. Cambridge, MA: Harvard University Press.

Sober, E., and D. S. Wilson. 1998. *Unto others: The evolution and psychology of unselfish behavior*. Cambridge, MA: Harvard University Press.

Stabler, E. 1998. "Acquiring languages with movement." *Syntax* 1: 72–97.

——— 2004. "Varieties of crossing dependencies." *Cognitive Science* 28: 699–720.

Stadler, P. F. 1992. "Correlation in landscapes of combinatorial optimization problems." *Europhys. Lett.* 20: 479–482.

——— 1999. "Fitness landscapes arising from the sequence-structure maps of biopolymers." *J. Mol. Struc.—Theochem* 463: 7–19.

Stadler, P. F., and P. Schuster. 1990. "Dynamics of small autocatalytic reaction networks: Bifurcations, permanence, and exclusion." *B. Math. Biol.* 52: 485–508.

Stadler, P. F., and P. Schuster. 1992. "Mutation in autocatalytic reaction networks: An analysis based on perturbation theory." *J. Math. Biol.* 30: 597–631.

Steels, L. 1997. "Self-organizing vocabularies." In C. G. Langton and K. Shimohara, eds., *Artificial life V: Proceedings of the fifth international workshop on the synthesis and simulation of living systems, 16–18 May 1996, Nara, Japan*, 179–184. Cambridge: MIT Press.

Stewart, F. M., and B. R. Levin. 1984. "The population biology of bacterial viruses: Why be temperate." *Theor. Popul. Biol.* 26: 93–117.

Strauss, B. S. 1998. "Hypermutability in carcinogenesis." *Genetics* 148: 1619–1626.

Strogatz, S. H. 1994. *Nonlinear dynamics and chaos: With applications to physics, biology, chemistry, and engineering*. Cambridge, MA: Perseus Books.

——— 2001. "Exploring complex networks." *Nature* 410: 268–276.

Sugden, R. 1986. *The economics of rights, co-operation and welfare*. Oxford: Blackwell.

Sugihara, G., and R. M. May. 1990. "Nonlinear forecasting as a way of distinguishing chaos from measurement error in time series." *Nature* 344: 734–741.

Swetina, J., and P. Schuster. 1982. "Self-replication with errors: A model for polynucleotide replication." *Biophys. Chem.* 16: 329–345.

Szabó, G., and C. Hauert. 2002. "Phase transitions and volunteering in spatial public goods games." *Phys. Rev. Lett.* 89: 118101 (1–4).

Szabó, G., and J. Vukov. 2004. "Cooperation for volunteering and partially random partnerships." *Phys. Rev. E* 69: 036107 (1–7).

Szathmáry, E., and L. Demeter. 1987. "Group selection of early replicators and the origin of life." *J. Theor. Biol.* 128: 463–486.

Taddei, F., M. Radman, J. Maynard Smith, B. Toupance, P. H. Gouyon, and B. Godelle. 1997. "Role of mutator alleles in adaptive evolution." *Nature* 387: 700–702.

Taubes, C. H. 2001. *Modeling differential equations in biology*. Upper Saddle River, NJ: Prentice Hall.

Taylor, C., D. Fudenberg, A. Sasaki, and M. A. Nowak. 2004. "Evolutionary game dynamics in finite populations." *B. Math. Biol.* 66: 1621–1644.

Taylor, P. D. 1989. "Evolutionary stability in one-parameter models under weak selection." *Theor. Popul. Biol.* 36: 125–143.

Taylor, P. D., and L. B. Jonker. 1978. "Evolutionary stable strategies and game dynamics." *Math. Biosci.* 40: 145–156.

Tesar, B., and P. Smolensky. 2000. *Learnability in optimality theory*. Cambridge: MIT Press.

Tilman, D., and P. Kareiva, eds. 1997. *Spatial ecology: The role of space in population dynamics and interspecific interactions*. Monographs in Population Biology. Princeton, NJ: Princeton University Press.

Tilman, D., R. M. May, C. L. Lehman, and M. A. Nowak. 1994. "Habitat destruction and the extinction debt." *Nature* 371: 65–66.

Toffoli, T., and N. Margolus. 1987. *Cellular automata machines*. Cambridge: MIT Press.

Tomasello, M. 1999. *The cultural origins of human cognition*. Cambridge, MA: Harvard University Press.

Tomlinson, I. P., M. R. Novelli, and W. F. Bodmer. 1996. "The mutation rate and cancer." *P. Natl. Acad. Sci. USA* 93: 14800–14803.

Tomlinson, I. P., P. Sasieni, and W. Bodmer. 2002. "How many mutations in a cancer?" *Am. J. Pathol.* 160: 755–758.

Trapa, P. E., and M. A. Nowak. 2000. "Nash equilibria for an evolutionary language game." *J. Math. Biol.* 41: 172–188.

Trivers, R. L. 1971. "The evolution of reciprocal altruism." *Q. Rev. Biol.* 46: 35–57.

——— 1983. "The evolution of sex." *Q. Rev. Biol.* 58: 62–67.

——— 2002. *Natural selection and social theory: Selected papers of Robert Trivers*. Oxford: Oxford University Press.

Turelli, M., N. H. Barton, and J. A. Coyne. 2001. "Theory and speciation." *Trends Ecol. Evol.* 16: 330–343.

Turing, A. M. 1936. "On computable numbers, with an application to the Entscheidungsproblem." *P. Lond. Math. Soc.* 42: 230–265.

——— 1950. "Computing machinery and intelligence." *Mind* 59: 433–460.

Turner, P. E., and L. Chao. 1999. "Prisoner's dilemma in an RNA virus." *Nature* 398: 441–443.

Valiant, L. G. 1984. "A theory of learnable." *Commun. ACM* 27: 436–445.

van Baalan, M. 2000. "Pair approximations for different spatial geometries." In U. Dieckmann, R. Law, and J. A. J. Metz, eds., *The geometry of ecological interactions: Simplifying spatial complexity*, 359–387. Cambridge: Cambridge University Press.

van Baalen, M., and D. A. Rand. 1998. "The unit of selection in viscous populations and the evolution of altruism." *J. Theor. Biol.* 193: 631–648.

Van Valen, L. 1973. "A new evolutionary law." *Evol. Theor.* 1: 1–30.

Vapnik, V. N. 1998. *Statistical learning theory*. Hoboken, NJ: John Wiley.

Vapnik, V. N., and A. Y. Chervonenkis. 1971. "On the uniform convergence of relative frequencies of events to their probabilities." *Theor. Probab. Appl.* 17: 264–280.

——— 1981. "The necessary and sufficient conditions for the uniform convergence of averages to their expected values." *Teor. Ver. Prim.* 26: 543–564.

Vargha-Khadem, F., K. E. Watkins, C. J. Price, J. Ashburner, K. J. Alcock, A. Connelly, R. S. J. Frackowiak, K. J. Friston, M. E. Pembrey, M. Mishkin, D. G. Gadian, and R. E. Passingham. 1998. "Neural basis of an inherited speech and language disorder." *P. Natl. Acad. Sci. USA* 95: 12695–12700.

Velicer, G. J., and Y. N. Yu. 2003. "Evolution of novel cooperative swarming in the bacterium *Myxococcus xanthus*." *Nature* 425: 75–78.

Vincent, T., and J. S. Brown. 2005. *Evolutionary game theory, natural selection, and Darwinian dynamics*. Cambridge: Cambridge University Press.

Vogelstein, B., and K. W. Kinzler, eds. 1998. *The genetic basis of human cancer*. Toronto: McGraw-Hill.

Volterra, V. 1926. "Variazioni e fluttuazioni del numero d'individui in specie animali conviventi. *Mem. Acad. Lincei.* 2: 31–113. (Translation in an appendix to R, N. Chapman, *Animal ecology*, New York, 1931.)

von Neumann, J. 1966. *Theory of self-reproducing automata*. Champaign: University of Illinois Press.

von Neumann, J., and O. Morgenstern. 1944. *Theory of games and economic behavior*. Princeton, NJ: Princeton University Press.

Wahl, L. M., and M. A. Nowak. 1999a. "The continuous prisoner's dilemma: I. Linear reactive strategies." *J. Theor. Biol.* 200: 307–321.

——— 1999b. "The continuous prisoner's dilemma: II. Linear reactive strategies with noise." *J. Theor. Biol.* 200: 323–338.

Wakeley, J. 2004. "Metapopulation models for historical inference." *Mol. Ecol.* 13: 865–875.

Wang, W. S.-Y. 1998. "Language and the evolution of modern humans." In K. Omoto and P. V. Tobias, eds., *The origins and past of modern humans: Towards reconciliation*, 247–262. Singapore: World Scientific.

Warnow, T., S. N. Evans, D. Ringe, and L. Nakhleh. 2006. "A stochastic model of language evolution that incorporates homoplasy and borrowing." In P. Forster and C. Renfrew, eds., *Phylogenetic methods and the prehistory of languages*. London: MacDonald Institute for Archaeological Research.

Watts, D. J. 1999. *Small worlds*. Princeton, NJ: Princeton University Press.

Watts, D. J., and S. H. Strogatz. 1998. "Collective dynamics of 'small-world' networks." *Nature* 393: 440–442.

Wedekind, C., and M. Milinski. 1996. "Human cooperation in the simultaneous and the alternating prisoner's dilemma: Pavlov versus generous tit-for-tat." *P. Natl. Acad. Sci. USA* 93: 2686–2689.

——— 2000. "Cooperation through image scoring in humans." *Science* 288: 850–852.

Wei, X., J. M. Decker, S. Wang, H. Hui, J. C. Kappes, X. Wu, J. F. Salazar-Gonzalez, M. G. Salazar, J. M. Kilby, M. S. Saag, N. L. Komarova, M. A. Nowak, B. H. Hahn, P. D. Kwong, and G. M. Shaw. 2003. "Antibody neutralization and escape by HIV-1." *Nature* 422: 307–312.

Wei, X., S. K. Ghosh, M. E. Taylor, V. A. Johnson, E. A. Emini, P. Deutsch, J. D. Lifson, S. Bonhoeffer, M. A. Nowak, B. H. Hahn, M. S. Saag, and G. M. Shaw. 1995. "Viral dynamics in human immunodeficiency virus type 1 infection." *Nature* 373: 117–122.

Weibull, J. W. 1995. *Evolutionary game theory*. Cambridge: MIT Press.

Weinberg, R. A. 1991. "Tumor suppressor genes." *Science* 254: 1138–1146.

Wexler, K., and P. Culicover. 1980. *Formal principles of language acquisition*. Cambridge: MIT Press.

Wheeler, J. M., N. E. Beck, H. C. Kim, I. P. Tomlinson, N. J. Mortensen, and W. F. Bodmer. 1999. "Mechanisms of inactivation of mismatch repair genes in human colorectal cancer cell lines: The predominant role of hMLH1." *P. Natl. Acad. Sci. USA* 96: 10296–10301.

Whitlock, M. 2003. "Fixation probability and time in subdivided populations." *Genetics* 164: 767–779.

Wilkinson, G. S. 1984. "Reciprocal food sharing in the vampire bat." *Nature* 308: 181–184.

Williams, G. C. 1966. *Adaptation and natural selection*. Princeton, NJ: Princeton University Press.

——— 1992. *Natural selection: Domains, levels, and challenges*. Oxford: Oxford University Press.

Wilson, D. S. 1980. *The natural selection of populations and communities*. Menlo Park, CA: Benjamin Cummings.

Wilson, D. S., G. B. Pollock, and L. A. Dugatkin. 1992. "Can altruism evolve in purely viscous populations?" *Evol. Ecol.* 6: 331–341.

Wilson, E. O. 1978. *On human nature*. Cambridge, MA: Harvard University Press.

——— 2000. *Sociobiology: The new synthesis*. Cambridge, MA: Harvard University Press.

Wodarz, D., and D. C. Krakauer. 2001. "Genetic instability and the evolution of angiogenic tumor cell lines." *Oncol. Rep.* 8: 1195–1201.

Wodarz, D., and M. A. Nowak. 1999. "Specific therapy regimes could lead to long-term immunological control of HIV." *P. Natl. Acad. Sci. USA* 96: 14464–14469.

Wolfram, S. 1984. "Cellular automata as models of complexity." *Nature* 311: 419–424.

——— 1994. *Cellular automata and complexity: Collected papers.* New York: Perseus Books.

——— 2002. *A new kind of science.* Champaign, IL: Wolfram Media.

Wright, S. 1931. "Evolution in Mendelian populations." *Genetics* 16: 97–159.

——— 1932. "The roles of mutation, inbreeding, crossbreeding and selection in evolution." In D. F. Jones, ed., *Proceedings of the sixth international congress of genetics, Ithaca, NY*, vol. 1, 356–366. Menasha, WI: Brooklyn Institute of Arts and Sciences Botanic Garden.

——— 1968. *Evolution and the genetics of populations 1: Genetics and biometric foundations.* Chicago: University of Chicago Press.

——— 1969. *Evolution and the genetics of populations 2: The theory of gene frequencies.* Chicago: University of Chicago Press.

Yamamura, N. 1993. "Vertical transmission and evolution of mutualism from parasitism." *Theor. Popul. Biol.* 52: 95–109.

Yang, C. 2002. *Knowledge and learning in natural language.* Oxford: Oxford University Press.

Yatabe, Y., S. Tavare, and D. Shibata. 2001. "Investigating stem cells in human colon by using methylation patterns." *P. Natl. Acad. Sci. USA* 98: 10839–10844.

Zeeman, E. C. 1980. "Population dynamics from game theory." In Z. H. Nitecki and R. C. Robinson, eds., *Global theory of dynamical systems: Proceedings of an international conference held at Northwestern University, Evanston, Illinois, June 18–22, 1979*, Lecture Notes in Mathematics, vol. 819. Berlin: Springer-Verlag.

Zheng, X., S. M. Wise, and V. Cristini. 2005. "Nonlinear simulation of tumor necrosis, neo-vascularization and tissue invasion via an adaptive finite-element/level-set method." *B. Math. Biol.* 67: 211–259.

INDEX

Further Reading page numbers are followed by *r*; figure page numbers are followed by *f*.